# Identification of Defects in Semiconductors

SEMICONDUCTORS
AND SEMIMETALS
Volume 51A

# Semiconductors and Semimetals

A Treatise

*Edited by R. K. Willardson*
CONSULTING PHYSICIST
SPOKANE, WASHINGTON

*Eicke R. Weber*
DEPARTMENT OF MATERIALS SCIENCE
AND MINERAL ENGINEERING
UNIVERSITY OF CALIFORNIA AT
BERKELEY

# Identification of Defects in Semiconductors

SEMICONDUCTORS
AND SEMIMETALS
Volume 51A

*Volume Editor*
MICHAEL STAVOLA

DEPARTMENT OF PHYSICS
LEHIGH UNIVERSITY
BETHLEHEM, PENNSYLVANIA

ACADEMIC PRESS
San Diego   London   Boston   New York
Sydney   Tokyo   Toronto

This book is printed on acid-free paper.

COPYRIGHT © 1998 BY ACADEMIC PRESS

All rights reserved.
NO PART OF THIS PUBLICATION MAY BE REPRODUCED OR TRANSMITTED IN ANY FORM OR BY ANY MEANS, ELECTRONIC OR MECHANICAL, INCLUDING PHOTOCOPY, RECORDING, OR ANY INFORMATION STORAGE AND RETRIEVAL SYSTEM, WITHOUT PERMISSION IN WRITING FROM THE PUBLISHER.
    The appearance of the code at the bottom of the first page of a chapter in this book indicates the Publisher's consent that copies of the chapter may be made for personal or internal use of specific clients. This consent is given on the condition, however, that the copier pay the stated per-copy fee through the Copyright Clearance Center, Inc. (222 Rosewood Drive, Danvers, Massachusetts 01923), for copying beyond that permitted by Sections 107 or 108 of the U.S. Copyright Law. This consent does not extend to other kinds of copying, such as copying for general distribution, for advertising or promotional purposes, for creating new collective works, or for resale. Copy fees for pre-1998 chapters are as shown on the title pages; if no fee code appears on the title page, the copy fee is the same as for current chapters. 0080-8784/98 $25.00

ACADEMIC PRESS
525 B Street, Suite 1900, San Diego, CA 92101-4495, USA
1300 Boylston Street, Chestnut Hill, Massachusetts 02167, USA

ACADEMIC PRESS LIMITED
24–28 Oval Road, London NW1 7DX, UK
http://www.hbuk.co.uk/ap/

International Standard Book Number: 0-12-752159-3
International Standard Serial Number: 0080-8784

PRINTED IN THE UNITED STATES OF AMERICA
98 99 00 01 02 BB 9 8 7 6 5 4 3 2 1

# Contents

PREFACE . . . . . . . . . . . . . . . . . . . . . . . . . . . . . . . . . ix
LIST OF CONTRIBUTORS . . . . . . . . . . . . . . . . . . . . . . . . . xiii

## Chapter 1  EPR and ENDOR Studies of Defects in Semiconductors
*George D. Watkins*

I. Introduction . . . . . . . . . . . . . . . . . . . . . . . . . . . 1
II. The EPR/ENDOR Experiment . . . . . . . . . . . . . . . . . . . 3
III. Theory of EPR and ENDOR . . . . . . . . . . . . . . . . . . . 4
    1. The g-Tensor . . . . . . . . . . . . . . . . . . . . . . . . . 4
    2. Hyperfine Interactions . . . . . . . . . . . . . . . . . . . . 7
    3. ENDOR . . . . . . . . . . . . . . . . . . . . . . . . . . . 12
    4. Fine-Structure Terms for $S > 1/2$ . . . . . . . . . . . . . 13
    5. Summary . . . . . . . . . . . . . . . . . . . . . . . . . . 15
IV. Additional Examples . . . . . . . . . . . . . . . . . . . . . . 16
    1. Defects in Irradiated Silicon . . . . . . . . . . . . . . . . 16
    2. Intrinsic Defects in Wide-Bandgap Semiconductors . . . . . . 24
    3. Transition Element Impurities . . . . . . . . . . . . . . . . 27
V. Auxiliary Techniques . . . . . . . . . . . . . . . . . . . . . . 28
    1. Applied Uniaxial Stress . . . . . . . . . . . . . . . . . . . 29
    2. Optical Illumination *In Situ* . . . . . . . . . . . . . . . 35
    3. Effect of Temperature . . . . . . . . . . . . . . . . . . . . 37
    4. Defect Production . . . . . . . . . . . . . . . . . . . . . . 39
    References . . . . . . . . . . . . . . . . . . . . . . . . . . . 41

## Chapter 2  Magneto-Optical and Electrical Detection of Paramagnetic Resonance in Semiconductors
*J.-M. Spaeth*

I. Introduction . . . . . . . . . . . . . . . . . . . . . . . . . . . 45
II. Magneto-Optical Detection of EPR and ENDOR . . . . . . . . . . 48
    1. EPR Detected with Magnetic Circular Dichroism of Absorption (MCDA) . 48

2. Examples for MCDA-detected EPR . . . . . . . . . . . . . . . . . . . 53
3. MCDA Excitation Spectra of ODEPR Lines (MCDA Tagged by EPR) . . 61
4. Spatially Resolved MCDA and ODEPR Spectra . . . . . . . . . . . . 66
5. Determination of the Spin State with the MCDA Method . . . . . . . . 68
6. ENDOR Detected with the MCDA Method . . . . . . . . . . . . . 70
III. Electrical Detection of EPR (EDEPR) . . . . . . . . . . . . . . . . . 73
1. Experimental Observaton of EDEPR . . . . . . . . . . . . . . . . 73
2. The Donor–Acceptor Recombination Model . . . . . . . . . . . . . 78
3. Concentration and Temperature Dependence of EDEPR Signals . . . . 82
IV. Electrical Detection of ENDOR (EDENDOR) . . . . . . . . . . . . . 84
V. New Possibilities . . . . . . . . . . . . . . . . . . . . . . . . . . . 86
References . . . . . . . . . . . . . . . . . . . . . . . . . . . . . . 90

## Chapter 3 Magnetic Resonance of Epitaxial Layers Detected by Photoluminescence

### T. A. Kennedy and E. R. Glaser

I. Introduction . . . . . . . . . . . . . . . . . . . . . . . . . . . . . 93
II. Fundamentals of ODMR and Epitaxy . . . . . . . . . . . . . . . . . 94
1. ODMR Detected on Photoluminescence . . . . . . . . . . . . . . 94
2. Epitaxy . . . . . . . . . . . . . . . . . . . . . . . . . . . . . 101
III. Illustrative Example: Bulk InP:Zn . . . . . . . . . . . . . . . . . . 104
1. Introduction . . . . . . . . . . . . . . . . . . . . . . . . . . . 104
2. ODMR of the Effective-Mass Donor . . . . . . . . . . . . . . . 105
3. ODMR of $P_{In}$ . . . . . . . . . . . . . . . . . . . . . . . . . 106
4. ODMR of the Effective-Mass Acceptor . . . . . . . . . . . . . . 110
5. Conclusion . . . . . . . . . . . . . . . . . . . . . . . . . . . 112
IV. Examples in Epilayers . . . . . . . . . . . . . . . . . . . . . . . 113
1. Effective-Mass Donors . . . . . . . . . . . . . . . . . . . . . 113
2. Deep Centers . . . . . . . . . . . . . . . . . . . . . . . . . . 122
3. Acceptors . . . . . . . . . . . . . . . . . . . . . . . . . . . 129
V. Summary and Future Directions . . . . . . . . . . . . . . . . . . . 133
References . . . . . . . . . . . . . . . . . . . . . . . . . . . . . 134

## Chapter 4 µSR on Muonium in Semiconductors and Its Relation to Hydrogen

### K. H. Chow, B. Hitti, and R. F. Kiefl

I. Introduction . . . . . . . . . . . . . . . . . . . . . . . . . . . . 138
II. Fundamentals of $\mu$SR in Semiconductors . . . . . . . . . . . . . . 143
1. Production and Decay of Spin-Polarized Muons . . . . . . . . . . 144
2. Calculation of the Muon Spin Polarization Function $\vec{P}(t)$ . . . . . 145
3. Effective-Field Approximation . . . . . . . . . . . . . . . . . . 152
4. Influence of Nuclear Spins . . . . . . . . . . . . . . . . . . . 153
III. Experimental Techniques and Examples . . . . . . . . . . . . . . . 155
1. Muon Spin Rotation in a Transverse Magnetic Field (TF-$\mu$SR) . . . . 157
2. Muon Level-Crossing Resonance ($\mu$LCR) . . . . . . . . . . . . . 167
3. Zero-Field Muon Spin Relaxation/Rotation (ZF-$\mu$SR) . . . . . . . 176
4. Muon Spin Relaxation in a Longitudinal Magnetic Field (LF-$\mu$SR) . . . . 180

|   |   |
|---|---|
| 5. Muon Spin Resonance in an RF Magnetic Field (RF-$\mu$SR) | 193 |
| 6. Comparison of Techniques and Facilities | 201 |
| IV. Summary | 202 |
| References | 204 |

## Chapter 5  Positron Annihilation Spectroscopy of Defects in Semiconductors
*Kimmo Saarinen, Pekka Hautojärvi, and Catherine Corbel*

|   |   |
|---|---|
| I. Introduction | 210 |
| II. Positrons in Solids | 211 |
| 1. Positrons for Bulk Studies | 212 |
| 2. Positrons for Layer Studies | 212 |
| 3. Positron Diffusion and Mobility | 213 |
| 4. Positron Wave Function and Positron States | 215 |
| 5. Annihilation Characteristics | 216 |
| III. Positron Trapping | 218 |
| 1. Trapping Rate and Trapping Coefficient | 218 |
| 2. Kinetic Trapping Model | 220 |
| IV. Experimental Techniques | 222 |
| 1. Positron Lifetime Spectroscopy | 222 |
| 2. Angular Correlation Spectroscopy | 225 |
| 3. Doppler Broadening Spectroscopy | 227 |
| V. Identification of Vacancies and Their Charge States | 232 |
| 1. Open Volume and Positron Lifetime | 232 |
| 2. Vacancy Charge State: Ga Vacancy in GaAs | 235 |
| 3. Identification of Vacancy Sublattice and Impurity Surroundings | 239 |
| VI. Negative Ions as Shallow Positron Traps | 243 |
| 1. Native Defects in As-grown GaAs as Negative Ions | 244 |
| 2. Negative Vacancies and Negative Ions in Electron-Irradiated GaAs | 246 |
| VII. Defects in Layers Studied by a Low-Energy Positron Beam | 250 |
| 1. Compensating Defects in Highly Si-doped GaAs Layers | 250 |
| 2. Analysis of Depth Profiles of Vacancies in Ion-Implanted GaAs | 254 |
| VIII. Investigation of Vacancy Ionization Levels | 258 |
| 1. Arsenic Vacancy in n-Type GaAs: Thermal Ionization | 258 |
| 2. Arsenic Vacancy in SI GaAs: Optical Transitions | 262 |
| IX. Investigation of the Atomic Structures of Metastable Defects | 268 |
| 1. As-grown SI GaAs: The Midgap Donor EL2 | 268 |
| 2. Metastability of Defects in SI GaAs After Electron Irradiation | 273 |
| 3. As-grown n-Type $Al_xGa_{1-x}As$ Layers: The Deep Donor Level DX | 275 |
| 4. The Atomic Structure of EL2 and DX | 279 |
| X. Summary | 282 |
| References | 282 |

## Chapter 6  The *Ab Initio* Cluster Method and the Dynamics of Defects in Semiconductors
*R. Jones and P. R. Briddon*

|   |   |
|---|---|
| I. Introduction | 288 |
| II. The Many-Body Problem | 290 |

|  |  |
|---|---|
| 1. The Born-Oppenheimer Approximation | 291 |
| 2. Hartree-Fock Theory | 292 |
| 3. The Homogeneous Electron Gas | 295 |
| 4. The Spin-Polarized Electron Gas | 297 |
| 5. Density Functional Theory | 300 |
| III. Pseudopotential Theory | 306 |
| IV. The Real-Space Cluster Method | 310 |
| 1. The Hartree Energy | 312 |
| 2. The Exchange-Correlation Energy | 314 |
| 3. Matrix Formulation | 317 |
| V. Self-Consistency and Atomic Forces | 319 |
| 1. Self-Consistency | 319 |
| 2. Evaluation of Forces | 323 |
| VI. Structural Optimization | 324 |
| 1. Unconstrained Relaxation | 324 |
| 2. Constrained Relaxation | 326 |
| VII. Determination of Vibrational Modes | 327 |
| 1. Energy Second Derivatives and Musgrave-Pople Potentials | 327 |
| 2. Effective Charges | 329 |
| 3. Resonant Modes | 331 |
| VIII. Practical Considerations | 332 |
| 1. Choice of Basis Sets | 332 |
| 2. The Construction of a Suitable Cluster | 334 |
| 3. Mulliken Populations | 336 |
| 4. Radiative Lifetimes | 337 |
| IX. Applications | 337 |
| 1. General | 337 |
| 2. Point Defects in Bulk Solids | 338 |
| 3. Line Defects | 343 |
| X. Summary | 345 |
| References | 346 |

| | |
|---|---|
| Index | 351 |
| Contents of Volumes in this Series | 361 |

# Preface

This is the first of two volumes on the identification of defects in semiconductors. The idea for these volumes originated with a graduate course that has been taught for several years at Lehigh University by George Watkins. In his course, Watkins lectures on the physics of defects in semiconductors and discusses the various characterization methods that are commonly used to reveal defect properties. Throughout the lectures, numerous examples are discussed to illustrate the physical principles and to expose students to the state-of-the-art in defect science. I have had the pleasure of attending this course twice since my arrival at Lehigh.

It was suggested to me by Robert Folk, also a member of the Lehigh Physics Department, that someone should write a book based on Watkins's lectures. This would have been a difficult task, I think even for Watkins, and Folk's idea was put aside. Not long afterwards, near the end of the 1994 Gordon Conference on Defects in Semiconductors, Eicke Weber suggested that I edit a volume for the *Semiconductors and Semimetals* series. A plan was developed to solicit contributions from leading researchers that would cover the various methods of identifying defects in semiconductors, and to assemble a volume that would be loosely patterned after Watkins's course. Authors were asked to discuss recent examples that would illustrate the capabilities of their measurement techniques and that would also provide a survey of many of the successes of defect studies. This volume and its companion are the result of that effort.

The identification of a defect means different things to different people, depending upon one's interests. In the context of these volumes, *identification* means identifying the atoms in the defect, their location in the semiconductor lattice, and their static and dynamic properties. Introduction mechanisms, reactions, diffusion coefficients, and so on, and the microscopic mechanisms for these processes complete the picture. Determining these fine details for the small concentrations of defects that are often of interest in semiconductors is a challenging task and often requires the application of a number of complementary experimental techniques and theory.

Many of the techniques used to identify defects and defect processes are discussed in these volumes. While the coverage is broad, it is not comprehensive, and I apologize to those researchers whose favorite measurement techniques have been omitted. It is hoped that these volumes will be useful to newcomers to the field of defects in semiconductors, and also to active researchers who want to broaden their horizons, as a means to learn the capabilities and limitations of many of the techniques that are used in semiconductor-defect science.

Some of the questions asked in this field are fundamental, while other questions arise from problems in microelectronics technology and are directly related to applications. However, in most cases, it is not easy to separate fundamental issues from technological concerns. Fundamental defect problems that were studied and understood (at least partially) years ago provide an important foundation for understanding technological problems that arise years later. One example is the study of oxygen in Si that was begun in the 1950s, followed by the widespread use of oxygen precipitates as intrinsic gettering sinks in semiconductor processing, beginning in the 1970s. Another recent example is the work that is being done to understand and control the effect of ion implantation on dopant diffusion during Si processing. In this case, ion implantation introduces interstitial defects that enhance dopant diffusion. The fundamental understanding of interstitial defects and defect reactions in Si that has evolved from many years of study can be directly applied to these implanation-related effects that are attracting much current attention.

In Chapter 1 of the first volume, Watkins introduces the powerful electron paramagnetic resonance (EPR) and electron-nuclear double-resonance (ENDOR) techniques. The hyperfine structure and local symmetry determined by magnetic resonance methods provide both chemical and structural information. The variety of examples that are discussed throughout emphasizes the important physics that can be learned.

In Chapter 2, Spaeth discusses EPR and ENDOR detected by magnetic circular dichroism of absorption (MCDA) and also by electrical detection methods. The optical detection methods, photoluminescence (PL) and MCDA, offer selectivity (i.e., the ability to measure the magnetic resonance associated with the optical spectrum associated with a particular defect. And from the magnetic resonance, rich structural information can be pulled out of the broad, overlapping optical bands that are often observed for deep-level defects. Electrical detection offers greatly enhanced sensitivity and the possibility to perform magnetic resonance for very small sample volumes and in device geometries. In Chapter 3, Kennedy and Glaser discuss the application of magnetic resonance spectroscopy to defects in epitaxial layers. In this case, the magnetic resonance is detected by its effect on PL, to obtain the sensitivity required to study defects in thin-layer structures. Kennedy and Glaser emphasize the important role that strain plays in the characterization of epitaxial structures.

In Chapter 4, Chow *et al.* discuss muon spin resonance spectroscopy in semiconductors. An implanted muon acts as a light, quasi-isotope of hydrogen and has provided unique information about the static and dynamic properties of isolated hydrogen in semiconductors, an important species that has been difficult to probe directly. In Chapter 5, Saarinen *et al.* discuss the use of positron annihilation spectroscopy as a probe of defects that contain open volume. The positron lifetime, the angular correlaton of the 511-keV photons emitted upon annihilation, and the Doppler broadening of these emitted photons all provide information about the center where the positron is trapped. Positrons are a particularly good probe of vacancy-like defects and also provide a means to probe metastable or bistable defects when there is a configurational change that leads to increased open volume.

This volume concludes with Chapter 6 on theoretical methods by Jones and Briddon, who discuss the first principles local density cluster method and its applications. Theory plays an increasingly important role in the interpretation of experimental results, predicting defect structures and properties, and in several important cases, explaining phenomena that cannot yet be probed by experiment. The companion to this volume will include chapters on absorption and luminescence spectroscopies, deep-level transient spectroscopy, local vibrational mode spectroscopy, perturbed angular correlation, transmission electron microscopy, and scanning tunneling microscopy.

I am happy with how Robert Folk's suggestion to write a book patterned after Watkins's course has been realized by these volumes and delighted by the quality of the contributions of the individual chapter authors, whom I thank for their efforts. However, I still highly recommend Watkins's lectures, for which there can be no substitute.

MICHAEL STAVOLA

# List of Contributors

Numbers in parenthesis indicate the pages on which the authors' contribution begins.

P. R. BRIDDON (287), *Department of Physics, University of Newcastle Upon Tyne, Newcastle Upon Tyne, NE1 7RU, United Kingdom*

K. H. CHOW (137), *Department of Physics, Lehigh University, Bethlehem, PA 18015*

CATHERINE CORBEL (209), *Institut National des Sciences et Techniques Nucléaires, Centre D'Etudes Nucléaires de Saclay, 91191 Gif-sur-Yvette Cedex, France*

E. R. GLASER (93), *Code 6877, Naval Research Laboratory, Washington, DC 20375*

PEKKA HAUTOJÄRVI (209), *Laboratory of Physics, Helsinki University of Technology, 02150 Espoo, Finland*

B. HITTI (137), *TRIUMF, Vancouver, B.C., Canada V6T 1Z1*

R. JONES (287), *Department of Physics, University of Exeter, Exeter, EX4 4QL, United Kingdom*

T. A. KENNEDY (93), *Code 6877, Naval Research Laboratory, Washington, DC 20375*

R. F. KIEFL (137), *Physics Department, University of British Columbia, Vancouver, B.C., Canada V6T 1Z1*

KIMMO SAARINEN (209), *Laboratory of Physics, Helsinki University of Technology, 02150 Espoo, Finland*

J.-M. SPAETH (45), *Fachbereich Physik, Universität-GH Paderborn, D-33095 Paderborn, Germany*

GEORGE D. WATKINS (1), *Department of Physics, Lehigh University, Bethlehem, PA 18015*

CHAPTER 1

# EPR and ENDOR Studies of Defects in Semiconductors

*George D. Watkins*

DEPARTMENT OF PHYSICS
LEHIGH UNIVERSITY
BETHLEHEM, PA

| | |
|---|---|
| I. INTRODUCTION | 1 |
| II. THE EPR/ENDOR EXPERIMENT | 3 |
| III. THEORY OF EPR AND ENDOR | 4 |
|    1. The g-Tensor | 4 |
|    2. Hyperfine Interactions | 7 |
|    3. ENDOR | 12 |
|    4. Fine-Structure Terms for $S > \frac{1}{2}$ | 13 |
|    5. Summary | 15 |
| IV. ADDITIONAL EXAMPLES | 16 |
|    1. Defects in Irradiated Silicon | 16 |
|    2. Intrinsic Defects in Wide-Bandgap Semiconductors | 24 |
|    3. Transition Element Impurities | 27 |
| V. AUXILIARY TECHNIQUES | 28 |
|    1. Applied Uniaxial Stress | 29 |
|    2. Optical Illumination In Situ | 35 |
|    3. Effect of Temperature | 37 |
|    4. Defect Production | 39 |
|    References | 41 |

## I. Introduction

Electron paramagnetic resonance (EPR) and electron-nuclear double-resonance (ENDOR) can be extremely powerful experimental tools for the study of point defects in semiconductors. Their magnetic resonance spectra, properly interpreted, often contain highly detailed microscopic information about the structure of a defect which cannot be obtained in any other way. In addition, a rich variety of complementary information is often available about a defect when auxiliary techniques are used in conjunction with the studies. Techniques such as applied uniaxial stress, temperature variations, optical illumination *in situ*, and so forth, offer different ways to perturb the defect, which reveal important dynamic and energetic features of the defect not available from the static spectra alone.

Not all defects are paramagnetic, however. In fact, in a semiconducting solid, electrons generally tend to be incorporated into the bonds between the atoms with their electron spins paired off and hence with their magnetic dipoles canceling. Defects with even numbers of electrons are, therefore, often (but not always) nonmagnetic and unobservable by EPR or ENDOR. This is not as serious as it sounds, however, because for semiconductors the defects of interest are usually the *electrically active* ones. This means that the defects can take on more than one charge state in the forbidden gap, usually presenting therefore at least one available charge state that involves an odd number of electrons. There is no way of assembling an odd number of electrons that will totally cancel their spins. Electrically active defects therefore should be paramagnetic in at least one of their charge states and, if everything else is favorable, magnetic resonance should be observable.

In this chapter, briefly outlined first will be the basic concepts of magnetic resonance and some of the principal interaction terms revealed in the EPR and ENDOR spectra. Several examples will be cited to illustrate how each of these show up in the spectra and how, in turn, they can be used to reveal the microscopic structure of a defect. Then some of the various auxiliary techniques used in conjunction with EPR and ENDOR will be enumerated, again using examples to demonstrate the rich variety of additional information that can be obtained. Throughout, I will take the opportunity, where appropriate, to emphasize the important physics that has been learned from the studies.

No attempt will be made to catalog the many semiconductor defects that have been studied by EPR and ENDOR. For this purpose, the interested reader may want to consult the Landolt-Börnstein (1982, 1989) tabulations that are available for many of the semiconductor materials, or browse some of the limited-area reviews (Ludwig and Woodbury, 1962; Lancaster, 1968; Title, 1967; Schneider, 1967, 1985; Watkins, 1975a; Kaufmann and Schneider, 1980a; Wilsey and Kennedy, 1985; Clerjaud, 1985; Ammerlaan *et al.*, 1985, Bourgoin *et. al.*, 1988; etc.).

The treatment in this chapter is elementary. It is not intended for the EPR or ENDOR experts, although even they might enjoy reading about the various powerful auxiliary techniques that have been used that they may or may not be familiar with. Instead it is aimed primarily at the semiconductor physicists or students interested in point defects and what can be learned of interest to them from the magnetic resonance techniques. Perhaps it will also interest them in adding EPR and ENDOR to *their* list of investigative tools. This chapter will serve to get them started. Before long they will want to learn more about the techniques than is included here. For this there are many good texts and reference books (Abragam and Bleaney, 1970; Pake and Estle, 1973; Slichter, 1990; Spaeth *et al.*, 1992; Weil *et al.*, 1994).

## II. The EPR/ENDOR Experiment

In an EPR experiment, the sample is placed in a magnetic field and transitions are induced between the Zeeman levels of the defect. For magnetic fields conveniently available in commercial electromagnets (3–10 kG), the Zeeman energy differences correspond to transitions in the microwave region of the electromagnetic spectrum ($\sim 10$–30 GHz). To maintain high sensitivity, it is common practice to keep the microwave frequency fixed so that the sample can be placed in a tuned resonant cavity where the microwave magnetic field can be highly concentrated. Resonance is then detected by either a loss in the Q of the cavity (absorption) or a slight shift in its resonance frequency (dispersion) when the magnetic field is adjusted so that the Zeeman splitting and the microwave frequency match. For ENDOR, an additional radio-frequency magnetic field is introduced by a coil wrapped around or contained within the cavity, and changes in the intensity of the EPR signal are monitored as the radio frequency is swept through the nuclear resonance frequencies of nearby atoms.

A simplified schematic diagram of an EPR/ENDOR spectrometer is shown in Fig. 1. Here resonance is monitored as a change in the power reflected from the cavity, which can be cooled to cryogenic temperatures (cryostat not shown). Either dispersion or absorption can be studied by adjusting the phase of the reference signal derived directly from the microwave source. To further increase the sensitivity, low-frequency magnetic field modulation is used. This serves to bring the sample in and out of resonance at the modulation frequency, and the corresponding AC signal is amplified, phase-sensitive-detected, and fed to a recorder or computer. For EPR, the magnetic field is swept through the resonance. (The shape of the resulting signal will vary depending upon the rate of spin-lattice relaxation for the defect. If it is fast with respect to the modulation frequency, derivative curves of absorption or dispersion result. If it is slow, absorption-like signals are often observed when monitoring dispersion. In the various figures of spectra to follow, examples of each of these different "passage cases" are present. The reader should not be concerned with these differences. Our interest will be on the spectral information contained, which is independent of the recorded shape of the resonances.) For ENDOR, the magnetic field is tuned to an EPR resonance and the radio frequency is

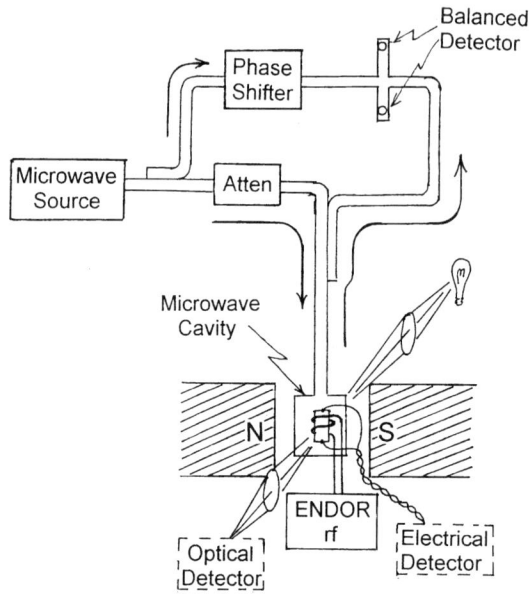

FIG. 1. Simplified block diagram of an EPR spectrometer with ENDOR capabilities. Also shown schematically (dashed) are optical (ODMR) and electrical (EDMR) detection methods.

swept. For study of the angular dependence of a spectrum, the magnet (or sample) is rotated.

It is sometimes possible to detect EPR or ENDOR also via its effect on optical absorption or luminescence (ODMR), or on the electrical properties (EDMR) of the sample. These techniques are also illustrated schematically in Fig. 1. Separate chapters in this volume will be devoted specifically to these important studies, so we will not consider them further here. They are noted, however, because they represent only different *detection* techniques for EPR and ENDOR, still sharing the common spectral information, which will be outlined.

### III. Theory of EPR and ENDOR

1. THE g-TENSOR

The Zeeman interaction of a paramagnetic defect is a magnetic field is given by

$$\mathcal{H} = -\mu \cdot \mathbf{B}, \tag{1}$$

where $\boldsymbol{\mu}$ is the electronic magnetic dipole moment associated with the defect. The magnetic moment will arise from both its electron spin ($\mathbf{S}$) and orbital ($\mathbf{L}$) angular momenta,

$$\boldsymbol{\mu} = -g_S \mu_B \mathbf{S} - g_L \mu_B \mathbf{L}. \tag{2}$$

Here, $\mu_B$ is the Bohr magneton ($eh/2mc$), with $g_S = 2.0023$ and $g_L = 1$.

For many of the defects encountered in solids, the orbital angular momentum is strongly quenched. Viewed physically, electrical charges of nearby atoms cause hills and valleys in the potential for the electrons associated with the defect, preventing their free circulation. For these, the total angular momentum and the magnetic dipole moment originate primarily, therefore, from the electron spin $\mathbf{S}$. However, the spin can still induce a small orbital magnetic moment into its environment (via its spin–orbit interaction $\lambda \mathbf{L} \cdot \mathbf{S}$) in much the same way that an electric dipole induces a dipole moment in a surrounding dielectric. Since this orbital contribution is induced by the spin, its contribution can be included by writing.

$$\boldsymbol{\mu} = -\mu_B \mathbf{S} \cdot \mathbf{g}, \tag{3}$$

where $\mathbf{g}$ is now a symmetric tensor

$$\mathbf{g} = 2.0023 + \Delta \mathbf{g}, \tag{4}$$

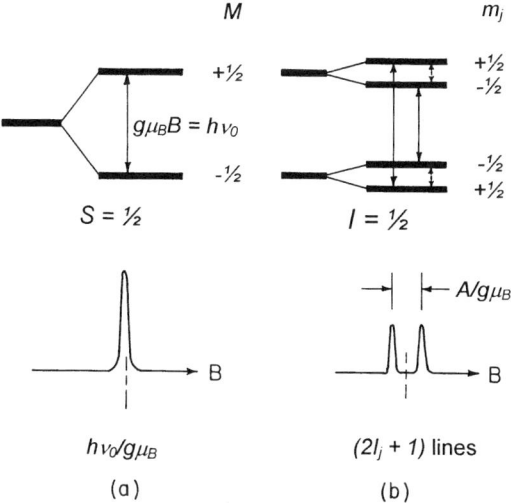

FIG. 2. Energy levels and EPR spectrum: (a) for an unpaired electron, $S = 1/2$; (b) plus hyperfine interaction with a single nucleus, $I_j = 1/2$. The $\Delta m_j = \pm 1$, $\Delta M = 0$ transitions detectable by ENDOR are also indicated.

and $\Delta\mathbf{g}$ reflects the small orbital magnetic contribution induced by the spin. It is a tensor because it is easier to induce the moment in some directions than others, reflecting the symmetry of the defect and its environment.

The Hamiltonian now becomes

$$\mathcal{H} = \mu_B \mathbf{S} \cdot \mathbf{g} \cdot \mathbf{B}, \tag{5}$$

and the levels split into $2S + 1$ equally spaced levels given by

$$E(M) = g\mu_B BM, \tag{6}$$

where the azimuthal spin quantum number $M$ takes on the $2S + 1$ values from $-S$ to $+S$. This is shown in Fig. 2a for the simple case of a single electrons, $S = \frac{1}{2}$. With a fixed microwave frequency $\nu_0$, the magnetic field is swept through "resonance" at

$$B = h\nu_0/g\mu_B. \tag{7}$$

The resonant field, detected and recorded by the EPR spectrometer, therefore serves as a direct measure of $g$, whose angular dependence, from the solution of Eq. (5), is given by

$$g = (g_1^2 n_1^2 + g_2^2 n_2^2 + g_3^2 n_3^2)^{1/2} \tag{8}$$

Here $g_1$, $g_2$, and $g_3$ are the principal values of $\mathbf{g}$, and $n_1$, $n_2$, $n_3$ are the direction cosines of $\mathbf{B}$ with respect to the 1,2,3 principal axes of the $g$-tensor. A study of the resonant field versus magnetic field orientation gives us directly the magnitude and orientation in the crystal lattice of the principal $g$-values. In Fig. 3, we show the characteristic angular rotation patterns ($\mathbf{B} \perp [0\bar{1}1]$) of $g$ for defects of each of the possible reduced-point group symmetries in a cubic ($T_d$) semiconductor lattice. The several branches in each result from the corresponding number of equally probable orientations of the defect in the lattice. The rotational pattern gives us our first bit of microscopic information about the defect, uniquely identifying its symmetry at a glance.

Including $\lambda \mathbf{L} \cdot \mathbf{S}$ in second-order perturbation theory leads to the expression

$$\Delta\mathbf{g} = -2\lambda \sum_n \frac{\langle 0|\mathbf{L}|n\rangle\langle n|\mathbf{L}|0\rangle}{E_n - E_0}, \tag{9}$$

where the sum is overall excited states $n$. With Eq. (9), the magnitude of the $g$-shifts can often also be related to the electronic structure of the defect and its excited states.

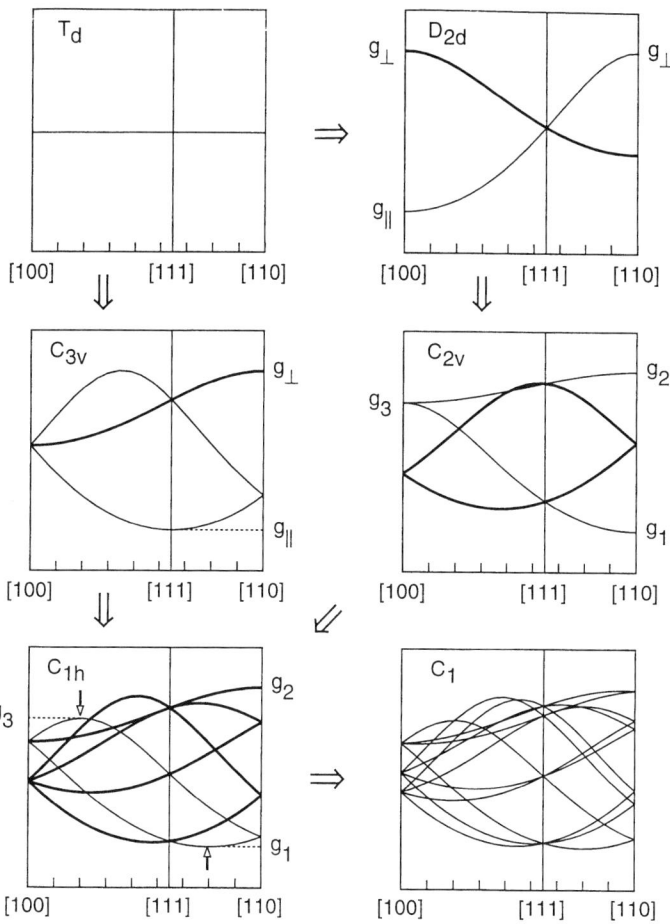

FIG. 3. Characteristic EPR spectrum angular dependences ($\mathbf{B} \perp [0\bar{1}1]$) for defects of sequentially reduced point group symmetry in a cubic semiconductor. The heavy lines arise from defect pairs, which are equivalent for rotation in this plane and are therefore of 2× intensity.

## 2. HYPERFINE INTERACTIONS

Strictly speaking, the magnetic field in Eq. (5) should have been the total field seen by the electron

$$\mathbf{B}_{\text{tot}} = \mathbf{B} + \mathbf{B}_{\text{loc}}, \tag{10}$$

where $B_{loc}$ can arise from weak stray magnetic fields of neighboring atomic nuclei that have magnetic moments associated with their nuclear spins. The local field can be written

$$\mu_B \mathbf{g} \cdot \mathbf{B}_{loc} = \sum_j \mathbf{A}_j \cdot \mathbf{I}_j, \qquad (11)$$

where the sum is over all nearby nuclei $j$ and the tensor $\mathbf{A}_j$ contains the magnitude of the nuclear dipole for each nucleus $j$, and the position and angular dependent parts of the field produced by it at the average position of the paramagnetic electron. The Hamiltonian now becomes

$$\mathcal{H} = \mu_B \mathbf{S} \cdot \mathbf{g} \cdot \mathbf{B} + \sum_j \mathbf{S} \cdot \mathbf{A}_j \cdot \mathbf{I}_j, \qquad (12)$$

which is the compact notation generally found in the literature. The quantum mechanical solution of Eq. (12) gives, for the allowed energy states (first-order perturbation theory),

$$E(M, m_j) \cong \left( g\mu_B B + \sum_j A_j m_j \right) M, \qquad (13)$$

and the EPR resonance condition now becomes ($\Delta M = \pm 1$, $\Delta m_j = 0$)

$$B = \frac{1}{g\mu_B} \left( h\nu_0 - \sum_j A_j m_j \right). \qquad (14)$$

Here $m_j$ is the nuclear azimuthal quantum number and, in analogy to the expression for $g$,

$$A_j = \sum_\alpha (A_{j\alpha}^2 n_{j\alpha}^2), \qquad (15)$$

where the $A_{j\alpha}$ and $n_{j\alpha}$ are, respectively, the principal values of $\mathbf{A}_j$ and the direction cosines of its corresponding principal axes with respect to the direction of $\mathbf{g} \cdot \mathbf{B}$ (the quantization direction for $\mathbf{S}$).

Figure 2b shows the EPR spectrum expected for the case of an $S = \frac{1}{2}$ electron interacting with a single nuclear spin, $I_j = \frac{1}{2}$. The resonance line splits symmetrically into two equally intense lines, separated by $A_j/g\mu_B$ (for a general spin $I_j$, it will split into $2I_j + 1$ lines). Angular dependence study of these splittings leads directly therefore to the determination of the principal values and axes of $A_j$, using Eq. (15).

An example of hyperfine interaction with a single nucleus is given in Fig.

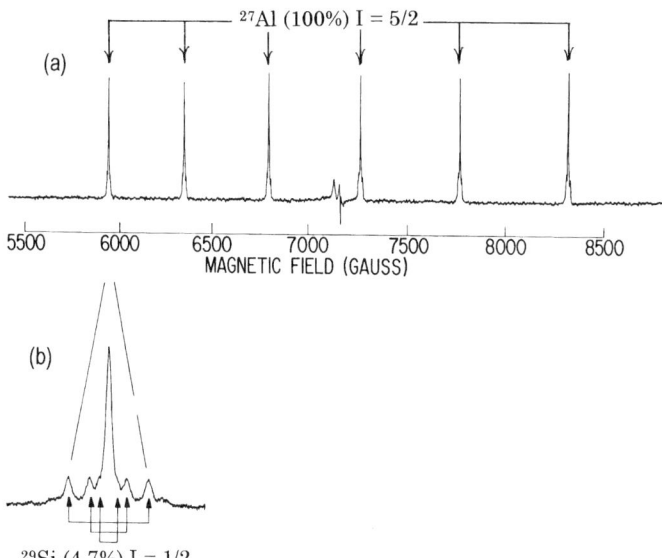

FIG. 4. (a) EPR spectrum of interstitial $Al_i^{2+}$ in silicon at 20.4 K, $v_0 \sim 20$ GHz, after *in situ* electron irradiation at 20.4 K. (b) Satellites, which arise from hyperfine interaction with nearby $^{29}Si$ nuclei, are revealed under higher magnetic field resolution.

4. This is an EPR spectrum produced by high-energy electron irradiation of aluminum-doped silicon (Watkins, 1965; Brower, 1970). The $2I + 1 = 6$ hyperfine components immediately identify it as arising from aluminum ($^{27}Al$, $I = \frac{5}{2}$, 100% abundant). The hyperfine splittings and the $g$-value are isotropic, showing no variation with magnetic field orientation, revealing, in turn, that the aluminum atom is in a high symmetry ($T_d$) position in the lattice.

Note also that each hyperfine line displays weak satellite structure when the magnetic field is expanded. This can be explained by additional weaker hyperfine interactions with neighboring silicon atoms, the interactions occurring with 4.7% abundant $^{29}Si$, $I = \frac{1}{2}$. Satellites can arise each time one of the near-neighbor silicon atoms happens to contain a $^{29}Si$ nucleus.

The additional hyperfine interaction with it causes a further splitting into $2I + 1 = 2$ lines symmetrically disposed around each of the unsplit lines. In Fig. 4, several pairs of satellites can be seen corresponding to interaction with a corresponding number of nonequivalent silicon neighbor sites. The satellite pairs are weak, reflecting the low (4.7%) abundance of $^{29}Si$. Angular dependence studies reveal that the satellite splittings are anisotropic, reflecting the symmetry of the defect wave function as viewed from the lower symmetry site of each nearby neighboring lattice atom. (Because of the identical form for the angular dependence of **g**, Eq. (8), and $\mathbf{A}_j$, Eq. (15), the hyperfine patterns for each symmetry class are similar to those of Fig. 2, supplying again the local symmetry by inspection).

Hyperfine interactions such as these obviously contain a great deal of information about a defect, which can be put on a semiquantitative basis as follows:

The classical dipole–dipole interaction between an electronic and nuclear magnetic dipole $\boldsymbol{\mu}_e$ and $\boldsymbol{\mu}_n$ separated by **r** is

$$\frac{\boldsymbol{\mu}_e \cdot \boldsymbol{\mu}_n}{r^3} - 3\frac{(\boldsymbol{\mu}_e \cdot \mathbf{r})(\mathbf{r} \cdot \boldsymbol{\mu}_n)}{r^5}. \tag{16}$$

To this must be added the "Fermi contact term"

$$-(8\pi/3)\boldsymbol{\mu}_e \cdot \boldsymbol{\mu}_n \delta(r), \tag{17}$$

which is applicable when there is no separation (**r** = 0).

Substituting $\boldsymbol{\mu}_e = -g\mu_B \mathbf{S}$ and $\boldsymbol{\mu}_n = +g_N \mu_N \mathbf{I}$, these can be written in the form

$$\mathbf{S} \cdot \mathbf{A} \cdot \mathbf{I}$$

where the tensor components of **A** are given by

$$A_{ij} = gg_N \mu_B \mu_N \left\{ \left\langle \frac{3x_i x_j}{r^5} - \delta_{ij}\frac{1}{r^3} \right\rangle + \frac{8\pi}{3}|\Psi(0)|^2 \delta_{ij} \right\}. \tag{18}$$

By the angular brackets, we mean an average taken over all of the electron positions in its orbit. This is therefore the expectation value, or matrix element, of the enclosed function in the ground electronic state. The term with $|\Psi(0)|^2$, the amplitude of the wave function at the nucleus, comes directly from the matrix element of the Dirac delta function $\delta(\mathbf{r})$.

Equation (18) gives a direct recipe for calculating the hyperfine interaction tensor once the electronic wave function is known. For an atomic s-function, the interaction with the central nucleus comes only from the $|\Psi(0)|^2$ term, the remaining terms averaging to zero over the spherically symmetric s-state. Therefore, for an s-function, **A** is isotropic with

$$A_{11}(s) = A_{22}(s) = A_{33}(s) = a = (8\pi/3)gg_N \mu_B \mu_N |\Psi(0)|^2. \tag{19}$$

For an atomic p-function, $|\Psi(0)|^2$ is zero. For the function $p_3$, with its lobe directed along the 3-axis, the hyperfine tensor is axially symmetric around the 3-axis with

$$A_{33}(p) = -2A_{11}(p) = -2A_{22}(p) = 2b = (4/5)gg_N\mu_B\mu_N \langle r^{-3} \rangle_p. \quad (20)$$

Tabulated Hartree-Fock wave functions for most atoms are available from which $|\Psi(0)|^2$ and $\langle r^{-3} \rangle_p$ have been estimated for the orbital of interest (Koh and Miller, 1985; Morton and Preston, 1978). In addition, accurate experimental estimates of $\langle r^{-3} \rangle_p$ can be obtained from atomic fine-structure splittings (Barnes and Smith, 1954).

Returning to the example in Fig. 4, the $^{27}$Al hyperfine interaction is isotropic, and we can conclude therefore that it arises from an s-function. The splitting $A/g\mu_B$ of 471 G corresponds to $a/h$ of 1321 MHz, which is ~one-third that expected for the free ion Al$^{2+}$3s function. This suggests that the charge state is Al$^{2+}$ (free ion configuration 1s$^2$2s$^2$2p$^6$3s$^1$), the reduction of $a$ from the free ion value indicating a spreading of the wave function into the surrounding lattice. Consistent with this is the sizable overlap with the neighboring silicon atoms as indicated by the $^{29}$Si hyperfine satellite structure in Fig. 4.

In analyzing such problems, where the electronic wave function overlaps many atom sites, it is often convenient to approximate the wave function for the unpaired electron as a molecular orbital made up of a linear combination of atomic orbitals (LCAO-MO) centered on each of the sites involved,

$$\Psi = \sum_j \eta_j \psi_j. \quad (21)$$

Here $\eta_j^2$ represents the fraction of the total wave function that is localized in the atomic orbital $\psi_j$ centered on each atomic site $j$. At site $j$, $\psi_j$ is further broken down into a linear combination of the various s, p, and so on, valence atomic orbitals on that site,

$$\psi_j = \alpha_j(\psi_s)_j + \beta_j(\psi_p)_j + \cdots. \quad (22)$$

In Eqs. (19) and (20), $|\Psi(0)|^2$ and $\langle r^{-3} \rangle_p$ strongly weight only those parts of the wave function very close to the nucleus. Therefore, the hyperfine interaction at site $j$ results primarily from $\psi_j$, the part of the total wave function that is atomic-like around site $j$. The hyperfine interaction at site $j$ therefore allows a direct estimate of $\eta_j^2$, the fraction of the total wave function at site $j$, as well as $\alpha_j^2$ from the isotropic part of the interaction, and $\beta_j^2$ from the anisotropic part, and so on. Such an analysis for the $^{29}$Si hyperfine satellites of the Al$^{2+}$ spectrum reveals that an additional $\sim 45\%$ of the wave function can be accounted for in terms of partial occupancy of 3s, 3p orbitals on three shells of neighboring silicon sites. From the symmetry of the hyperfine angular dependence, these sites are deduced to be those of the four nearest, the six next-nearest, and the four fourth-nearest neighbors surrounding the high symmetry $T_d$ *interstitial* site. They are not consistent with the $T_d$ substitutional site. In this example, therefore, a highly detailed and unambiguous picture of the defect has been determined — that of interstitial aluminum produced as a product of the electron irradiation.

## 3. ENDOR

Hyperfine interactions with more distant neighbors also exist, of course, due to the more extended parts of a defect wave function. They may not be resolved in the EPR spectrum but can often be studied by ENDOR (electron-nuclear double-resonance). A $\Delta m_j = \pm 1, \Delta M = 0$, nuclear ENDOR resonance transition is also shown in Fig. 2b.

Because of the increase resolution of ENDOR (and the difference in selection rules), the approximations leading to Eq. (14) are no longer accurate enough. To the spin-Hamiltonian of Eq. (12) must be added

$$+ \sum_j (-g_{Nj}\mu_N \mathbf{I}_j \cdot \mathbf{B} + \mathbf{I}_j \cdot \mathbf{Q}_j \cdot \mathbf{I}_j), \tag{23}$$

where the first term is the direct interaction of the nucleus with the external field and the second is the nuclear quadrupole interaction (present if $I_j > \frac{1}{2}$, and the symmetry at the nucleus is lower than $T_d$).

To first order in $A/g\mu_B B$, this leads to an effective Hamiltonian for nucleus $j$ of

$$\mathcal{H}_{Nj} \cong -g_{Nj}\mu_N \mathbf{I}_j \cdot \left\{ \mathbf{B} - \mathbf{A}_j \cdot \frac{\mathbf{g} \cdot \mathbf{B}}{g_{Nj}\mu_N g B} M \right\} + \mathbf{I}_j \cdot \mathbf{Q}_j \cdot \mathbf{I}_j, \tag{24}$$

where the term in curly brackets defines a net magnetic field, $\mathbf{B}_{\text{eff}}$, seen by the nucleus. In the often encountered case $|Q_j/g_{nj}\mu_N B_{\text{eff}}| \ll 1$, Eq. (24) leads for the $\Delta m_j = \pm 1, \Delta M = 0$, ENDOR transitions to

$$h\nu(m_j \to m_j - 1) \cong g_{Nj}\mu_N B_{\text{eff}} \pm \frac{3}{2}(2m_j - 1)\left(\sum_\beta Q_{j\beta}^2 n_{j\beta}^2\right), \tag{25}$$

where $Q_{j\beta}$ and $n_{j\beta}$ are, respectively, the principal values of $\mathbf{Q}_j$, and their direction cosines with respect to the direction of $\mathbf{B}_{\text{eff}}$. The transitions are centered at $g_{nj}\mu_N B_{\text{eff}}$, and their angular dependence provides a precision determination of $\mathbf{A}_j$. For $I_j > \frac{1}{2}$, the second term provides splittings from which $\mathbf{Q}_j$ can be determined, which supplies, in turn, important information relating to the total electronic charge distribution around atom $j$. Another important bonus of ENDOR studies comes from the direct nuclear interaction with the external field. It produces a linear shift in the ENDOR resonance versus the applied field $B$, from which $g_{Nj}$ can be directly determined, providing a conclusive chemical identification of atom $j$, otherwise often ambiguous. Still another important bonus is the $M$-dependence for $\mathbf{B}_{\text{eff}}$ in Eq. (24). There are as many sets of transitions for each nucleus as there are values of $M$, $(2S + 1)$, providing a direct determination of $S$, often not apparent from EPR alone.

ENDOR has been applied to the $\text{Al}_i^{2+}$ defect and has probed several additional shells of silicon neighbors as well as made more precise the near-neighbor interactions (Brower, 1970; Niklas et al., 1985). The model of the defect as *interstitial* $\text{Al}^{2+}$ with $S = \frac{1}{2}$ has been unambiguously confirmed.

## 4. Fine-Structure Terms for $S > \frac{1}{2}$

So far we have considered only terms linear in the spin **S**. This is all that is required when the magnetism arises from a single unpaired electron with $S = \frac{1}{2}$. Many defect problems fall into this class, and for them, Eq. (14) is usually sufficient to completely describe the EPR spectrum. Sometimes, however, a defect has two or more paramagnetic electrons coupled together to give a resultant spin $S > \frac{1}{2}$. For such a defect, it may be necessary to include higher order terms in **S**. These can cause additional structure in the spectrum and are called fine-structure terms.

The most important of these terms is usually the quadratic one, which introduces an additional term into the Hamiltonian of the form

$$\mathbf{S} \cdot \mathbf{D} \cdot \mathbf{S}, \tag{26}$$

if the symmetry of the defect is lower than cubic. If $D \ll g\mu_B B$, its effect is to add an $M^2$-dependent term to the energy levels of Eq. (13)

$$\frac{3}{2} M^2 \sum_\gamma D_\gamma n_\gamma^2, \tag{27}$$

giving for the EPR transitions now

$$B(M \to M - 1) \cong \frac{1}{g\mu_B}\left(h\nu_0 - \sum_\alpha A_\alpha^2 n_\alpha^2 - \frac{3}{2}(2M - 1)\sum_\gamma D_\gamma n_\gamma^2\right). \tag{28}$$

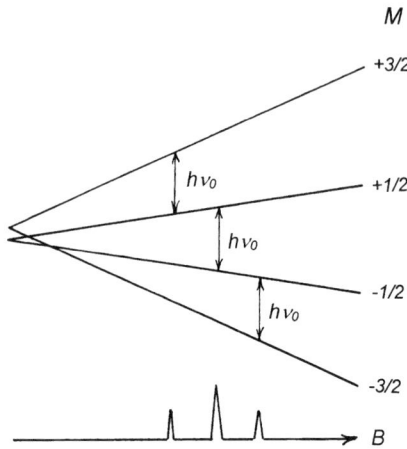

FIG. 5. Energy levels and EPR spectrum for S = 3/2, with a fine-structure term $\mathbf{S} \cdot \mathbf{D} \cdot \mathbf{S}$.

Here, $D_\gamma$ and $n_\gamma$ are, respectively, the principal values of the traceless tensor $\mathbf{D}$ and the direction cosines of its principal axes with respect to $\mathbf{g} \cdot \mathbf{B}$.

The effect of these terms is to destroy the equal energy spacings produced by the $\mu_B \mathbf{S} \cdot \mathbf{g} \cdot \mathbf{B}$ term, producing now $2S$ equally spaced groups of transitions (one for each $M \to M - 1$ transition and its hyperfine structure). This is illustrated in Fig. 5, for $S = \frac{3}{2}$. (The relative intensities of the transitions are given by the binomial coefficients, 1:2:1 for $S = \frac{3}{2}$, 1:3:3:1 for $S = 2$, etc.). Angular dependence studies of the splittings give the principal values and axes of $\mathbf{D}$. Both $\mathbf{D}$ and $\mathbf{g}$ reflect the overall symmetry of the defect and therefore, for $C_{3v}$, $D_{2d}$, and $C_{2v}$ symmetries, will have identical principal axes. However, for $C_{1h}$, they must share only the axis perpendicular to reflection plane, and for $C_1$ symmetry defects, all three axes can in general be different.

$\mathbf{D}$ commonly arises from spin–spin and spin–orbit-induced interactions between the electrons averaged over their spatial distribution. In addition to reflecting very sensitively the symmetry of the defect, it can also often serve therefore as a useful diagnostic measure of either an average separation of the spins, or the relationship to $\mathbf{g}$, which also reflects the role of spin–orbit interactions, as in Eq. (9). In addition, a very important bonus of the interaction, when present, is that simply counting the number of transitions ($2S$) immediately determines the spin $S$.

In cubic symmetry, $D_1 = D_2 = D_3 = 0$. Weaker terms reflecting cubic symmetry and of the form

$$S_x^3 B_x + S_y^3 By + S_z^3 B_z, \; S_x^3 I_x + S_y^3 I_y + S_z^3 I_z, \; S_x^4 + S_y^4 + S_z^4, \ldots \quad (29)$$

can also exist in the Hamiltonian, but it is easy to show that these cause splittings only if $S \geq \frac{3}{2}$ for the first two, $S \geq 2$ for the third, and so on. In ENDOR, the preceding middle term and terms of the form $S^2 I^2$ may also be important (van Wezep and Ammerlaan, 1988).

## 5. Summary

For most defects, orbital angular momentum is quenched in first order by the crystalline electric field. For these, the EPR and ENDOR spectra can be described by the spin-Hamiltonian

$$\mathcal{H} = \mu_B \mathbf{S} \cdot \mathbf{g} \cdot \mathbf{B} + \mathbf{S} \cdot \mathbf{D} \cdot \mathbf{S} + O(S^3 B; S^3 I, S^4, \ldots)$$
$$+ \sum_j (\mathbf{S} \cdot \mathbf{A}_j \cdot \mathbf{I}_j - \mu_N g_{Nj} \mathbf{B} \cdot \mathbf{I}_j + \mathbf{I}_j \cdot \mathbf{Q}_j \cdot \mathbf{I}_j). \tag{30}$$

1. The first term is the electron-spin Zeeman term. The departure of the $g$-tensor from the spin-only value 2.0023 arises from the small orbital angular magnetic moment induced into the surroundings via spin–orbit interaction at the atomic cores.
2. The next two terms are fine-structure terms.
3. The remaining term contains the magnetic hyperfine interaction with neighboring nuclei, plus their direct Zeeman interaction with the applied field and their quadrupole interaction.

The selection rule for EPR transitions is $\Delta M = \pm 1$, $\Delta m_j = 0$. From a study of the angular dependence of the EPR spectrum in a single crystal, $\mathbf{g}$, $\mathbf{D}$, and $\mathbf{A}_j$ can be determined; $\mathbf{g}$ and $\mathbf{D}$ reflect the overall character and symmetry of the defect, and, from the number of fine-structure lines, the spin $S$ can be determined. The $\mathbf{A}_j$ reflects the symmetry and character of the spin-containing electronic wave function at each of the $j$ atom sites in the immediate vicinity of the defect. The intensity of the hyperfine satellites determines the isotopic abundance of the $j$ nucleus involved. With this microscopic information, a detailed model for the defect can often be constructed.

ENDOR involves nuclear transitions $\Delta m_j = \pm 1$, $\Delta M = 0$. Hyperfine interactions with distant nuclei, not resolved by EPR, can often be resolved and studied in ENDOR. The nuclear Zeeman term provides an unambiguous chemical identification of the $j$-th atom involved, and its quadrupole interaction, normally not seen in EPR studies, can also be measured. The quadrupole interaction probes the total electron charge density surrounding the $j$-th atom, providing useful information concerning its bonding arrangement.

Occasionally one finds cases in which the orbital angular momentum is not quenched. This is usually true for rare earth element impurities and also sometimes for the 3d transition element ions. It is also true for shallow hole states in cubic semiconductors. In this case the total angular momentum $J = L + S$ replaces $S$ in the spin-Hamiltonian described previously. The analysis is the same as before, but the microscopic origin of each of the spin-Hamiltonian parameters, $g$, $D$, $A_j$, and so on, must be considered separately.

In the preceding presentation I have given simple first-order treatments for the various terms in the spin-Hamiltonian to demonstrate their general effect upon the spectra and how the appropriate parameters can be extracted experimentally. Of course, in the actual analysis of a spectrum, one either carries the treatments to high enough order in perturbation theory for the accuracy required, or instead diagonalizes the relevant parts of the Hamiltonian. With modern personal computers, the latter approach is routinely used more and more, while simultaneously computer-fitting to the stored experimental data. Even so, the analysis of magnetic resonance data by perturbation theory remains an invaluable means to verify that the computer-generated fits yield parameters that make physical sense.

## IV. Additional Examples

1. DEFECTS IN IRRADIATED SILICON

*a. Annealing of Interstitial Aluminum in Silicon*

Described earlier was the interstitial aluminum spectrum produced by radiation damage in silicon (see Fig. 4). Annealing at 200–250°C produces interesting changes in the spectrum (Watkins, 1965, 1969), which give further insight into the complex processes involved in the radiation-damage behavior of this material. A new spectrum emerges (Fig. 6) that still displays a large six-line hyperfine splitting characteristic of interstitial $Al_i^{2+}$, but each "line" is further split into six closely spaced ones indicating the presence of a nearby second aluminum atom. This weaker "superhyperfine" interaction is anisotropic, reflecting a $\langle 111 \rangle$ axis of the crystal. From this it can be concluded that in this annealing stage, interstitial aluminum has diffused through the lattice until trapped by substitutional aluminum ($Al_s$), not yet converted to interstitial aluminum by the irradiation, to form $Al_i^{2+} \cdot Al_s^-$ pairs. The $\langle 111 \rangle$ axis reflects the direction from the substitutional $Al_s^-$ to the adjacent interstitial site occupied by $Al_i^{2+}$. This defect has also been

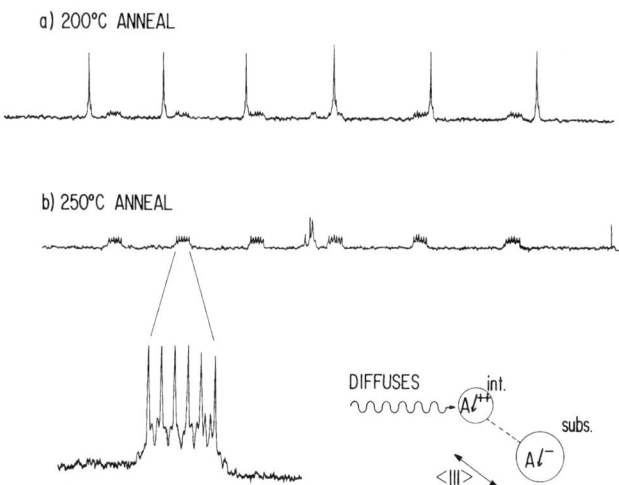

FIG. 6. Loss of interstitial $Al_i^{2+}$ and emergence of $Al_i^{2+} \cdot Al_s^-$ pairs upon annealing.

studied by ENDOR with $^{29}$Si hyperfine interactions determined for several shells around the defect (Niklas et al., 1985).

### b. The Silicon Lattice Vacancy

The spectrum of $Al_i^{2+}$ shown in Fig. 4 was actually taken directly after *in situ* 1.5-MeV electron irradiation at 20.4 K. We note additional weak structure near 7200 G, not associated with the $Al_i^{2+}$ spectrum. This structure can be greatly enhanced by shining light on the sample and is found to be composed of two separate spectra, which have been identified as two different charge states of the isolated lattice vacancy (Watkins, 1963, 1975b, 1986). These are shown with expanded magnetic field scales in Figs. 7 and 8.

The spectrum in Fig. 7a is identified as arising from the isolated lattice vacancy in its singly positive charge state ($V^+$). Its angular dependence is shown in Fig. 7b, which, from the strong central group, reveals the defect symmetry to be $D_{2d}$, (see Fig. 3). The satellites, which arise from hyperfine interaction with $^{29}$Si, split for each central line into the 1:2:1 intensity pattern characteristic of $C_{3v}$ symmetry, revealing hyperfine interaction with four equivalent Si neighbors but each with a different $\langle 111 \rangle$ axis of symmetry. The magnitude of the interaction accounts for most ($\sim 60\%$) of the defect wave function on these four silicon atoms. This has led to the model shown in the inset in Fig. 7, the unpaired electron spread equally over

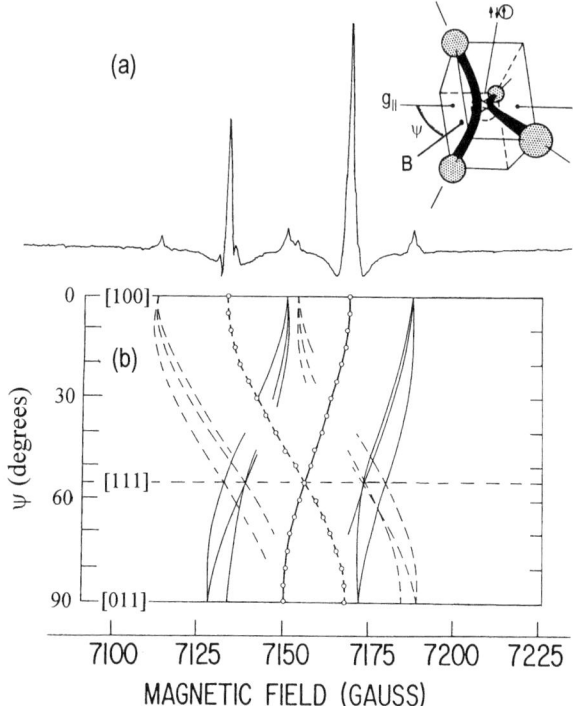

FIG. 7. (a) EPR spectrum of the V$^+$ lattice vacancy in silicon at 4.2 K, $\nu_0 \sim 20$ GHz, $B \parallel [100]$, and the model deduced for its structure. (b) Angular dependence of the spectrum, the dashed lines arising from the specific defect orientation depicted in (a). The spectrum has been produced in p-type material by illumination with $h\nu < 0.35$ eV.

the four silicon neighbors surrounding the vacancy. The pairing by twos shown in the figure is detected in the $D_{2d}$ symmetry of the g-tensor as well as in slight tilting of the $^{29}$Si hyperfine tensor axes. The spectrum is optimized by light with $h\nu < 0.35$ eV. Such light, in this p-type material, produces only free holes. This is consistent with the charge stage identification, the neutral vacancy trapping a hole to become V$^+$.

Bandgap light, on the other hand, can produce free electrons, which when trapped by the neutral vacancy can give V$^-$. The spectrum for V$^-$ generated in this way is shown in Fig. 8a. Again, the $^{29}$Si satellites serve as the main diagnostic element, the defect in this case taking on the interesting configuration as shown, in which the unpaired electron sloshes to one side and is spread between only two atoms. The defect therefore has $C_{2v}$ symmetry, as evidenced in Fig. 8b by the angular dependence of the central lines. The

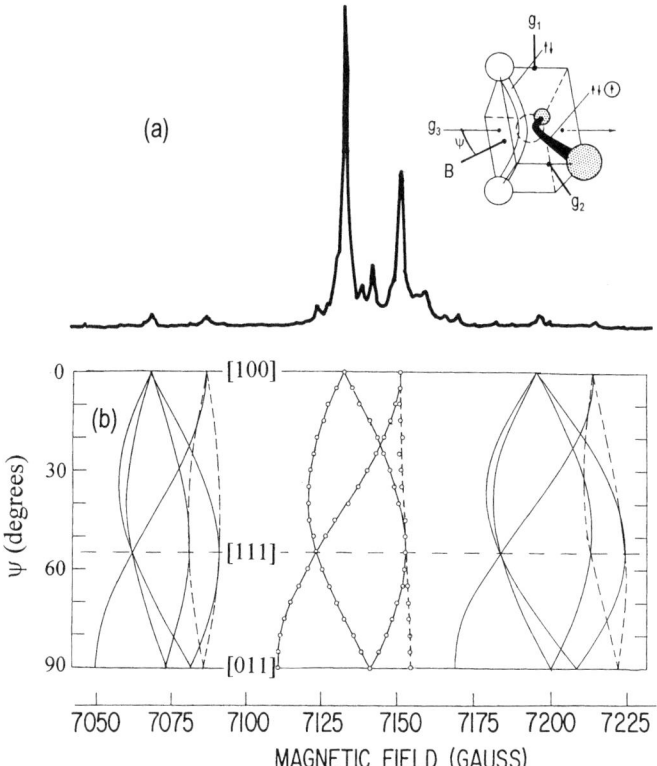

FIG. 8. (a) EPR spectrum of $V^-$ in silicon at 20.4 K, $v_0 \sim 20$ GHz, $\mathbf{B} \parallel [100]$, and the model deduced for its structure. (b) Angular dependence of the spectrum, the dashed lines arising from the specific defect orientation depicted in (a). The spectrum has been generated in p-type material by illumination with $hv > 1.0$ eV.

two-silicon character is revealed directly from the angular dependence of the satellites, which split into two for each central branch. Combining the satellite angular dependence for all branches again reveals very close to $C_{3v}$ symmetry for each of the two silicon hyperfine interactions. [ENDOR measurements on this center (Sprenger *et al.*, 1983) have resolved hyperfine interactions for 27 shells of silicon neighbors, containing 73 atoms! The results fully confirm the assignment to $V^-$.]

The EPR studies have led to a very simple LCAO-MO model for understanding the electronic and lattice structure of the various charge states of the vacancy. This is summarized in Fig. 9. Here, the atomic orbitals are the "broken bonds" on the four (a, b, c, d) atoms surrounding the

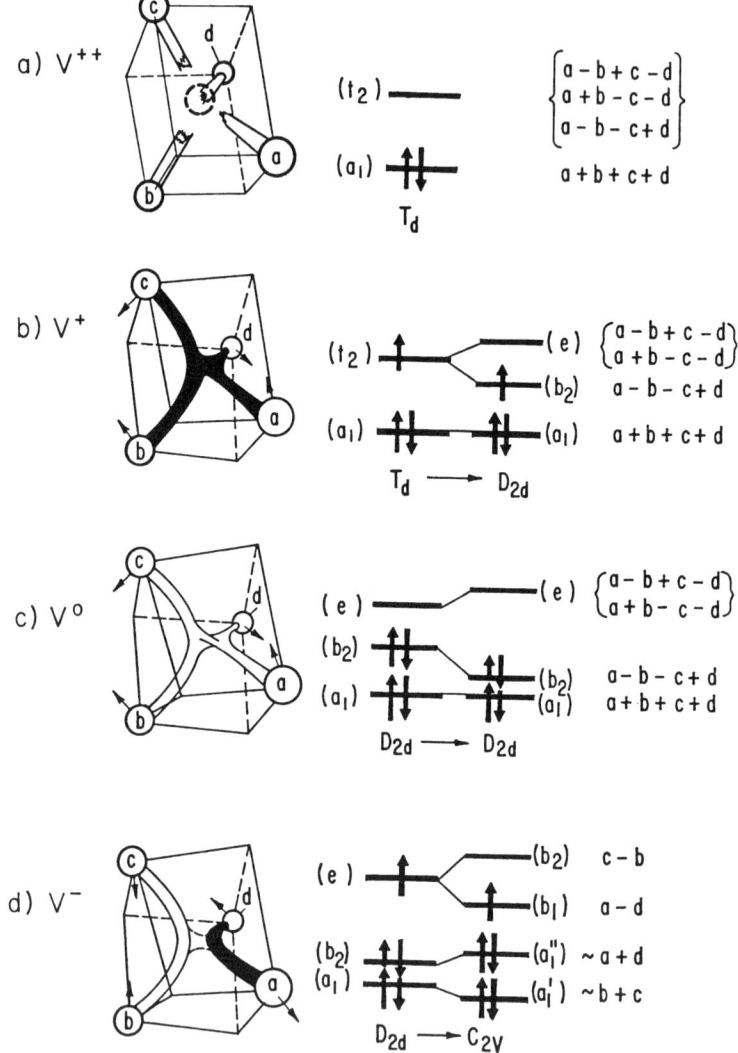

FIG. 9. LCAO model for the various charge states of the vacancy in silicon deduced from the EPR studies.

vacancy. In the symmetry $T_d$ of the undistorted vacancy, a singlet ($a_1$) and triplet ($t_2$) sets of molecular orbitals are formed. For $V^{2+}$, (see Fig. 9a), two electrons go into the $a_1$ orbital paired off, the defect is nonparamagnetic, and no EPR is observed. For $V^+$, the third electron goes into the $t_2$ orbital. Because of the degeneracy associated with the orbital, a Jahn-Teller distortion occurs, as shown, lowering the symmetry to $D_{2d}$. The resulting orbital containing the unpaired electron ($b_2$) is spread equally over the four atoms, explaining the $^{29}$Si hyperfine interactions observed in the V$^+$ spectrum, and the tetragonal distortion accounts for the axial $\langle 100 \rangle$-anisotropy of **g**.

In forming V$^0$, (see Fig. 9c), the next electron also goes into the $b_2$ orbital, paired off, further enhancing the tetragonal Jahn-Teller distortion. The defect is again nonparamagnetic and no resonance is observed.

For V$^-$, (see Fig. 9d), the fifth electron goes into the degenerate e orbital and an additional Jahn-Teller distortion occurs, reducing the symmetry to $C_{2v}$. Atoms b and c pull together, a and d separate slightly, and the unpaired electron ends up on only two of the neighbors, a and d, providing a simple logical explanation for the observed $^{29}$Si hyperfine structure. The success of these simple one-electron models has served to reveal the important physical fact that the electron–electron interactions that normally favor parallel spins (Hund's rule) are small with respect to the relevant crystal field energies in this case. We fill the $a_1$ level before going to the $t_2$. Then, in partially filling the $t_2$ level, a Jahn-Teller distortion occurs, which imposes a strong enough new crystal field to decouple the electrons again. This was not anticipated and was a surprise at the time.

### c. *Defect–Vacancy Complexes in Silicon*

Being able to produce and freeze in isolated lattice vacancies, EPR can in turn be used to monitor the vacancy migration as it is trapped and forms new defect complexes. Figure 10 provides a dramatic example of hyperfine interactions to reveal vacancy trapping by a substitutional Ge impurity in the silicon lattice (Watkins, 1969) after annealing at 200°K. The spectrum shown for (V·Ge)$^-$ is similar to that of V$^-$, but under higher gain the resolved structure reveals hyperfine interaction now with only one $^{29}$Si atom plus the $2I + 1 = 10$ satellites expected from interaction with one 7.6% abundant $^{73}$Ge ($I = \frac{9}{2}$). From the angular dependence studies, the model shown in Fig. 10 can be deduced with the electron shared between the germanium and one of the silicon neighbors. (V·Ge)$^+$ can also be observed giving, in that case, a spectrum similar to V$^+$. For it, one of the four $^{29}$Si hf interactions is replaced by a single one from $^{73}$Ge.

Many other trapped vacancy EPR spectra have been observed and identified. In addition to the V·Ge defects, these include (V·O)$^-$ (Bemski, 1959; Watkins and Corbett, 1961), (V·Al)$^-$ (Watkins, 1967), (V·B)$^0$ (Watkins, 1976), (V·P)$^0$ (Watkins and Corbett, 1964), (V·As)$^0$ and (V·Sb)$^0$ (Elkin and Watkins, 1968), (V·Sn)$^0$ (Watkins, 1975c), and vacancy trapping to form divacancies, (V·V)$^{+,-}$ (Watkins and Corbett, 1965; Corbett and

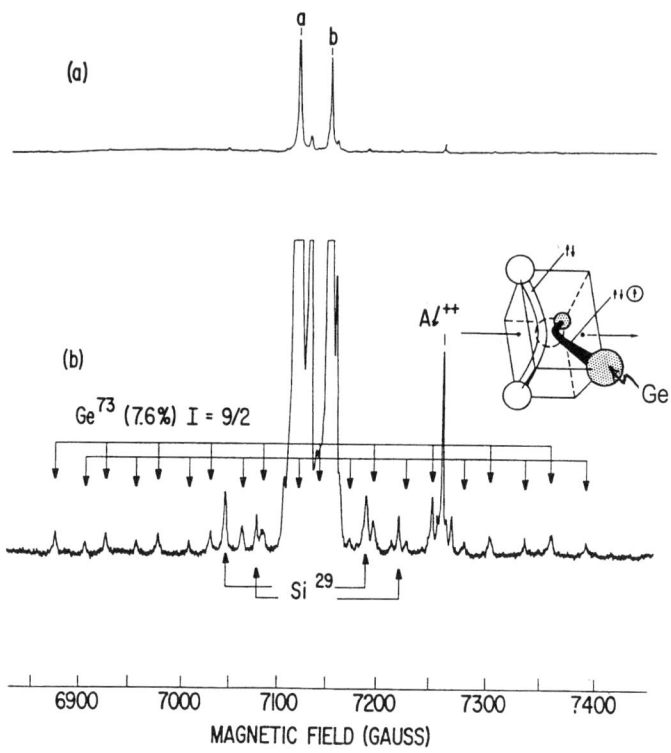

FIG. 10. EPR spectrum of the (V·Ge)⁻ pair in silicon at 4.2 K, **B** ∥ [100], $v_0 \sim 20$ GHz. (a) Gain ×1; (b) gain ×25.

Watkins, 1965). ENDOR has been performed on many of them in these original studies, or subsequently for $(V \cdot O)^-$ (van Kemp et al., 1986), $(V \cdot B)^0$ (Sprenger et al., 1985), $(V \cdot V)^+$ (de Wit et al., 1976), and $(V \cdot V)^-$ (Sieverts et al., 1978), with results in each case consistent with the detailed microscopic models deduced from the EPR studies. In all cases, the structures can again be understood in terms of the isolated vacancy one-electron molecular orbital models as simply perturbed by the presence of the other defect. Again, many electron effects are found to be unimportant.

### d. The Divacancy in Silicon

Figure 11 shows the spectrum and the angular dependence of the strong central lines for the single positive $(V \cdot V)^+$ charge state of the divacancy in silicon. As shown in the model, the $C_{1h}$ symmetry (evident from its angular

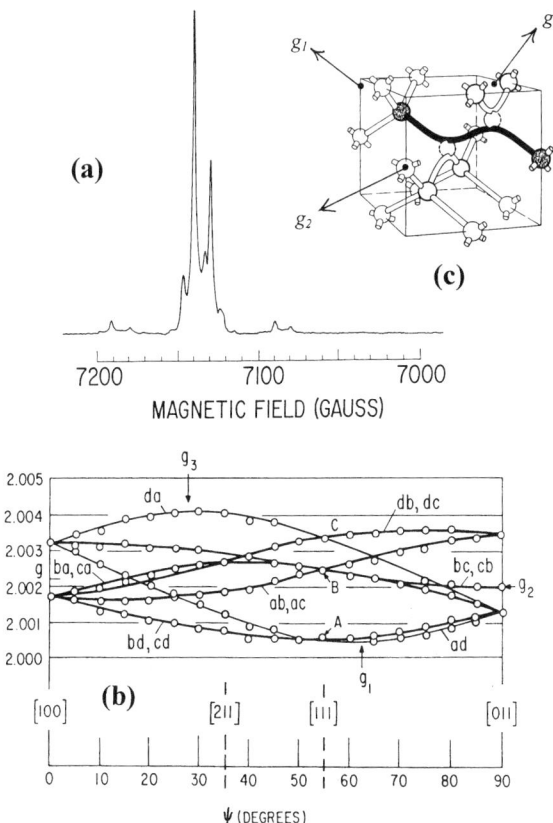

FIG. 11. (a) EPR of the positive charge state of the divacancy $V \cdot V^+$ in silicon at 20.4 K, $B \parallel [100]$, $v_0 \sim 20$ GHz. (b) The $C_{1h}$ angular dependence of the strong central lines. (c) The model.

dependence; see Fig. 3) results as four of the silicon neighbors pull together by pairs to form weak reconstructed bonds, leaving the unpaired electron to divide between the dangling bonds on the two silicon atoms at the opposite extremes of the defect. The $^{29}$Si hyperfine satellites evident in the spectrum result from these two silicon atoms, which remain fully equivalent at all orientations of **B**. A very similar spectrum is observed for the single negative charge state, $(V \cdot V)^-$, for which three electrons (i.e., one $S = \frac{1}{2}$ hole) are accommodated between the two atoms. A simple one-electron molecular orbital model for the divacancy reveals the bond reconstruction by pairs again as that of a Jahn-Teller distortion, necessary to relieve degeneracy in the $C_{3v}$ symmetry of the undistorted $\langle 111 \rangle$ vacancy–vacancy pair. In Section V, where we discuss the variety of interesting new things that can often be learned by the use of auxiliary techniques with EPR, the divacancy will serve as a rich example.

## 2. INTRINSIC DEFECTS IN WIDE-BANDGAP SEMICONDUCTORS

### a. The Gallium Vacancy in GaP

Figure 12 shows the EPR spectrum of the neutral gallium vacancy ($V_{Ga}^0$) in GaP (Kennedy and Wilsey, 1978; Kennedy et al., 1983). Produced by electron irradiation, it is identified by the hyperfine pattern of the four $^{31}$P ($I = \frac{1}{2}$, 100% abundant) neighbors surrounding the vacancy. When $\mathbf{B} \parallel [100]$, all four hyperfine splittings are equal producing the characteristic 1:4:6:4:1-intensity, equally spaced structure shown in Fig. 12. For other orientations of $\mathbf{B}$, the lines partially split, as shown, revealing a different $\langle 111 \rangle$-axis hyperfine anisotropy for each of the neighbors. Analysis of the

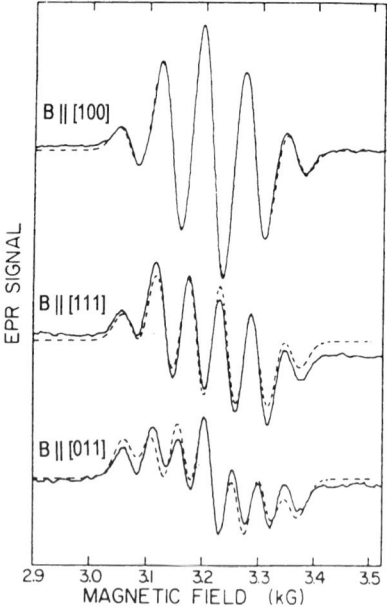

FIG. 12. EPR spectrum of the neutral gallium vacancy in GaP at 77 K, $v_0 \sim 10$ GHz. The dashed lines represent computer simulations of the spectra from which the ligand hyperfine parameters were determined. With permission from T. A. Kennedy and N. D. Wilsey. *Phys. Rev. Lett.* **41**, 977 (1978).

hyperfine interactions reveals ~86% of the wave function on the four nearest phosphorus neighbors, the orbitals on each being mostly p-like (92% p, 8% s). The central field does not change with orientation of **B**, revealing an isotropic $g$-value.

The molecular orbital models for the silicon vacancy (see Fig. 9), are essentially those only of symmetry and should provide an equally valid starting point therefore for $V_{Ga}^0$ in GaP, where the a, b, c, d atomic orbitals are taken as those of the P neighbors. $V_{Ga}^0$ has five electrons to distribute between the $a_1$ and $t_2$ orbitals of the unrelaxed $T_d$ state, and is therefore isoelectronic to $V_{Si}^-$ in silicon. However, unlike $V_{Si}^-$, $V_{Ga}^0$ remain *undistorted*, with full $T_d$ symmetry. A surprise when first observed, we now know that the three electrons in the $t_2$ state have *parallel spins*, $S = \frac{3}{2}$, providing a half-closed shell (i.e., a $^4A_2$ state with no degeneracy). A proof for this has come from ENDOR, from which, in addition to confirming the isolated $V_{Ga}$ identification by probing surrounding neighbor shells, the fact that $S = \frac{3}{2}$ has been directly and unambiguously confirmed (Hage et al., 1986). In Section V.1.e, a different proof will be described, actually the first, involving the application of uniaxial stress to the sample. [The absence of the D fine-structure splittings in $T_d$ symmetry makes the EPR spectrum insensitive to $S$, and what seemed to be a reasonable interpretation as an $S = \frac{1}{2}$ system in the early studies led to much confusion (Kennedy and Wilsey, 1978).]

This was a very important discovery, revealing that in this system the many-electron effects leading to Hund's rules are stronger and dominate in the competition with Jahn-Teller effects. Having gotten used to the reverse situation for the much studied defects in silicon, it has forced a new appraisal for what we might find, as we are only recently turning toward the wider bandgap semiconductors.

### b. *The $P_{Ga}$ Antisite in GaP*

In Fig. 13, we show the $S = \frac{1}{2}$ spectrum of the $P_{Ga}^+$ antisite (Kaufmann et al., 1976; Kaufmann and Schneider, 1980a; Kaufmann and Kennedy, 1981), which is the singly ionized state of a phosphorus atom double donor substituted for Ga in GaP. The identity here is provided by the isotropic hyperfine splitting due to the central phosphorus atom ($^{31}P$, 100%, $I = \frac{1}{2}$). Otherwise, it is very similar to that for the isolated vacancy (see Fig. 12), with superhyperfine structure, reflecting again the 1:4:6:4:1 structure of the four equivalent P neighbors for **B** ∥ [100], as shown. Angular dependence of the structure again reveals a different ⟨111⟩ hyperfine axis for each of the neighbors, and analysis in this case gives ~26% of the wave function in an

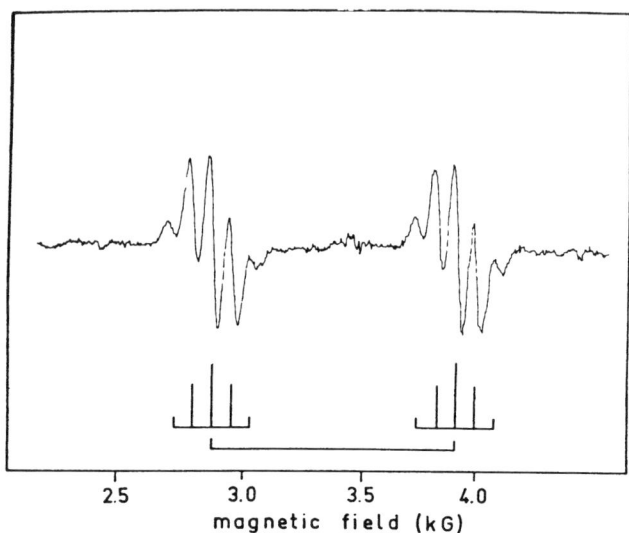

FIG. 13. EPR spectrum of the $P_{Ga}^+$ antisite in GaP at 20 K, $v_0 \sim 10\,\mathrm{GHz}$, $\mathbf{B} \parallel [100]$. Decomposition into the central $^{31}$P hyperfine splitting and that of the four $^{31}$P neighbors is shown. With permission from U. Kaufmann and T. A. Kennedy. *Journal of Electronic Materials* **10**, 347 (1981).

s-orbital on the central atom and 66% on the four neighbors, the neighbor orbitals now having somewhat less p-character (88% p, 12% s).

### c. The Lattice Vacancy in Diamond

EPR of the singly negative charged vacancy (V$^-$) in diamond has also been identified (Isoya *et al.*, 1992). Again, the spectrum is isotropic, revealing undistorted $T_d$ symmetry, and $^{13}$C ($I = \frac{1}{2}$, 1.1% natural abundance) hyperfine interactions with the four nearest and twelve next-nearest carbon neighbors have been observed both by EPR and, in a 10% abundant enriched $^{13}$C diamond, by ENDOR. From the ENDOR study, it has been established unambiguously that $S = \frac{3}{2}$. So again, in this very wide bandgap semiconductor, many-electron effects within the $t_2$ state dominate over Jahn-Teller effects, leading to a ground $^4A_2$ state for the $a_1^2 t_2^3$ configuration of V$^-$. [The EPR spectrum was actually first observed in 1963 (Baldwin), but was assumed at the time and in subsequent studies (Loubser and van Wyk, 1978) to be the $S = \frac{1}{2}$ state of V$^+$, using the silicon vacancy observations as a guide.]

## 3. Transition Element Impurities

The study by EPR of the 3d transition element impurities in semiconductors has proven to be a very active and productive field, involving many research groups. There are several reasons. The impurities have reasonable solubilities; are electrically active, affecting both electrical and optical properties; and their paramagnetic partially filled inner d-shells generally provide a surefire EPR probe for the study. Here I will only briefly summarize a few of the important general features that have been established.

In the compound semiconductors (II–VI, III–V), the transition element impurity primarily substitutes on a metal atom site. Only a few cases of interstitial incorporation have been documented (van Gisbergen *et al.*, 1991). Few surprises have been found, in that the neutral defect d-configuration has almost universally been found to correspond to that of the 2+ or 3+ free ion when substituting for the II or III metal ion, respectively, as one might expect from an ionic picture, and, for the other charge states, the added or removed electrons generally come from the inner d-shell. Hund's rule applies for the occupancy of the d-orbitals, which are split into e and $t_2$ orbitals as expected by the $T_d$ crystal field. In the cases in which orbital degeneracy remains, Jahn-Teller effects occur, removing the degeneracy. A good example is afforded by chromium, which in ZnSe (Baranowski, 1986) has been observed to follow this recipe accurately as $d^3$ in its ionized donor state $[Cr]^+$ (Rai *et al.*, 1969)), $d^4$ in its neutral state $[Cr]^0$ (Title, 1964; Vallin and Watkins, 1974), and $d^5$ in its ionized acceptor state $[Cr]^-$ (Röppischer *et al.*, 1975). Conversely, in GaAs (Allen, 1986), $[Cr]^+$ is $d^2$ (Strauss *et al.*, 1980; Kaufmann and Schneider, 1980b; Krebs and Strauss, 1982), $[Cr]^0$ is $d^3$ (Krebs and Stauss, 1977a), and $[Cr]^-$ is $d^4$ (Krebs and Stauss, 1977b). [There is one interesting exception so far to this simple recipe, and that is for neutral manganese in GaAs. Instead of being $d^4$, EPR has revealed it to be $d^5$ exchange coupled to a shallow bound hole (Schneider *et al.*, 1987).]

Following this simple ionic picture, it is not so obvious what to expect for the configuration of the ion in the completely covalent group-IV elemental semiconductors. Actually, as far as the substitutional impurities are concerned, it turns out that if one blindly follows this trend and considers the neutral impurity configuration to be that of its 4+ free ion, you get the right result, at least for the so far observed cases. In silicon, for example, $[Cr]^0$ and $[Mn]^+$ have been established to be $d^2$, and $[Mn]^{2-}$ to be $d^5$ (Ludwig and Woodbury, 1962), and $[Ni]^-$ to be $d^7$ (Vlasenko *et al.*, 1990; Son *et al.*, 1991). For the first three, the Hund's rule filling of the $T_d$ crystal field split d-levels also applies, but in the case of $[Ni]^-$, an orthorhombic Jahn-Teller distortion overcomes Hund's rules, giving an $S = \frac{1}{2}$ defect, in a manner very similar to that observed for the isolated vacancy. This and similar observations for the 4d$^7$ and 5d$^7$ counterparts, [Pd]$^-$, and [Pt]$^-$, has led to a "vacancy" model for substitutional transition elements near the end of each series, the t$_2^3$ gap orbitals for the impurities in silicon being primarily on the four silicon neighbors, with little central ion d-character, leading to greatly reduced many-electron effects as for the vacancy (Watkins, 1983; Watkins and Williams, 1995).

Clearly, there is something unsettling, however, about this ionic picture. The assignment, for example, of 1+ for the metal in an alkali halide seems clearly reasonable, but we know that in going to the II–VIs, to the III–Vs, to the group IVs, the materials are becoming progressively less ionic, not more. The inconsistency of the ionic picture only becomes really apparent when we make the last move from the III–Vs to the IVs. It was the pioneering EPR studies of Ludwig and Woodbury (1962) in silicon that set the physics straight. They argued that the missing four electrons were actually involved in the covalent bonds of the impurity to its neighbors. If we work backward from this starting point, we need three electrons to complete the bonds at the III atom site in a III–V semiconductor, two in a II–VI, and so on. This is a much more physically reasonable way to look at it, particularly for the substantially covalent tetrahedrally bonded semiconductors, and one that leads to the correct configuration throughout.

Another major difference in silicon discovered by the EPR studies (Ludwig and Woodbury, 1962) is that, except for the elements at the very end of each series, the ions prefer to be incorporated as interstitials. Several charge states of the various interstitial transition element ions have been observed and they fall again into a simple pattern. For them, their valence electrons are found in the 3d-shell, as might be expected, because the ions are not bonded. For example, interstitial manganese has been observed in four charge states: Mn$^{2+}$(d$^5$), Mn$^+$(d$^6$), Mn$^0$(d$^7$), and Mn$^-$(d$^8$). For these and all other interstitials, Hund's rule again applies for the occupancy of the d-orbitals, which are split by a T$_d$ crystal field of opposite sign to that for the substitutionals.

Finally, the transition elements have proven to be a powerful probe for studying defect pair and complex formation in semiconductors (Ludwig and Woodbury, 1962; van Kooten *et al.*, 1984; Kreissl *et al.*, 1992; Gehlhoff *et al.*, 1993; Irmscher *et al.*, 1994).

## V. Auxiliary Techniques

So far, we have considered only the information that can be obtained from analysis of the EPR (or ENDOR) spectrum itself. We have shown, by way of selected examples, that this information is often sufficient to identify the defect and, at the same time, reveal considerable microscopic detail about its structure.

However, the power of EPR (or ENDOR) can often be greatly enhanced by the use of auxiliary techniques in conjunction with them. In this section, some of the techniques are briefly discussed. To illustrate the wealth of additional information obtainable, I will describe their use on the defects described in the previous sections.

1. APPLIED UNIAXIAL STRESS

Perhaps the most powerful single addition to an EPR spectrometer is the facility to apply uniaxial stress to the sample *in situ*. Some of the properties that can be studied using stress are as follows.

*a. Jahn-Teller Alignment*

When a defect undergoes a Jahn-Teller distortion, there are always several equivalent such distortions, each of which produces a different equivalent orientation of the distorted defect in the lattice. Consider the divacancy in silicon (Watkins and Corbett, 1965; Corbett and Watkins, 1965) as an example. The distortion illustrated in Fig. 11 is just one of three possible distortions that could result for the particular $\langle 111 \rangle$-oriented vacancy–vacancy pair in the figure. Any two pairs of the six silicon atoms neighboring the two vacancies could have pulled together, leaving the unpaired electron on the remaining pair. The three distortions are equally probable and equally energetic, and, as a result, the EPR spectrum reveals all three, equally intense.

The application of uniaxial stress will destroy the equivalence of the different Jahn-Teller orientations. For some orientations, the stress will aid the distortion, further lowering the energy of the defect. For others, the stress will oppose the distortion, raising the energy. If the defects can reorient at the temperature of the experiment, they will do so, and this can be observed directly by a change in the relative intensities of the corresponding lines in the EPR spectrum. This is illustrated for $(V \cdot V)^+$ in Fig. 14.

Such an experiment immediately reveals several important bits of information about the defect. First, it demonstrates convincingly that the lowered symmetry ($C_{1h}$) of the defect is indeed the result of a Jahn-Teller distortion: The fact that this reorientation can occur at such a low temperature (30 K) argues against the alternative possibility that the reduced symmetry could have been due to a nearby defect locked in the lattice. This could not have been ruled out from the static spectrum alone. Second, it confirms the *sign* of the distortion (i.e., that the two pairs of silicon atoms *pull together*), as

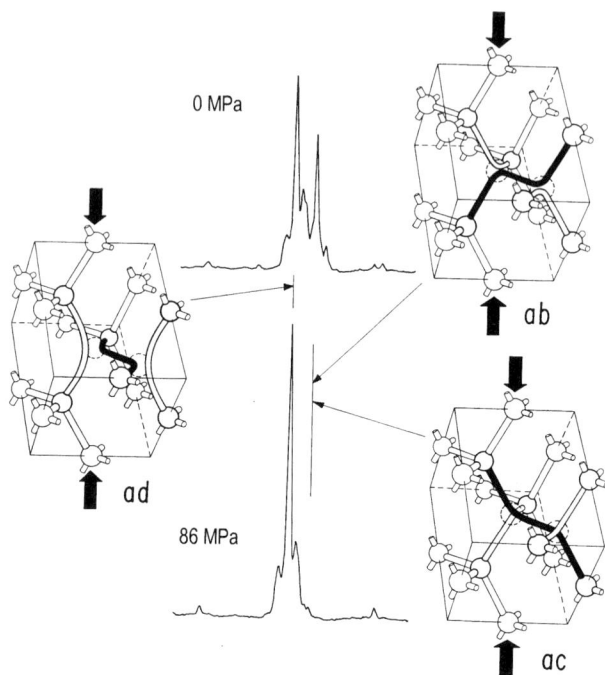

FIG. 14. Jahn-Teller alignment of $(V \cdot V)^+$ in silicon produced by $[0\bar{1}1]$ compressional stress at 30°K, $\mathbf{B} \parallel [011]$. The insets show a typical defect orientation for each line in the spectrum.

shown by the increase in intensity of the orientation labeled *ad* in Fig. 14. Third, monitoring the recovery rate versus temperature allows a direct measurement of the kinetics of the reorientation process. The results of such measurements for the two charge states of the divacancy are shown in Fig. 15 for the range $T \leqslant 30$ K, where the transient recovery could be conveniently monitored over the range from a few seconds to $\sim 24$ hours. (The other points shown in Fig. 15 for much shorter time constants have been determined by a different technique, which will be described later.)

The degree of alignment per unit stress, measured accurately from the amplitudes of the spectral components, also gives a direct *quantitative* measure of the coupling of the defect to the stress field. For this, the difference in energy $E_1 - E_2$ for two defect orientations is reflected in their relative amplitudes, $n_1$ and $n_2$, via a Boltzman distribution function

$$n_1/n_2 = \exp[-(E_1 - E_2)/kT]. \tag{31}$$

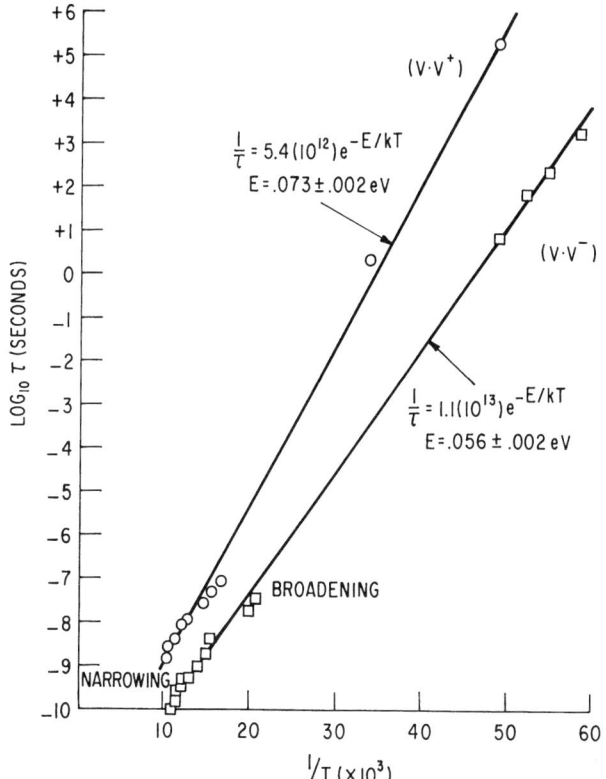

FIG. 15. Relaxation time for the Jahn-Teller reorientation of the two charge states of the silicon divacancy versus temperature, as determined from stress alignment and linewidth studies.

By applying stress in different crystallographic directions, a complete characterization of the coupling between the defect and applied stress can be achieved. Such studies, in turn, allow an approximate estimate of the magnitude of the Jahn-Teller distortion, a quantity not easily available by any other technique. For the V·V pair, the Jahn-Teller relaxation energy was estimated in this way to be ~1.3 eV for $(V \cdot V)^+$, and ~2.4 eV for $(V \cdot V)^-$.

Similar studies have also been performed on the isolated vacancy in silicon as well as for most of the impurity–vacancy pairs. Again, the Jahn-Teller energy estimates range from ~0.4 to ~>2.0 eV. In the case of the vacancy, there is an independent estimate available from the effect of the distortions on the vacancy energy level positions in the gap (Baraff *et al.*, 1980), which suggests that the estimates from the stress analysis may be a factor of ~2 too large. Still, even with this correction, these remain large Jahn-Teller energies, which can have serious consequences for the properties of the defects. Stress studies with EPR provide a unique way of probing these quantities directly.

### b. *Defect Alignment*

At higher temperatures, stress may also induce alignment that involves atomic rearrangements in the lattice. For example, for the divacancy, a preferential alignment of its vacancy–vacancy $\langle 111 \rangle$-axis has been observed directly in the EPR spectrum by stressing at 160°C and then cooling down to cryogenic temperatures for the EPR measurement. As shown in Fig. 16, the changes in the intensities of the lines reveal a substantial alignment of the defect frozen in by this process.

Again, monitoring the decay of the EPR alignment versus subsequent annealing allows a direct study of the kinetics of the reorientation process, in this case involving *atomic* rearrangements. This is particularly interesting

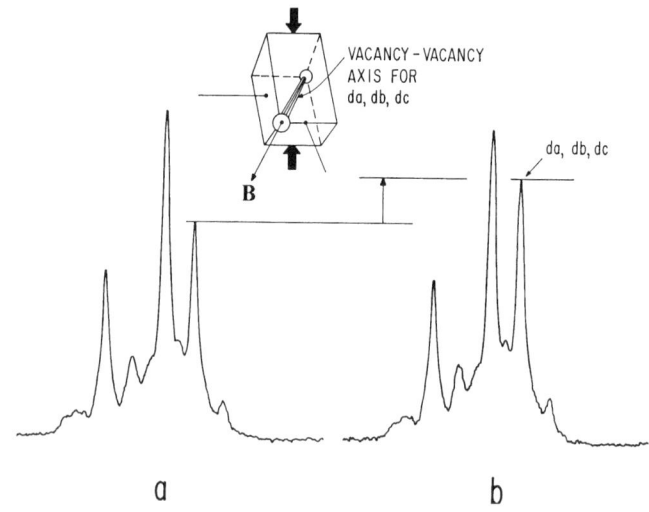

FIG. 16. Alignment of the vacancy–vacancy axis observed in the $(V \cdot V)^+$ spectrum, $\mathbf{B} \parallel [111]$, (a) before and (b) after 183 MPa $[0\bar{1}1]$ compressional stress at 160°C. As shown, the transition to the right (lower field) of each spectrum arises from the defects with their axis along $\mathbf{B}$.

for the divacancy because reorientation necessarily involves displacement of the defect from one pair of atom sites to another. For it, therefore, the activation energy for reorientation (estimated from the EPR study to be $\sim 1.3\,\text{eV}$) must also be the activation energy for *migration*, an important property measured unambiguously and for the first time by this simple procedure. For the group V atom–vacancy pairs, on the other hand, whose stress-induced alignments have also been studied by this technique, migration through the lattice requires both reorientation and substitutional impurity atom–vacancy interchange. In this case, additional information from annealing studies (via EPR or other techniques) could be combined to unravel the separate processes (Elkin and Watkins, 1968).

*c. Correlation with Optical Dichroism*

Optical transitions associated with an anisotropic point defect often have well-defined polarization properties with respect to the defect axes. As a result, stress-induced alignment of the defect can produce a difference in the optical absorption or luminescence intensity with the electric vector of light parallel and perpendicular to the stress direction (dichroism). By combining such studies with stress-induced alignment studies observed directly in EPR, a direct identification of optical transitions associated with a specific defect is possible.

Again the divacancy provides a good example. Optical absorption bands at 3.3, 1.8, and 3.9 μm have been directly assigned to electronic excitations of different charge states of the divacancy using this technique (Cheng *et al.*, 1966). By correlation of the magnitude and sign of the dichroism to the corresponding EPR results for both the low-temperature Jahn-Teller and the higher temperature divacancy axis alignments, and for stress along several crystal directions, an unambiguous assignment could be made, and the orientation of the transition dipole moment with respect to the defect axes determined for each of the bands. With this information, tentative assignments for the transitions could be made within the molecular orbital models for the divacancy.

*d. Electrical-Level Determination*

In general, the electrical-level position of a defect in the bandgap will shift under applied stress. For an anisotropic defect, the shift will be different depending upon the orientation of the defect with respect to the stress, some going up, others down. If the defect level is only partially occupied (Fermi level locked to the defect level), and if, at the temperature of observation, emission of an electron (or hole) to the band proceeds at a measurable rate but the defect itself is frozen in with no reorientation occurring, apparent alignment may still be observed in the EPR. This results because the paramagnetic electrons (or holes) redistribute among the defects, seeking out the defects with the lower (or higher) energy. For such a case, the rate of the process is a direct measure of the electron (or hole) emission rate, providing, in turn, a direct measure of the energy-level position. In effect, this can be considered the ideal "deep-level transient spectroscopy (DLTS) experiment", because it measures the emission rate in neutral material (no Poole-Frenkel effect, which is a problem in the capacitance measurements) and directly on an identified defect. This potentially powerful technique has seen only limited use so far, a successful example being for the oxygen–vacancy pair in silicon (Watkins and Corbett, 1961), which has an acceptor level 0.16 eV below the conduction band. It is potentially useful for a wide range of defects, however, so long as they are shallow enough so that emission to the nearest band is occurring on a convenient time scale at a temperature at which the EPR can be measured.

In addition, the changes in the amplitudes of the lines in this type of experiment can again, via Eq. (31), be used to give a detailed quantitative description of the interaction of the electrical level position and the applied stress tensor. Stress-induced-level position shifts have also been detected directly by electrical measurement techniques using conventional DLTS (Meese *et al.*, 1983; Benton *et al.*, 1985), supplying yet another direct way of correlating a defect studied by EPR with its electrical-level structure.

*e. Miscellaneous*

As described earlier, a defect with $S > \frac{1}{2}$ experiences no axial fine-structure splittings in cubic ($T_d$) symmetry. It is often difficult, therefore, to distinguish the spin of a center in cubic symmetry by EPR. Uniaxial stress, however, in lowering the symmetry, can produce splittings if $S > \frac{1}{2}$, and by counting the number of components, the spin can be determined. This information is often critical in determining the electronic structure of the defect, its charge state, or even its identity. A good example here is the gallium vacancy in GaP. It was not until uniaxial stress was applied (Kennedy *et al.*, 1983) that it was realized that $S = \frac{3}{2}$ (not $\frac{1}{2}$ as originally presumed). As discussed in Section IV.2.a, this information was essential in unraveling the electronic structure of the defect.

As mentioned earlier, orbital angular momentum is not always quenched for a defect. In these cases, uniaxial stress can be used to lift this orbital degeneracy. An important example is the shallow acceptor states in cubic semiconductors, which retain the orbital degeneracy of the top of the valence band and are therefore usually so broadened by random internal strains that they are unobservable in EPR. Uniaxial stress has been successful in some cases in overcoming the internal strains, allowing direct observation and study by EPR (Feher *et al.*, 1960; Mehran *et al.*, 1972).

2. OPTICAL ILLUMINATION *IN SITU*

The ability to introduce light into the microwave cavity *in situ* opens up many additional experimental opportunities. Excited states of defects may be generated for study; new EPR centers can be produced, and others removed, as charge states are changed by photoconductive processes; and, defect alignment can be produced with polarized light.

*a. Excited States*

The vacancy in diamond provides an interesting example of a photo-induced excited state with lifetime long enough for EPR and ENDOR study. The ground state of the *neutral* defect is diamagnetic, $S = 0$, and gives no EPR. However, illumination with ultraviolet light produces a new $S = 2$ EPR center that has been identified as the excited $^5A_2$ ($a_1t_2^3$) state of the neutral defect (van Wyk *et al.*, 1995). For this center, $^{13}C$ hyperfine interactions have been determined for the four nearest and twelve next nearest neighbors from both EPR and ENDOR studies, confirming the identification. The important fact was that $S = 2$ was not evident from the EPR, but required the ENDOR study. Although a great deal of information concerning the ground and optically accessible excited states of the neutral vacancy have been extracted from study of its GR1 optical band (Davies and Foy, 1980; Davies, 1982), this represents the first experimental information on this important Hund's rule excited state.

*b. Defect Charge-State Changes*

At cryogenic temperatures, where EPR is often studied, electronic equilibrium times in semiconductors can be extremely long. As a result, different charge states of defects can be produced and studied by selective generation and bleaching with monochromatic light. From the wavelength dependence of these processes, the electrical-level positions of defects can be estimated and their role in complex photoconductive processes unraveled.

Many examples exist in the literature. A convenient example is the case of the $P_{Ga}$ antisite in GaP. By monitoring the generation of its $P_{Ga}^+$ EPR signal versus photon wavelength in p-type material,

$$P_{Ga}^{2+} + h\nu \rightarrow P_{Ga}^+ + h^+, \qquad (32)$$

Kaufmann and Schneider (1980a) have estimated the second donor level $(+/++)$ to be located at $E_V + 1.1\,\text{eV}$.

In this example, the desired charge state was being optically generated because the Fermi level was in the wrong position. A different case is when the desired charge state is unstable, independent of Fermi-level position. An example is the isolated vacancy in silicon, which has been shown to have negative-U properties, with regions of Fermi-level position where $V^0$ or $V^{2+}$ are stable, but none for $V^+$. The optical generation of the $V^+$ signal, described in section IV.1.b, was actually producing a *metastable* charge state, not accessible in any other way. It was the EPR studies of these optically and thermally induced charge state changes that made it possible to establish the negative-U properties of the vacancy (Watkins, 1984), as earlier predicted by theory (Baraff *et al.*, 1980).

### c. *Optical Alignment*

Illumination with *polarized light* into an absorption band characteristic of a defect can often produce alignment of the defect that can be directly monitored in the EPR. This can result, for instance, if easy reorientation occurs in the excited state to which the defect is carried in the optical transition. The net effect in this case is to bleach orientations that couple most strongly to the polarized light, with a corresponding increase in the population of the others. Two examples of this can be found in studies of the divacancy in silicon (Ammerlaan and Watkins, 1972; van der Linde and Ammerlaan, 1979). Alignment can also occur via selective photogeneration or annihilation of the EPR active charge state. An example of this is available for the $V \cdot O$ pair in silicon (Lee *et al.*, 1976). In either case, such experiments unambiguously identify the optical bands associated with the defect because alignment can result only if the transition is a local one occurring at the defect site. From the sense of the alignment, the selection rules for the transition can also be determined directly.

1  EPR AND ENDOR STUDIES OF DEFECTS IN SEMICONDUCTORS        37

3. EFFECT OF TEMPERATURE

*a. Annealing*

Complex annealing processes in solids involving long-range diffusion of defects can sometimes be unraveled unambiguously by EPR studies, because the defects involved can often be followed throughout the whole process, in both the initial and the final configurations. An example of this has been given for the conversion of interstitial $Al_i^{2+}$ to $Al_i^{2+} \cdot Al_s^-$ pairs (see Fig. 6). Another important example is the isolated lattice vacancy in silicon, with its subsequent migration to form divacancies, and impurity–vacancy pairs. An example of this is the trapping by a germanium impurity (see Fig. 10). In these studies, accurate kinetic studies have allowed the determination of the activation energy for vacancy migration in silicon and its dependence upon charge state (Watkins, 1975c, 1986), important quantities unavailable by other techniques.

For annealing studies, changes in the EPR spectrum are usually monitored at a fixed cryogenic temperature at periodic intervals between isochronal or isothermal higher temperature anneals. There are, however, other interesting effects that must be observed at temperature. These involve changes in the character of the spectrum itself.

*b. Linewidth*

An example of linewidth changes versus temperature is illustrated in Fig. 17, for the positive charge state of the divacancy in the temperature region 60–150 K (Watkins and Corbett, 1965). Some of the lines broaden and disappear as the temperature is raised, and at higher temperatures, a simplified spectrum emerges.

What is being observed is the effect of thermally activated reorientation from one Jahn-Teller distortion direction to another. At these temperatures, the rate becomes so fast that it begins to exceed the natural width of the lines themselves (measured in frequency), and the lines become lifetime-broadened directly by the hopping process. At higher temperatures, a phenomenon known as "motional narrowing" sets in. Here, the hopping rate is occurring so rapidly that the resonance begins to take on properties associated with the average over each of the Jahn-Teller distortions. In effect, the hopping is too fast for the experiment to distingish whether the defect is actually reorienting or simply in a single configuration that is spread equally between the various distortions.

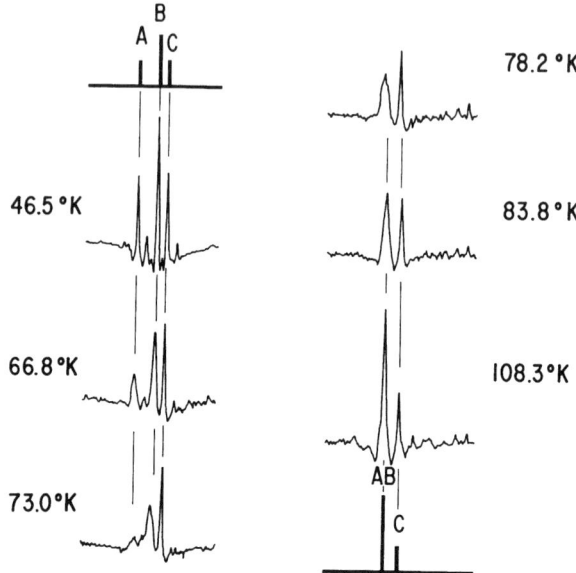

FIG. 17. EPR spectrum of $(V \cdot V)^+$ versus temperature, $\mathbf{B} \parallel [111]$, showing the effect of thermally activated reorientation from one Jahn-Teller distortion to another. Line C (defects with vacancy–vacancy axis along $\mathbf{B}$) remains unaffected, while lines A and B broaden and disappear, eventually reappearing in their motionally averaged positions AB.

A quantitative theory has been given for this effect that permits direct extraction of the lifetime of the orientation process from the measured linewidth in this temperature region (Gutowsky and Saika, 1953). The results of such an analysis for the two charge states of the divacancy are included in Fig. 15, along with the results of the stress alignment recovery studies discussed earlier. By combining these two studies, we have points spanning over 13 decades for a single process, allowing unprecedented accuracy in the determination of the kinetics!

The stress alignment and linewidth studies provide essentially the same information about the reorientation kinetics. Each, however, also gives somewhat unique information as well, making it worthwhile to study a defect by both techniques when possible. In the stress alignment studies, the sign and magnitude of the Jahn-Teller coupling coefficients are also determined. Studies in the high-temperature, motionally averaged state also give an important piece of information not available by stress studies: $^{29}$Si hyperfine satellites also exist for the motionally average spectrum. A study of their angular dependence reveals that the "averaged" center is now a *six*-silicon center, the wave function uniformly distributed over six silicon atoms. This confirms unambiguously that the hopping is "electronic" between the three pairs of equivalent silicon sites neighboring the divacancy, as predicted by the model of Fig. 14. [From the stress alignment studies alone, one could not rule out the possibility that a silicon atom pair might be hopping from one distorted position to another, carrying the unpaired electron with it. An example of this type of motion, also at very low temperatures, has been observed by ODMR for a different radiation-produced center involving carbon in silicon (Lee *et al.*, 1982; O'Donnell *et al.*, 1983)].

4. DEFECT PRODUCTION

A simple and unambiguous way of introducing point defects into crystalline solids is by high-energy particle irradiation. Because of the high mobility often found for some of the simpler defects, the facility to irradiate *in situ* in an EPR apparatus at cryogenic temperatures greatly enhances the power of EPR and ENDOR in these studies.

For covalent materials, where damage is produced only by atomic collisions with the irradiation particles, high-energy electrons have a distinct advantage. In the first place, relatively simple damage (isolated simple vacancies, interstitials, etc.) tends to be produced because the light electron does not transfer energy efficiently to the atoms. In addition, because the electrons are characterized by a well-defined energy and momentum direction, several unique experiments can be done with EPR.

An example is afforded by the divacancy in silicon (Corbett and Watkins, 1965):

1. By comparing the intensity of the divacancy EPR spectrum with that for the oxygen-vacancy pair (a monitor of single-vacancy formation) versus electron-beam energy, the expected higher threshold energy for formation of the divacancy has been confirmed.
2. A study of the effect of incident beam direction has revealed the interesting result shown in Fig. 18. Here, the intensity of the line on the right of each recording measures the number of divacancies with their vacancy–vacancy axis along the magnetic field direction. The two recordings demonstrate clearly that a substantial preferential alignment of the divacancies along the incident beam direction has been frozen in. This alignment is a strong function of bombarding energy, being greatest near threshold. These studies have been used to construct a detailed microscopic picture of the damage event.

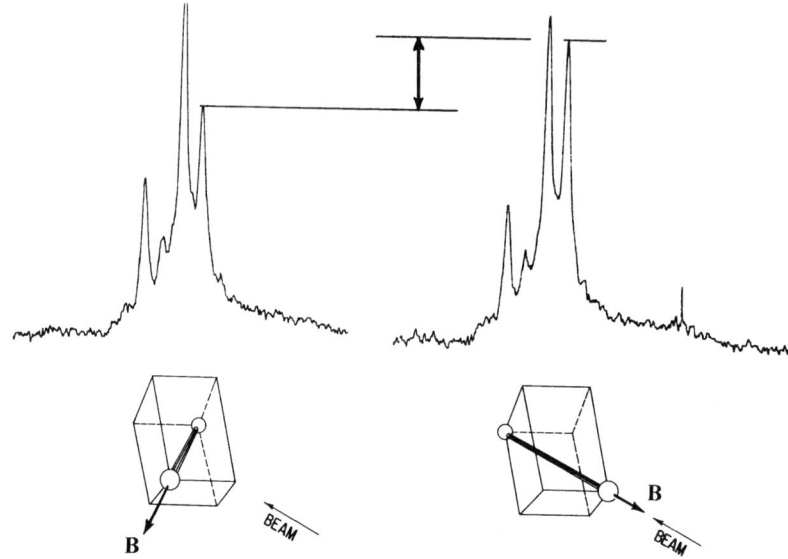

FIG. 18. EPR detection of divacancy alignment produced by 1.0-MeV electrons incident along a $\langle 111 \rangle$ direction. The transition to the right (lower field) in each spectrum of $(V \cdot V)^+$ measures the number of divacancies with their vacancy–vacancy axis along **B**.

Another interesting example is in ZnSe (Watkins, 1974, 1975d). There, *in situ* irradiations at 20.4 K produced zinc vacancies with Zn interstitials locked in the lattice nearby in a direction correlated with the incoming beam direction. These pristine Frenkel pairs could be monitored directly by EPR shedding light on the microscopic damage event in a compound semiconductor that has two sublattices.

Ion implantation represents an additional technique for producing defects for study. In addition to producing damage, it also offers a unique way to introduce selective impurities and isotopes. A good example is the EPR observation of interstatial bond-centered hydrogen after proton implantation at 77 K (Gorelkinskii and Nevinnyi, 1991).

### Acknowledgment

The preparation of this chapter was made possible through the support of National Science Foundation Grant No. DMR-92-04114 and U.S. Office of Naval Research Grant No. N0;0014-94-1-0117.

## References

Abragam, A. and Bleaney, B. (1970). *Electron Paramagnetic Resonance of Transition Ions.* Claredon, Oxford.
Allen, J. (1986). *Deep Centers in Semiconductors*, edited by S. T. Pantelides. Gordon and Breach, New York, Chapter 9.
Ammerlaan, C. A. J. and Watkins, G. D. (1972). *Phys. Rev. B* **5**, 3988.
Ammerlaan, C. A. J., Sprenger, M., van Kemp, R., and van Wezep, D. A. (1985). *Microscopic Identification of Electronic Defects in Semiconductors,* edited by N. M. Johnson, S. G. Bishop, and G. D. Watkins. MRS Vol. 46, Pittsburgh, p. 227.
Baldwin, J. A., Jr. (1963). *Phys. Rev. Lett.* **10**, 220.
Baraff, G. A., Kane, E. O., and Schlüter, M. (1980). *Phys. Rev. B* **21**, 5662.
Baranowski, J. (1986). *Deep Centers in Semiconductors*, edited by S. T. Pantelides, Gordon and Breach, New York, Chapter 10.
Barnes, R. G. and Smith, W. V. (1954). *Phys. Rev.* **93**, 95.
Bemski, G. (1959). *J. Appl. Phys.* **30**, 1195.
Benton, J. L., Lee, K. M., Freeland, P. E., and Kimerling, L. C. (1985). *J. Elect. Mat.* **14a**, 647.
Bourgoin, J. C., von Bardeleben, H. J., and Stievenard, D. (1988). *J. Appl. Phys.* **64**, R65.
Brower, K. L. (1970). *Phys. Rev. B* **1**, 1908.
Cheng, L. J., Corelli, J. C., Corbett, J. W., and Watkins, G. D. (1966). *Phys. Rev.* **152**, 761.
Clerjaud, B. (1985). *J. Phys. C: Solid State Phys.* **18**, 3615.
Corbett, J. W. and Watkins, G. D. (1965). *Phys. Rev.* **138**, A555.
Davies, G. and Foy, C. P. (1980). *J. Phys. C: Solid State Phys.* **13**, 2203.
Davies, G. (1982). *J. Phys. C: Solid State Phys.* **15**, L149.
De Wit, J. G., Sieverts, E. G., and Ammerlaan, C. A. J. (1976). *Phys. Rev. B*, **14**, 3494.
Elkin, E. L. and Watkins, G. D. (1968). *Phys. Rev.* **174**, 881.
Feher, G., Hensel, J. C., and Gere, E. A. (1960). *Phys. Rev. Lett.* **5**, 309.
Gehlhoff, W., Emanuelsson, P., Omling, P., and Grimmeiss, H. G. (1993). *Phys. Rev. B* **47**, 7025.
Gorelkinskii, Yu, V. and Nevinnyi, N. N. (1991). *Physica B* **170**, 155.
Gutowsky, H. S. and Saika, A. (1953). *J. Chem. Phys.* **21**, 1688.
Hage, J., Niklas, J. R., and Spaeth, J.-M. (1986). *Mat. Sci. Forum* **10–12**, 259.
Irmscher, K., Kind, T., and Gehlhoff, W. (1994). *Phys. Rev. B* **49**, 7964.
Isoya, J. Kanda, H., Uchida, Y., Lawson, S. C., Yamasaki, S., Itoh, H., and Morita, Y. (1992). *Phys. Rev. B* **45**, 1436.
Kaufmann, U., Schneider, J., and Räuber, A. (1976). *Appl. Phys. Lett.* **29**, 312.
Kaufmann, U. and Schneider, J. (1980a). *Festkörperprobleme (Advances in Solid State Physics) Vol. XX*, edited by J. Treusch. Viewig, Braunschweig, p. 87.
Kaufmann, U. and Schneider (1980b). *Appl. Phys. Lett.* **36**, 747.
Kaufmann, U. and Kennedy, T. A. (1981). *J. Electronic Mat.* **10**, 347.
Kennedy, T. A. and Wilsey, N. D. (1978). *Phys. Rev. Lett.* **41**, 977.
Kennedy, T. A., Wilsey, N. D., Krebs, J. J., and Stauss, G. H. (1983). *Phys. Rev. Lett.* **50**, 1281.
Koh, A. K. and Miller, D. J. (1985). *Atomic Data and Nuclear Tables* **33**, 235.
Krebs, J. J. and Stauss, G. H. (1977a). *Phys. Rev. B* **15**, 17.
Krebs, J. J. and Stauss, G. H. (1977b). *Phys. Rev. B* **16**, 971.
Krebs, J. J. and Stauss, G. H. (1982). *Phys. Rev. B* **26**, 2296.
Kreissl, J., Ulrici, W., Rehse, U., and Gehlhoff, W. (1992). *Phys. Rev. B* **45**, 4113.
Lancaster, G. (1968). *Electron Spin Resonance in Semiconductors.* Plenum, New York.
Landoldt-Börnstein (1982). *Vol. III/17b Semiconductors: Physics of the II–VI and I–VII*

Compounds, Semimagnetic Semiconductors, edited by O. Madelung. Springer-Verlag, Berlin.
Landoldt-Börnstein (1989). Vol. III/22b Semiconductors: Impurities and Defects in Group IV and III-V Compounds, edited by M. Schulz. Springer-Verlag, Berlin.
Lee, Y. H., Corelli, J. C., and Corbett, J. W. (1976). Phys. Lett. **59A**, 238.
Lee, K. M., O'Donnell, K. P., Weber, J., Cavenett, B. C., and Watkins, G. D. (1982). Phys. Rev. Lett. **48**, 37.
Loubser, J. H. N. and van Wyk, J. A. (1978). Rep. Prog. Phys. **41**, 1201.
Ludwig, G. W. and Woodbury, H. H. (1962). Solid State Physics, Vol. 13, edited by F. Seitz and D. Turnbull. Academic, New York, p. 223.
Meese, J. M., Farmer, J. W., and Lamp, C. D. (1983). Phys. Rev. Lett. **51**, 1286.
Mehran, F., Morgan, T. N., Title, R. S., and Blum, S. E. (1972). J. Magn. Res. **6**, 620.
Morton, J. R. and Preston, K. F. (1978). J. Magn. Res. **30**, 577.
Niklas, J. R., Spaeth, J.-M., and Watkins, G. D. (1985). Microscopic Identification of Electronic Defects in Semiconductors, edited by N. M. Johnson, S. G. Bishop, and G. D. Watkins. MRS Vol. 46, Pittsburgh, p. 237.
O'Donnell, K. P., Lee, K. M., and Watkins, G. D. (1983). Physica **116B**, 258.
Pake, G. E. and Estle, T. L. (1973). The Physical Principles of Electron Paramagnetic Resonance. Benjamin, New York.
Rai, R., Savard, J. Y., and Tousignuon, B. (1969). Can. J. Phys. **47**, 1147.
Röppischer, H., Elssner, W., and Böttner, H. (1975). Phys. Stat. Solidi (a) **27**, 375.
Schneider, J. (1967). II-VI Semiconducting Compounds, edited by D. J. Thomas Benjamin, New York, p. 40.
Schneider, J. (1985). Microscopic Identification of Electron Defects in Semiconductors, edited by N. M. Johnson, S. G. Bishop, and G. D. Watkins. MRS Vol. 46, Pittsburgh, p. 13.
Schneider, J., Kaufmann, U., Wilkening, W., Baeumler, M., and Köhl, F. (1987). Phys. Rev. Lett. **59**, 240.
Sieverts, E. G., Muller, S. H., and Ammerlaan, C. A. J. (1978). Phys. Rev. B **18**, 6834.
Slichter, C. P. (1990). Principles of Magnetic Resonance. Springer-Verlag, New York.
Son, N. T., van Oosten, A. B., and Ammerlaan, C. A. J. (1991). Solid State Commun. **80**, 439.
Spaeth, J.-M., Niklas, J. R., and Bartram, R. H. (1992). Structural Analysis of Point Defects in Solids. Springer-Verlag, New York.
Sprenger, M., Muller, S. H., and Ammerlaan, C. A. J. (1983). Physica **116B**, 224.
Sprenger, M., van Kemp, R., Sieverts, E. G., and Ammerlaan, C. A. J. (1985). J. Elect. Mat. **14a**, 815.
Stauss, G. H., Krebs, J. J., Lee, S. H., and Swiggard, E. M. (1980). Phys. Rev. B **22**, 3141.
Title, R. S. (1964). Phys. Rev. A **133**, 1613.
Title, R. S. (1967). Physics and Chemistry of II-VI Compounds, edited by M. Aven and J. S. Prener. North-Holland, Amsterdam, Chapter 6.
Vallin, J. T. and Watkins, G. D. (1974). Phys. Rev. B **9**, 2051.
van der Linde, R. H. and Ammerlaan, C. A. J. (1979). Defects and Radiation Effects in Semiconductors, 1978, edited by J. H. Albany. Inst. Phys. Conf. Ser. 46, London, p.242.
van Gisbergen, S. J. C. H. M., Godlewski, M., Gregorkiewicz, T., and Ammerlaan, C. A. J. (1991). Phys. Rev. B **44**, 3012.
van Kemp, R., Sieverts, E. G., Sprenger, M., and Ammerlaan, C. A. J. (1986). Mat. Sci. Forum **10-12**, 875.
van Kooten, J. J., Weller, G. A., and Ammerlaan, C. A. J. (1984). Phys. Rev. B **30**, 4564.
van Wezep, D. A. and Amerlaan, C. A. J. (1988). Phys. Rev. B **37**, 7628.
van Wyk, J. A., Tucker, O. D., Newton, M. E., Baker, J. M., Woods, G. S., and Spear, P. (1995). Phys. Rev. B **52**, 12657.

Vlasenko, L. S., Son, N. T., van Oosten, A. B., Ammerlaan, C. A. J., Lebedev, A. A., Taptygov, E. S., and Khramtsov, V. A. (1990). *Solid State Commun.* **73,** 393.
Watkins, G. D. and Corbett, J. W. (1961). *Phys. Rev.* **121,** 1001.
Watkins, G. D. (1963). *J. Phys. Soc. Jpn.* **18,** Suppl. II, 22.
Watkins, G. D. and Corbett, J. W. (1964). *Phys. Rev.* **134,** A1359.
Watkins, G. D. (1965). *Radiation Damage in Semiconductors,* edited by P. Baruch. Dunoid, Paris, p. 97.
Watkins, G. D. and Corbett, J. W. (1965). *Phys. Rev.* **138,** A543.
Watkins, G. D. (1967). *Phys. Rev.* **155,** 802.
Watkins, G. D. (1969). *IEEE Trans. Nucl. Sci.* **NS-16,** 13.
Watkins, G. D. (1974). *Phys. Rev. Lett.* **33,** 223.
Watkins, G. D. (1975a). *Point Defects in Solids, Vol. 2,* edited by J. H. Crawford, Jr., and L. M. Slifkin. Plenum, New York, Chapter 4.
Watkins, G. D. (1975b). *Lattice Defects in Semiconductors 1974,* edited by F. A. Huntley. Inst. Phys. Conf. Ser. 23, London, p. 1.
Watkins, G. D. (1975c). *Phys. Rev. B* **12,** 4383.
Watkins, G. D., (1975d). *Lattice Defects in Semiconductors 1974,* edited by F. A. Huntley. Inst. Phys. Conf. Ser. 23, London, p. 338.
Watkins, G. D. (1976). *Phys. Rev. B* **13,** 2511.
Watkins, G. D. (1983). *Physica* **117B/118B,**9.
Watkins, G. D. (1984). *Festkörperprobleme (Advances in Solid State Physics) Vol. XXIV,* edited by P. Grosse. Viewig, Braunschweig, p. 163.
Watkins, G. D. (1986). *Deep Centers in Semiconductors,* edited by S. Pantelides. Gordon and Breach, New York, Chapter 3.
Watkins, G. D. and Williams, P. M. (1995). *Phys. Rev. B* **52,** 16575.
Weil, J. A., Bolton, J. R., and Wertz, J. E. (1994). *Electron Paramagnetic Resonance.* Wiley, New York.
Wilsey, N. D. and Kennedy, T. A. (1985). *Microscopic Identification of Electronic Defects in Semiconductors,* edited by N. M. Johnson, S. G. Bishop, and G. D. Watkins. MRS Vol. 46, Pittsburgh, p. 309.

CHAPTER 2

# Magneto-Optical and Electrical Detection of Paramagnetic Resonance in Semiconductors

*J.-M. Spaeth*

FACHBEREICH PHYSIK
UNIVERSITÄT-GH PADERBORN
PADERBORN, GERMANY

|     |                                                                      |    |
| --- | -------------------------------------------------------------------- | -- |
| I.  | INTRODUCTION                                                         | 45 |
| II. | MAGNETO-OPTICAL DETECTION OF EPR AND ENDOR                           | 48 |
|     | 1. EPR Detected with Magnetic Circular Dichroism of Absorption (MCDA) | 48 |
|     | 2. Examples for MCDA-detected EPR                                    | 53 |
|     | 3. MCDA Excitation Spectra of ODEPR Lines (MCDA Tagged by EPR)       | 61 |
|     | 4. Spatially Resolved MCDA and ODEPR Spectra                         | 66 |
|     | 5. Determination of the Spin State with the MCDA Method              | 68 |
|     | 6. ENDOR Detected with the MCDA Method                               | 70 |
| III.| ELECTRICAL DETECTION OF EPR (EDEPR)                                  | 73 |
|     | 1. Experimental Observation of EDEPR                                 | 73 |
|     | 2. The Donor–Acceptor Recombination Model                            | 78 |
|     | 3. Concentration and Temperature Dependence of EDEPR Signals         | 82 |
| IV. | ELECTRICAL DETECTION OF ENDOR (EDENDOR)                              | 84 |
| V.  | NEW POSSIBILITIES                                                    | 86 |
|     | References                                                           | 90 |

## I. Introduction

Electron paramagnetic resonance (EPR) is a powerful tool for the study of the microscopic and electronic structures of point defects in nonmetallic solids. It can be successfully applied to investigate dopants, unwanted impurities, or intrinsic point defects in semiconductors, provided the number of point defects is sufficiently large, which, in general, requires a bulk sample and a defect concentration between $10^{12}$ and $10^{15}$ defects/cm$^3$, depending, of course, on the linewidth of the EPR lines. In the conventional detection of EPR, upon induction of a magnetic dipole transition between the Zeeman levels of the defect, a microwave absorption is measured in a microwave bridge. In a good commercial spectrometer, the sensitivity of conventional

EPR is about $5 \cdot 10^{10}$ spins per Gauss linewidth. Higher resolution of hyperfine (hf) and superhyperfine (shf) interactions is achieved by electron-nuclear double-resonance (ENDOR), where nuclear magnetic resonance (NMR) transitions are detected via an EPR transition, (i.e., ENDOR is a double-resonance method). Usually in semiconductors, the ENDOR effect is, at most, about a few percentages of the EPR effect, which requires, accordingly, a higher number of defects. For details of the experimental method and the analysis of the spectra, the reader is referred to, for example, Spaeth et al. (1992), and for an elementary account of EPR and ENDOR, to the chapter by G. D. Watkins in this book (Chapter 1, with further references therein).

In modern semiconductor physics, however, very often bulk samples are not available, and it is important to study a low concentration of defects. For example, GaN can be grown only as epitaxial layers, and defects in devices such as diodes are restricted to a very small semiconductor volume. Conventional EPR and ENDOR are mostly not sensitive enough. It is therefore necessary to look for more sensitive ways to detect EPR and ENDOR.

Optical detection of EPR (ODEPR) differs from conventional detection, basically, in that a microwave-induced repopulation of Zeeman levels is detected indirectly by a change of some properties of light, which is either absorbed or emitted by the defect under study. The light properties are polarizations or intensities, which are measured to detect EPR. The optical detection of EPR has a number of interesting new features. By virtue of the quantum transformation for detecting the signals from $10^{10}$ to $10^{15}$ Hz, there is an enormous gain in sensitivity by several orders of magnitude. Originally this sensitivity enhancement was used to study sparsely populated, excited defect states (see, e.g., Geschwind, 1972), but it soon became evident that application of optical techniques is also very useful for the study of ground states. Clearly, in this way optical properties can be associated directly with particular defects and their structure.

Another interesting feature connected with ODEPR is the possibility of studying defects with a high selectivity. One of the problems for EPR can be the simultaneous presence of many paramagnetic defects, which renders the spectra very complicated. With optical detection, every defect can be investigated separately, except for the rare situation of different defects having indistinguishable optical properties.

In ODEPR, the fact that magnetic sublevels can be selectively populated either by choosing appropriate experimental conditions or by physical mechanisms, such as spin selection rules for radiative or nonradiative transitions, plays an important role. This is the basis of the application of different optical techniques to detect EPR. Basically, one can use the optical absorption or the fluorescence or phosphorescence emission of a defect. Determining which of the techniques to apply depends on the system and the kind of problem one wants to study.

Optical absorption and emission bands of defects in solids are usually rather broad, with typical half-widths of 0.1 to 0.3 eV. This width is due to electron–phonon interactions. The Zeeman splittings of both ground and excited states are only of the order of $10^{-4}$ eV. The reason that one can observe ODEPR in the polarization of optical bands is that the polarization properties of the absorption or emission bands are practically not changed by the electron–phonon interaction. One can therefore use optical spectrometers with rather low optical resolution unless special features of very sharp optical transitions, such as zero-phonon lines, are of interest. It is because of this property of the electron–phonon interaction that spin selection rules are not affected much by it, which makes the observation of ODEPR in the broad optical bands of defects in solids possible.

The absorption-detected ODEPR method (magneto-optical method) of studying defect ground states and related techniques, such as EPR excitation spectroscopy (the so-called tagging) to correlate the optical transitions belonging to one defect with its EPR spectrum, as well as the measurements of the spin state and spatial resolution are discussed in Section II. The examples that illustrate the methods are taken from work done at the University of Paderborn. For the ODEPR techniques using optical emission, in particular by observing the radiative donor-acceptor pair recombinations in semiconductors, the reader is referred to the chapter by T. A. Kennedy and E. R. Glaser in this book (Chapter 3) and, for example, to Spaeth *et al.*, (1992).

The magneto-optical method via the magnetic circular dichroism of the optical absorption (MCDA) usually requires layer thicknesses of about 50 to 100 $\mu$m or more, and an optical transition associated with the defect is needed, while the detection via donor–acceptor–recombination luminescence requires the donor–acceptor recombination to be radiative, which is often not the case, in particular, in indirect semiconductors such as silicon. There are only very few ENDOR experiments via luminescence-detected EPR.

As an alternative approach for the study of semiconductor devices, there were attempts to measure EPR as microwave-induced changes of the electrical conductivity. This was, in principle, successful, but the experiments on Si diodes at room temperature gave, in most cases, not much information, except for a structureless line at $g = 2$ (Schmidt and Solomon, 1966; Stich *et al.*, 1995 and many references therein). Only in a few cases was an hf structure resolved (Kurylev and Karyagin, 1974; Christmann *et al.*, 1992; Stich *et al.*, 1993), allowing the identification of the defect involved.

Most authors explained their observations by assuming a spin-dependent recombination (SDR) of excess charge carriers at a paramagnetic recombination defect as the responsible process (Schmidt and Solomon, 1966; Solomon, 1976; Lépine and Prejean, 1970; Lépine, 1972; Mendz and Hanemann, 1978; Mendz et al., 1979; Vlasenko et al.,1986), and a number of specific mechanisms have been proposed (Lépine, 1972; Rong et al., 1991, 1992; Kaplan et al., 1978; Honig, 1966; Lannoo et al., 1993). The understanding of the experimental observations and the conditions for observing electrical detection of EPR (EDEPR) remained quite poor, such that no EDEPR spectroscopy has been developed. Only recently has some progress been made, via a systematic study of the conditions for observing EDEPR of shallow and deep-level defects in silicon for low or moderate concentrations of such defects, using microwave-induced changes of the photoconductivity. It was shown that a donor–acceptor pair recombination process can explain satisfactorily most experimental observations (Stich et al., 1995) and that ENDOR also can be detected via photoconductivity (Stich et al., 1996). In Sections III and IV, these results are reviewed and current understanding discussed. Further, some additional features for the study of point defects that are not available with conventional detection of EPR/ENDOR are pointed out (Section V).

## II. Magneto-Optical Detection of EPR and ENDOR

### 1. EPR Detected with Magnetic Circular Dichroism of Absorption (MCDA)

The MCD of the absorption is the differential absorption of right and left circularly polarized light, where the light is propagating along the direction of an externally applied static magnetic field $\mathbf{B}_0$. One is, therefore, concerned with the circularly polarized transitions of the Zeeman effect. As a measure of the MCDA, the quantity

$$\varepsilon = \frac{\omega d}{2c}(k_r - k_l) \tag{1}$$

is defined, where $k_r$ and $k_l$ are the absorption coefficients for right and left circularly polarized light, respectively; $d$ is the thickness of the crystal; and $\omega$ is the light frequency. Only the energy-dependent absorption constants $\alpha_{r,l}(E)$ are measurable. With the relation

$$k(E) = \frac{\hbar c}{2E}\alpha(E), \tag{2}$$

it follows with $E = \hbar\omega$ that

$$\varepsilon = \frac{1}{4}d(\alpha_r - \alpha_1). \tag{3}$$

In the experiment one measures the intensities $I_{r,1}$ that pass through the sample:

$$I_{r,1} = I_0 \exp(-\alpha_{r,1} d), \tag{4}$$

where $I_0$ is the light intensity incident on the sample. It follows from Eq. (4) that

$$\alpha_{r,1} = \frac{1}{d} \ln \frac{I_0}{I_{r,1}}. \tag{5}$$

From Eq. (3) one obtains a relation that depends only on the measurable quantities $\Delta I$ and $I_a$, where $I_a$ is the average intensity, $I_a = \frac{1}{2}(I_r + I_1)$, and $\Delta I$ is the difference in intensity, $\Delta I = I_r - I_1$. Since $\Delta I \ll I_a$, [i.e., $d(\alpha_r - \alpha_1) \ll 1$], one obtains the simple relation

$$\varepsilon = \frac{\Delta I}{4 I_a}. \tag{6}$$

The quantity $\varepsilon$ can be measured very precisely with the help of a stress modulator and lock-in techniques (Spaeth et al., 1992). According to the selection rules for circularly polarized electric dipole transitions $(x \pm iy)$ in a magnetic field, the MCDA signal $\varepsilon$ contains two parts (Mollenauer and Pan, 1972):

$$\varepsilon = \varepsilon_p(P) + \varepsilon_d(B_0), \tag{7}$$

where $\varepsilon_p$ is the paramagnetic part, which depends on the spin polarization $P$ of the ground state, while $\varepsilon_d$ is the diamagnetic part proportional to $B_0$ and arises from the unresolved Zeeman splittings in the optically excited states. For the detection of EPR, we are only concerned with the paramagnetic part. For $S = \frac{1}{2}$, one obtains for the paramagnetic term $\varepsilon_p$, since $P = (n_- - n_+)/(n_+ + n_-)$, where $n_+$ and $n_-$ are the occupation numbers of the $m_s = \pm \frac{1}{2}$ states:

$$\varepsilon_p(P) \propto P = \frac{n_- - n_+}{n_+ + n_-} = \tanh(g_e \mu_B B_0 / 2kT) \tag{8}$$

For $S > \frac{1}{2}$, $P$ is given by the Brillouin function.

The paramagnetic part $\varepsilon_p$ is temperature- and field-dependent according to Eq. (8). By measuring $\varepsilon_p$ as a function of temperature, it can easily be distinguished from nonmagnetic circular dichroisms of the sample. $\varepsilon_d$ is not temperature-dependent. The principle of EPR detection with the MCDA method is easily seen from Eq. (8). The equilibrium ground-state spin polarization $P$ can be changed by microwave-induced magnetic dipole transitions, provided the transition rate exceeds that of the spin lattice relaxation and the light intensity is weak enough, so that the populations are not influenced appreciably by the optical transitions by optical-pumping effects (Geschwind, 1972). The microwave transitions reduce $\varepsilon_p$. This reduc-

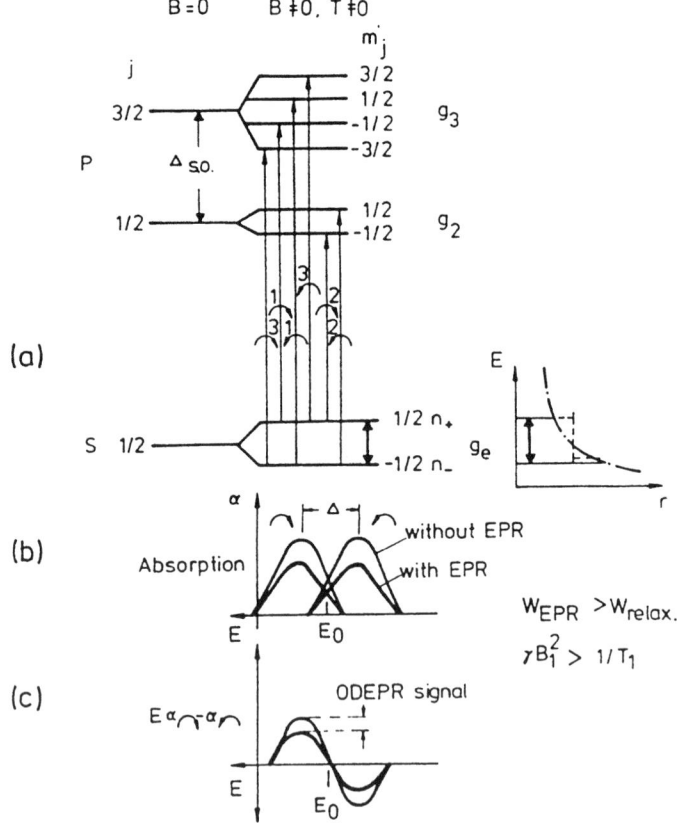

FIG. 1. Simple atomic model to explain the magnetic circular dichroism method of the absorption (MCDA) to detect EPR. (a) Level scheme and circularly polarized optical absorption transitions. The numbers represent the relative transition matrix elements. (b) Absorption band for right and left circularly polarized light, assuming strong electron–phonon coupling. (c) Magnetic circular dichroism and its change due to EPR transitions.

tion or decrease of $\varepsilon_p$ is monitored to detect the EPR transitions. In principle, one can obtain a 100% effect if the EPR transition is fully saturated (i.e., $n_+ = n_-$).

$\varepsilon_p$ is often very small. The detection limit is approximately $10^{-5}$ for an optical density of the absorption band of $\sim 1$. Accordingly, the EPR-induced changes in $\varepsilon_p$ are often difficult to measure. Therefore, samples having a large (diamagnetic) dichroism due to crystal structure (e.g., due to birefringence) or possessing large stress-induced dichroisms are not suitable for this technique to measure EPR.

To illustrate both the method and the experiment in more detail, let us discuss the MCDA of an atomic $s \to p$ transition such as that occurring in a free alkali atom. The level scheme is shown in Fig. 1a. In a magnetic field, the ground-state Kramers doublet is split into Zeeman levels $m_s = \pm\frac{1}{2}$, the excited $p$ states are split by the spin–orbit interaction into $j = \frac{1}{2}$ and $j = \frac{3}{2}$ states with the magnetic substates $m_j$. The absorption transitions for right and left circularly polarized light are indicated by arrows according to the quantum-mechanical selection rules with their relative transition probabilities (see, e.g., Schiff 1949). The equilibrium occupation of the ground-state levels $n_+$ and $n_-$ for finite temperatures is indicated. For $T = 0$ only $m_s = -\frac{1}{2}$ is occupied. If the small Zeeman splitting in the excited state is neglected, an absorption band for right polarization to $j = \frac{1}{2}$ and one with left polarization to $j = \frac{3}{2}$, both with equal intensities (the former with intensity 2 and the latter with intensity $3 - 1 = 2$) are observed.

The–spin orbit splitting $\Delta_{so}$ is often small compared to the absorption bandwidth. In this case, it is convenient to measure the dichroism $\varepsilon$, being the difference of the two absorptions $\alpha_r$ and $\alpha_1$, which can be approximated by a derivative:

$$\varepsilon(\omega) = \frac{\omega d}{2c}(k_r - k_1) \approx \frac{\omega d}{2c}\frac{dk}{d\omega}\Delta\omega$$

$$\approx \frac{Ed}{2\hbar c}\frac{dk(E)}{dE}\Delta E. \tag{9}$$

With Eq. (2) and the assumption that the optical band shape is Gaussian according to

$$\alpha(E) = \alpha_0 e^{-(E-E_0)^2/W^2}, \quad W_{1/2} = 2\sqrt{\ln 2}\, W, \tag{10}$$

one obtains for the extrema of $\varepsilon(E)$ (see Fig. 1c)

$$\varepsilon_{\text{extr}} = \pm\sqrt{\frac{\ln 2}{2e}}\frac{\alpha_0 d}{W_{1/2}}\Delta E, \tag{11}$$

with

$$\Delta E = 2\left\{\frac{\mu_B B_0(g_2 + 5g_3)}{3} - \frac{\Delta_{so}}{3}\tanh\frac{g_e\mu_B B_0}{2kT}\right\}. \qquad (12)$$

where $g_2$ and $g_3$ are g-factors of the excited state, $\alpha_0$ is the absorption coefficient in the peak of the band, $W_{1/2}$ is its half-width, and $\Delta_{so}$ is the spin–orbit splitting of the excited $j = \frac{1}{2}$ and $j = \frac{3}{2}$ states (Paus, 1980). The first term in Eq. (12) is the diamagnetic part, and the second term is the paramagnetic part of the MCDA. For the observation of EPR transitions, it is thus convenient to measure the change of the MCDA at the wavelength of the extrema of the MCDA. For $\Delta_{so} \gg W_{1/2}$, there are two separate bands for $\varepsilon$, one with a positive and the other with a negative sign.

To measure EPR with the MCDA method, it is not necessary to understand the optical transitions in detail as long as a paramagnetic MCDA signal can be measured. Figure 1 is, indeed, a special case, approximately observed for $F$ centers in alkali halides. In general, the MCDA spectra look quite different.

For a two-level system, the paramagnetic MCDA can be described in a more general way with the following expression:

$$\varepsilon_p(E) \propto \alpha_0(E_0)d\,\frac{\sigma_r(E) - \sigma_l(E)}{\sigma_r(E) + \sigma_l(E)}\,\frac{n_+ - n_-}{n_+ + n_-} \qquad (13)$$

where $\sigma_r$, and $\sigma_l$ are the cross-sections or transition probabilities for right and left circularly polarized light, respectively.

The cross-sections $\sigma_r(E)$ and $\sigma_l(E)$ are the sums over all transitions with right and left circular polarization, respectively, arising from the two magnetic sublevels, whose small energy difference ($10^{-4}$ eV) is being neglected compared to the optical transition energies. $\alpha_0(E_0)$ is the absorption constant at the peak of the absorption band, and $d$ is the crystal thickness. From Eq. (13) it is seen that there is no $\varepsilon_p$, if the ground state polarization is zero or if there is no difference between $\sigma_r(E)$ and $\sigma_l(E)$. This difference depends on the properties of the excited states. As a general rule, the difference is large for a large spin–orbit interaction at the defect, such as that observed for heavy impurity atoms. The difference is also influenced by the crystal field splitting in the excited states. If both are large, one observes separated bands for right- and left-polarized light as a function of photon energy. If $\alpha_0$ is too small, $\varepsilon_p$ is also only a small effect and hard to measure.

For a generalization to a multilevel ground-state system for $S > \frac{1}{2}$, the last term in Eq. (13) is to be replaced by the spin polarization $P$, which is then determined by the Brillouin function rather than the Langevin function, which determined the last term in Eq. (13) for $S = \frac{1}{2}$. Microwave transitions between fine-structure levels will also change the MCD of the absorption.

In Eqs. (11) and (13), $\alpha_0$ and therefore the oscillator strength of the optical transition enters. The oscillator strength is usually not known and is hard to calculate in a reliable fashion. Therefore, the measurement of the MCDA cannot be used to determine the concentration of defects quantitatively, unless the oscillator strength is known. The EPR spectra measured via MCDA cannot be calibrated with some other defect in a known concentration either, since the size of the spin–orbit interaction in the excited state enters in a crucial and usually unknown way. Although the MCDA method is very sensitive and selective, it has the disadvantage that a quantitative determination of defect numbers is not possible directly, unless the oscillator strength and the spin–orbit splittings are known or, by some other processes, the MCDA can be changed in a controllable manner. This is, for instance, possible if, by a charge transfer transition, the MCDA of the defect under study can be enhanced or decreased in a way in which the number of defects in the charge state from which the MCDA is built up or decreased can be determined. If, on the other hand, the number of defects is known, the MCDA measurement can yield valuable information on the properties of the excited states.

In some cases, both conventional EPR and ODEPR can be measured in the same sample. Then, by calibrating the conventional EPR signal, one can calibrate the MCD of the absorption.

2. Examples for MCDA-detected EPR

The very high sensitivity of the method compared to conventional EPR and conventional optical absorption spectroscopy can be demonstrated by experiments on the singly ionized midgap defect EL2 in semi-insulating as-grown GaAs single crystals grown by the liquid encapsulated Czochralski (LEC) method. Figure 2a shows the optical absorption spectrum in the near-infrared region for photon energies below the gap energy of 1.52 eV. Only a weak band at 1.18 eV, caused by an intracenter transition of the diamagnetic midgap defect EL2, is detectable. However, the MCDA reveals the existence of further absorption bands caused by paramagnetic defects, which turned out to be the singly ionized state of the midgap EL2 defect (Meyer et al., 1984). The ODEPR spectrum, measured at 1350 nm at 24 GHz (Fig. 3), shows a quartet structure due to the hyperfine (hf) splitting of $^{75}$As having $I = \frac{3}{2}$ and being 100% abundant. In Fig. 3, the trivial MCDA effect due to the tanh $(g_e \mu_B B_0 / 2kT)$ is already subtracted. Only the

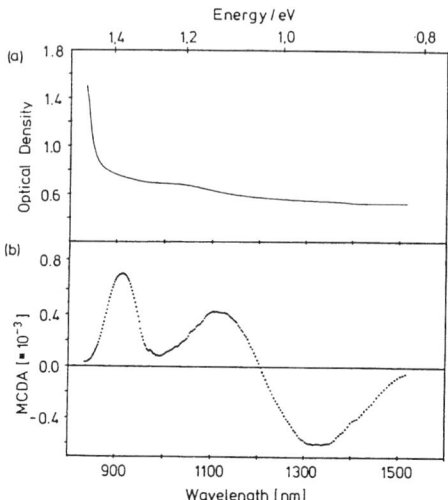

FIG. 2. (a) Optical absorption of as-grown semi-insulating GaAs at 1.4 K (crystal thickness, 0.3 mm). (b) Magnetic circular dichroism of the absorption of Fig. 2a. $T = 4.2$ K, $B = 2$ T.

microwave-induced changes of the MCDA are shown as the ODEPR spectrum. In the conventional EPR spectrum, the hf structure is hardly recognized and the signal-to-noise ratio observed for a defect concentration of $\sim 5 \times 10^{15}$, shown in Fig. 3, is not much greater than 1. The weakness of the conventional EPR spectrum is caused by the quartet splitting and the

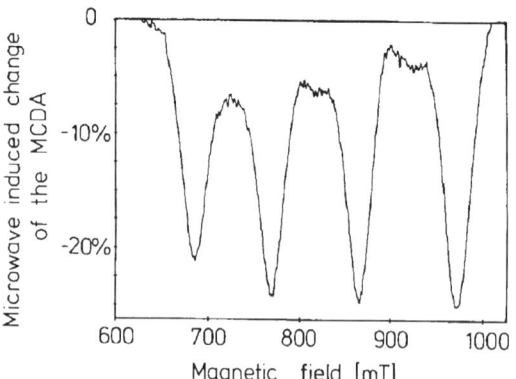

FIG. 3. EPR spectrum of paramagnetic EL2 defects in semi-insulating as-grown GaAs measured as microwave-induced decrease of the MCDA (see Fig. 2b) at 1350 nm. The spectrum shows an hf quartet structure due to $^{75}$As ($I = \frac{3}{2}$) ($K$-band, $T = 1.5$ K).

the large linewidth of each line ($\sim 30$ mT). The signal-to-noise ratio of the ODEPR spectrum is about two orders of magnitude higher. The defect structure was revealed by ODENDOR to be an As-antisite As-interstitial pair defect (Meyer et al., 1987). As seen from Fig. 2b, a sensitivity enhancement was also found for the detection of the optical absorption bands of this defect, which could not be measured in a conventional absorption experiment. Thus, the MCDA is also measurable for very weak absorptions as a consequence of the fact that one applies a sensitive modulation spectroscopy. Therefore, one also measures ODEPR spectra, even if the absorption spectrum does not show clear absorption bands.

MCDA-detected EPR can also be measured in zero phonon lines (ZPLs). This is illustrated in measurements of $V$ impurities in 6H-SiC single crystals. In 6H-SiC there are inequivalent lattice sites in both the Si and C sublattices: one hexagonal site and two quasi-cubic sites. Figure 4a shows the optical absorption (Schneider et al., 1990), and Fig. 4b shows the MCDA of $V_{Si}^{4+}$ measured at 1.5 K (Reinke et al., 1993a). The ZPL labeled $\alpha$ had been assigned to $V_{si}^{4+}$ on a hexagonal site, while those labeled $\beta$ and $\gamma$ had been assigned to the two quasi-cubic sites. The MCDA spectrum has lines with different line shapes for all three groups: $\alpha$, $\beta$, and $\gamma$. There is a positive MCDA for the $\alpha$ line, derivative-like lines for the $\beta$ line set, and a negative MCDA for the $\gamma$ line. At about 6 K an additional weak MCDA line at 7630 cm$^{-1}$ appears (at the low-energy side of the strong $\alpha$ transition). The MCDA lines show a field and temperature dependence and thus originate in paramagnetic ground states. Their temperature dependence is quite complicated. It was studied experimentally in detail and a theoretical interpretation of the $\alpha$ line was presented in the framework of crystal field theory (Reinke et al., 1993b). Basically, the optical transition is a $^2E \to {}^2T_2$ transition of $V^{4+}$ ($3d^1$).

Figure 5, trace a, shows the MCDA-detected EPR spectrum measured in the $\alpha$ line for $B_0 \parallel$ to the $c$-axis of the crystal; Figure 5, trace b, shows the spectrum for $B_0$ oriented 5° off the $\bar{c}$-axis. $^{51}V$ has $I = \frac{7}{2}$ with 100% natural abundance, from which eight hf lines (marked by bars in Fig. 5) are expected. The lines in between in Fig. 5, trace a, are forbidden EPR transitions ($\Delta m_s = \pm 1$, $\Delta m_I = \pm 1$), which are less pronounced in Fig. 5, trace b. That the MCDA-detected EPR of Fig. 5 is indeed due to $V_{Si}^{4+}$ at the hexagonal site could be confirmed by measuring the angular dependence of the MCDA-detected EPR lines. Within experimental error, the same lines were obtained as those published by Maier et al. (1992) from conventional EPR. When measuring in the MCDA lines $\beta$ and $\gamma$ one observes the EPR lines assigned to the two quasi-cubic sites (Maier et al., 1990). Figure 6 shows that the same EPR spectrum is measured in the two $\beta$ lines at 7397 cm$^{-1}$ (trace a) and 7420 cm$^{-1}$ (trace b). Hence, both $\beta$ lines belong to the same defect. The effective spin was determined to be $S = \frac{1}{2}$ for all three

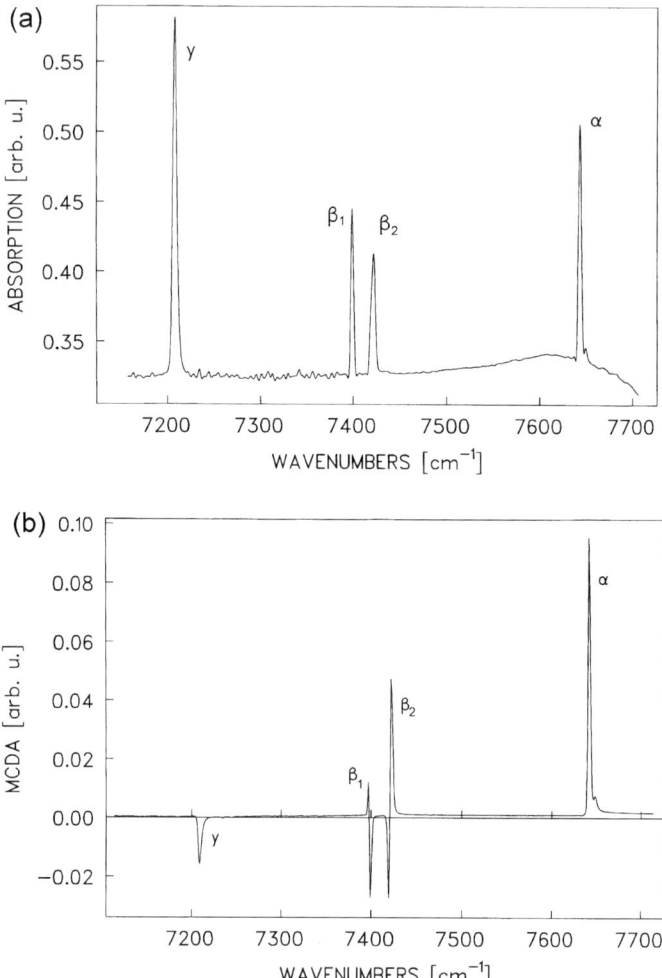

FIG. 4. (a) Absorption spectrum of a vanadium-containing 6H-SiC crystal measured at $T = 1.5$ K. The absorption lines are labeled after Schneider et al. (1990) and Reinke et al. (1993), according to whom, $\alpha$ was assigned to the hexagonal $V_{Si}^{4+}$-defect site and $\beta$ and $\gamma$ to the two quasi-cubic sites. (b) MCDA spectrum of the absorption spectrum of (a) measured at $T = 1.5$ K and $B = 2$ T.

lattice sites (see section II.5), confirming that the charge state is $V^{4+}(d^1)$. Figure 7 shows the $\alpha$-line MCDA measured towards higher energy where pronounced phonon side bands are apparent, extending from 7600 to 9000 cm$^{-1}$. All MCDA lines could be assigned to intrinsic lattice phonon replica of the strong $\alpha$ lines of the hexagonal defect. The EPR spectrum of

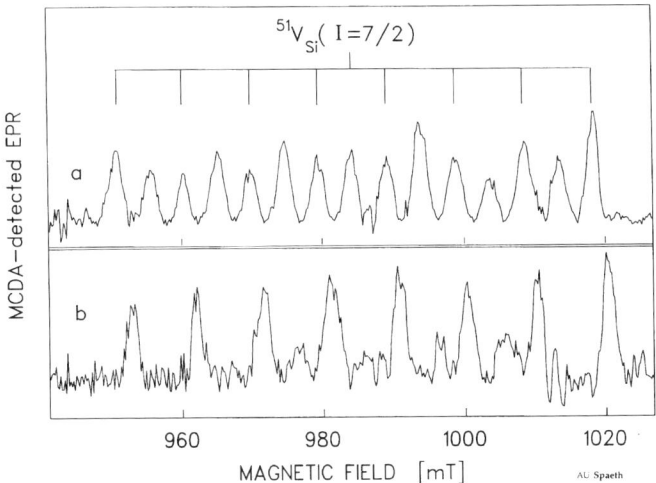

FIG. 5. MCDA-detected EPR of $V_{Si}^{4+}$ in 6H-SiC measured at $T = 1.5$ K in the $K$-band ($\nu_{EPR} = 24.2$ GHz) with the magnetic field $B_0 \parallel \vec{c}$-axis of the crystal (trace $a$) and for $B_0 + 5°$ off the $\vec{c}$-axis (trace $b$). The bars above trace a show the expected EPR line positions for $V^{51}$ ($I = \frac{7}{2}$) on the hexagonal site with the hf interaction and $g$-values (($g_\parallel = 1.749$ and $g_\perp = 0$) from Maier et al. (1990).

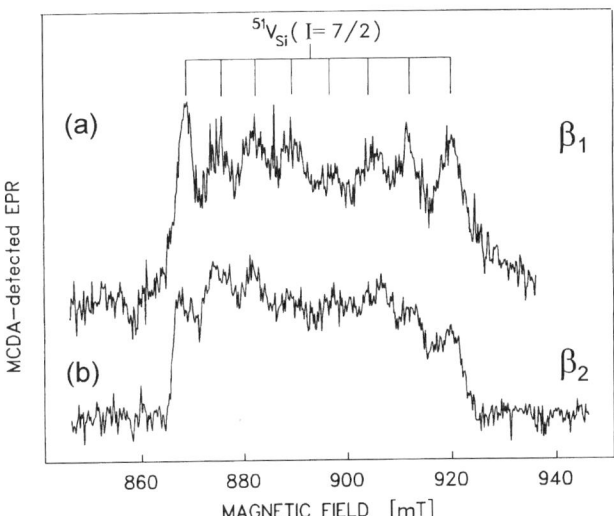

FIG. 6. MCDA-detected EPR of a quasi-cubic site of $V_{Si}^{4+}$ in 6H-SiC measured in the $\beta_1$-MCDA line (at 7397 cm$^{-1}$, trace $a$) and in the $\beta_2$-MCDA line (7420 cm$^{-1}$, trace $b$) at $T = 1.5$ K in the $K$-band ($\nu_{EPR} = 24.2$ GHz) with a magnetic field $B_0 \parallel \vec{c}$-axis of the crystal. The $g$-values ($g_\parallel = 1.9463$, $g_\perp = 1.937$) and hf interaction are identical within experimental error to those given by Maier et al. (1990).

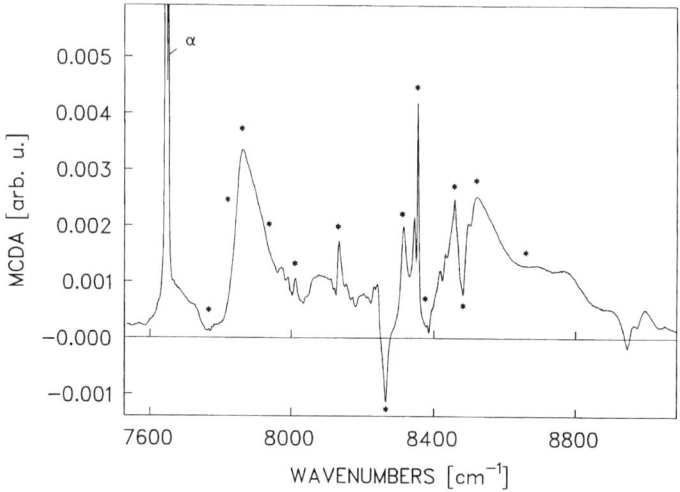

FIG. 7. MCDA spectrum measured in $V$ containing 6H-SiC in the spectral range between 7600 and 9000 cm$^{-1}$ at $T = 1.5$ K. (From Reinke et al., 1993a, with permission.)

the hexagonal site $V_{Si}^{4+}$ was seen in the replica at many photon energies (indicated by asterisks in Fig. 7). The spectra of the quasi-cubic site defects were not seen in the phonon side bands (no phonon side bands could be detected for the quasi-cubic site $V$-defects) (Reinke et al., 1993a). The latter experiments show that the MCDA technique can be used to identify phonon side bands belonging to a specific defect and, of course, to assign optical properties (the ZPLs) to specific EPR spectra.

Another nice example is the MCDA-detected EPR spectra of the shallow Al acceptors in the two polytypes 4H- and 6H-SiC. The MCDA in both polytypes consists of two broad bands, which were measured between 0.8 and 3.2 eV and assigned to a photoionization transition to the valence band and possibly a $V^{4+}$ impurity-related transition from Al$^0$ to the $V^{4+/5+}$ level (Reinke et al., 1996). Figure 8 shows the MCDA-detected EPR spectra for B $\|\bar{c}$-axis: Fig. 8a for 4H-SiC and Fig. 8b for 6H-SiC. As expected, in 4H-SiC there are two sites resolved (one hexagonal and one quasi-cubic), while in 6H-SiC there are three (one hexagonal, two quasi-cubic sites). The hf interaction with $^{27}$Al ($I = \frac{5}{2}$, 100% abundance) is resolved for this field orientation, where for all sites sextets are measured. Neither with conventional EPR nor with luminescence-detected EPR has this hf structure been resolved previously. Only the MCDA-detected EPR spectra could unambiguously show the involvement of the Al in the spectra. All spectra show axial symmetry about the $\bar{c}$-axis. Therefore, from the angular dependence of

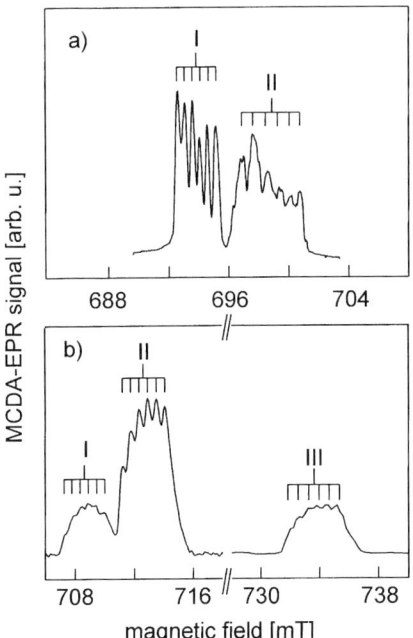

FIG. 8. (a) MCDA-EPR spectrum measured in the MCDA of Al-doped 4H-SiC at $T = 1.5$ K. (b) MCDA-EPR spectrum measured in the MCDA of Al-doped 6H-SiC ($T = 1.5$ K). In both measurements, $B_0$ was $\parallel$ to the $\vec{c}$-axis, $K$-band. (From Reinke et al., 1996, with permission.)

the spectra, no assignment can be made to specific lattice sites (Reinke et al., 1996). In the case of 6H-SiC, it could be argued from the intensities that spectra I and III belong to the quasi-cubic sites and II to the hexagonal site.

So far, examples have been discussed for which bulk samples were available. However, it has also been possible to apply the MCDA technique to thin layers. One example is the detection of an MCDA and MCDA-detected EPR spectra of low-temperature MBE-grown GaAs, grown at 200 and 300°C. The layer thickness was 2 μm. Because of the high concentration of the arsenic antisite-related defects of $10^{19}$ to $10^{20}$ cm$^{-3}$, spectra with very good signal-to-noise ratios could be measured. For growth at 200°C, it was found that the MCDA and MCDA-EPR spectra were very similar to those of EL2$^+$ defects (see Figs. 2 and 3), except that the bleaching properties and the lattice relaxation times were different. The ZPL of EL2$^0$ also was not found. These results were explained as being due to the high concentration (Krambrock et al., 1992a).

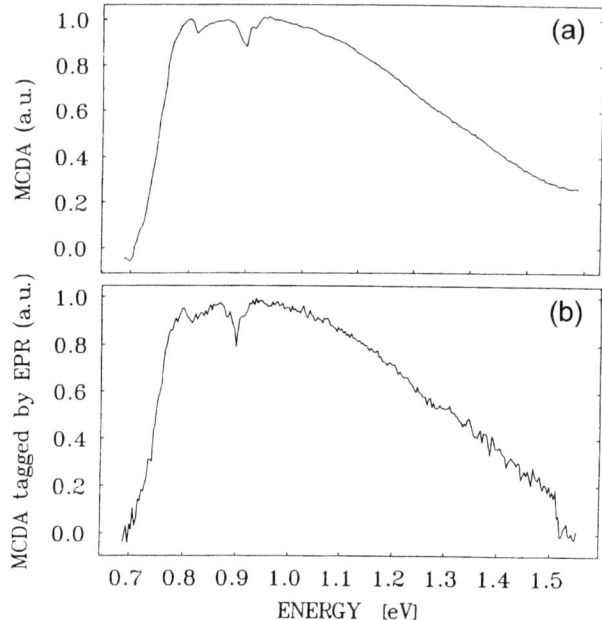

FIG. 9. (a) MCDA of an Sn-doped $Al_{0.39}Ga_{0.61}As$ layer measured at 2 T and 1.5 K after photoionization of DX centers. The substrate was removed. The two dips at 0.8 and 0.9 eV are experimental artifacts. (b) Excitation spectrum of the MCDA-detected EPR at 870 mT, $v_{EPR} = 23.95$ GHz (MCDA tagged by EPR). (From Fockele et al., 1990, with permission.)

Figure 9a shows, as another "layer" example, the MCDA measured after photoionization of DX centers in $Al_xGa_{1-x}As$ ($x = 0.35$) doped with Sn. The epitaxial layer was about 100 μm thick and removed from the substrate. Sn substitutes for Ga and is expected to be a shallow donor ($Sn^0$). In the dark, no EPR is detected. After photoionization of the DX centers with a threshold photon energy of 0.9 eV, the MCDA of Fig. 9 and an EPR spectrum measured in this MCDA (Fig. 10) are observed (Fockele et al., 1990). The EPR spectrum shows a broad central line with $g = 1.97$ and a half-width of 51 mT and two hf doublets due to the two Sn isotopes $^{117}Sn$ ($I = \frac{1}{2}$, 7.75% abundance) and $^{119}Sn$ ($I = \frac{1}{2}$, 8.58% abundance). The isotropic hf interaction constant a/h was determined to be 10.1 GHz for $^{119}Sn$. The sample contained approximately $3.10^{18}\,Sn^0$ donors/cm³. From these experiments it could be concluded that the DX center is diamagnetic in its ground state, accommodating two electrons with opposite spins. Upon photoionization of the DX center, a deep paramagnetic defect with tetrahedral symmetry is formed. The large isotropic hf interaction shows that about

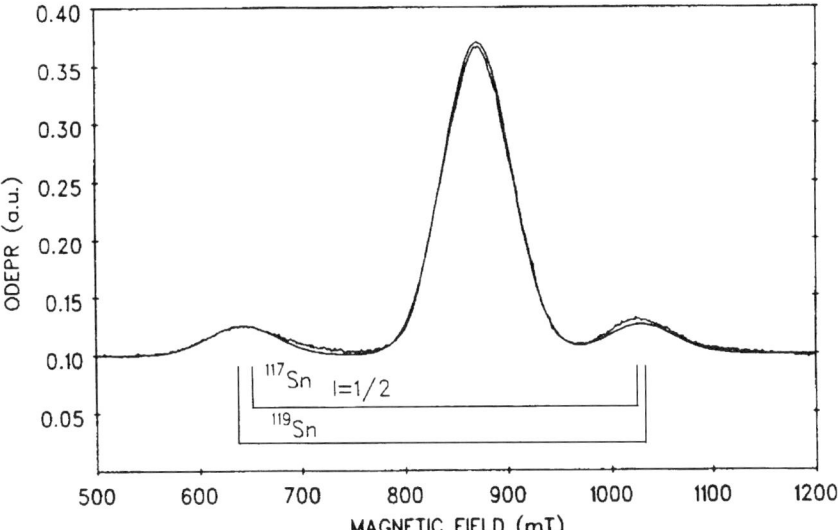

Fig. 10. MCDA-detected EPR spectrum of an Sn-doped layer ($x = 0.35$) with additional ionization of below-bandgap light at 1.24 eV (50 $\mu$W). The GaAs substrate was removed. The stick diagram indicates the resonance field positions of the two magnetic Sn isotopes, $^{117}$Sn and $^{119}$Sn. The smooth line is a simulation of the spectrum with the hf data. (From Fockele et al. 1990, with permission.)

13% of $s$-like spin densities are localized at the Sn nucleus. (For further discussion, see Spaeth and Linde, 1994).

### 3. MCDA Excitation Spectra of ODEPR Lines (MCDA Tagged by EPR)

By conventional EPR and optical spectroscopy, it is very difficult to unambiguously correlate the optical absorption spectra of a defect with its EPR spectrum, since they are both measured separately and the presence of several defects can never be excluded. By ODEPR, a direct correlation is possible.

One can measure a kind of excitation spectrum of the ODEPR lines. In Fig. 1, a simple scheme was discussed to explain the MCDA method to measure the ground-state EPR of a defect. Such a scheme applies to all optical transitions of a defect; that is, the EPR spectra appear in the MCDA of all absorption transitions, provided the condition expressed by Eq. (13)

is fulfilled, in particular, that $\sigma_r(E) - \sigma_l(E) \neq 0$ for the absorption transitions used for the experiments. The ground-state polarization is, of course, the same for all transitions. Therefore, the EPR resonance conditions can be set to a particular EPR line, the optical wavelength can be varied, and the EPR signal intensity can be monitored as the microwave-induced change of the MCDA. Thus, from the total MCDA of a sample, one can measure in this way only that part which belongs to the EPR line. Therefore, with this method, one distinguishes the optical properties of different defects and also of different defect orientations, if the EPR spectra are anisotropic. One thus has the possibility to study the polarization of optical transitions for different defect orientations. Such a measurement is particularly useful when a superposition of optical absorptions caused by several defects is present. This situation is typical for radiation damage or impurity problems.

Experiments have shown that ZPLs and sharp spin-forbidden transitions can be measured as tagged MCDA spectra and unambiguously assigned to a particular defect. Figure 11 shows the absorption spectrum of high-resistivity GaAs doped with $V$ measured with high resolution in a Fourier transform optical absorption spectrometer. The spectrum is attributed to $V^{3+}$ defects, where $V^{3+}$ substitutes for Ga. The broad feature is a $^3A_2 \rightarrow {}^3T_1$ transition into excited Jahn-Teller states. The sharp features are the ZPLs and spin-forbidden transitions (Clerjaud et al., 1985). The optically detected EPR spectrum of the $V^{3+}$ defects measured in the broad band (see Fig. 12a) agrees with the previously conventionally measured one and has an unre-

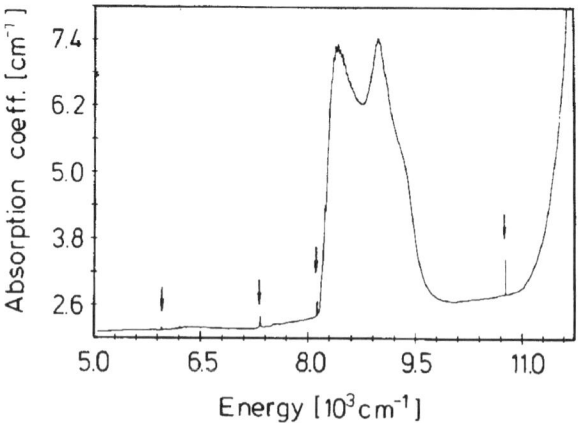

FIG. 11. Optical absorption spectrum of a semi-insulating GaAs:V sample measured with a Fourier transform spectrometer: Zero phonon lines are observed close to $5900\,\text{cm}^{-1}$, at 7333, 8131, and $10773\,\text{cm}^{-1}$. All transitions correspond to $V^{3+}$ internal transitions.

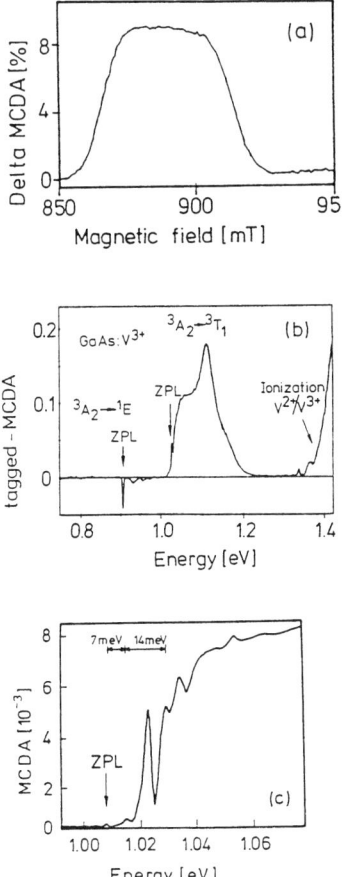

FIG. 12. (a) MCDA-detected EPR spectrum of $V_{Ga}^{3+}$ centers in high-resistivity (p-type) GaAs measured at 1.1 eV ($g = 1.96 \pm 0.01$, $v_{EPR} = 24.41$ GHz, $T = 1.5$ K). (b) MCDA of $V_{Ga}^{3}$ centers in high-resistivity GaAs (p-type) at $T = 1.6$ K tagged by an EPR spectrum of Fig. 12a. (c) Part of the MCDA spectrum of $V_{Ga}^{3+}$ centers in GaAs measured with high spectral resolution. (From Görger et al., 1988). (Transition elements in III-V-Semiconductors—A study with optically detected magnetic resonance. In: *Semi-insulating III-V Materials*, edited by G. Grossman and L. Lebedo. Adam Hilger, Bristol, with permission of IOL Publishing Ltd.)

solved hf splitting due to $^{51}V(I = \frac{7}{2})$, which is responsible for the unusual line shape. The MCDA tagged by this EPR spectrum (ODMR MCDA) (Fig. 12b) shows the same ZPLs and spin-forbidden sharp transitions as the conventionally measured absorption spectrum, including some additional features. It was verified in this way that the structure in the optical absorption spectrum indeed belong to this one defect, $V^{3+}$, which was not possible on the basis of optical measurements alone. The crystal also contained $V^{2+}$ defects and $V$-related complexes (Görger et al., 1988). Figure 12c shows details of the MCDA spectrum around the ZPL at 1.008 eV, which could not be detected conventionally.

In the tagged MCDA spectrum at $1.37 \pm 0.002$ eV, the onset of the ionization transition $V^{3+} \rightarrow V^{2+}$ + hole (in the valence band) is observed in agreement with the known energy position of the $V^{3+/2+}$ level at $E_c$-0.15 eV. This is a remarkable observation because it implies that the ODEPR spectrum can be observed in an ionizing transition, in this case to the valence band. Since this first observation (Görger et al., 1988), this phenomenon has been observed several times for a number of transition metal ions (Ti, Mn, Fe) in GaAs and GaP (Görger et al., 1988, Baeumler et al., 1989). From the foregoing discussion, such an observation is not expected, since localized states were always assumed in the mechanism to explain ODEPR. Possibly, during the time needed to induce an optical transition of the order of $10^{-15}$ s, defect-induced states resonant with the bands exist, allowing the MCDA and ODEPR observation before lattice phonons destroy them and the electrons or holes become thermalized within the bands. However, a theory for explaining the more recent observations is not available. That the electrons and holes are thermalized after the transition is demonstrated by the observation that no optical pumping was possible.

From a practical point of view, the possibility to observe the EPR in ionizing transitions is very important, since such defect transitions are always present in semiconductors, while there may not be intracenter transitions. It seems, therefore, that the MCDA method to investigate defects in semiconductors can be applied in a very general way.

Another example for tagging is seen in Fig. 9b, where the excitation spectrum of the $Sn^0$ spectrum was measured and shown to be identical with the MCDA measured after photoionization of the DX centers (Fockele et al., 1990).

In GaAs, a particular difficulty arose in that several different arsenic antisite-related defects showed almost undistinguishable EPR spectra, which were only characterized by the quartet structure caused by the hf splitting of the $^{75}As_{Ga}$ nucleus ($I = \frac{3}{2}$, 100% abundance) (see also Fig. 3). Figure 13 shows the MCDA-detected spectra that were measured after the following treatment of GaAs: trace (a) is observed after electron irradiation of semi-insulating GaAs at low temperature, when the sample is kept below and up to 77 K (Krambrock et al., 1992b); trace (b) is the MCDA-detected EPR of $EL2^+$ defects; trace (c) is measured after electron irradiation of all types of GaAs at low temperature and subsequent annealing to near room tempera-

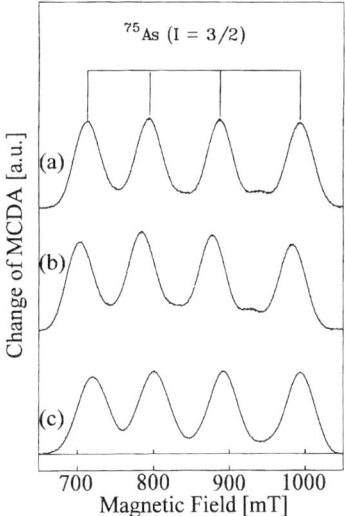

FIG. 13. MCDA-detected EPR spectra of three different As antisite-related defects. (a) The isolated $As_{Ga}$ defect, (b) the EL2 defect, and (c) the next nearest antistructure pair measured at 1.5 K. See also text. (From Spaeth and Krambrock 1993, with permission.)

ture or directly after electron irradiation at room temperature. The three EPR spectra are almost identical, with the exception of small differences in the inhomogenously broadened linewidth. In particular, the hf splitting of the quartets is practically identical for all of those spectra ($a/h \approx 2650$ MHz). The tagged MCDA spectra clearly show that three different defects are involved (Fig. 14). The MCDA spectra reflect the very different optical properties, which must originate in different microscopic structures. Their models have been established by MCDA-detected ENDOR: Fig. 14, trace a, comes from isolated arsenic antisite defects ($As_{Ga}$) that are not stable at room temperature (Krambrock et al., 1992b); Fig. 14, trace b, is due to the $EL2^+$ defect (an $As_{Ga}^-As_i$ pair defect) (Meyer et al., 1987, Meyer, 1988); Fig. 14, trace c, comes from an antistructure pair ($As_{Ga}^+ - Ga_{As}^{(nnn)}$), in which a Ga antisite is the next nearest neighbor to an As antisite (Krambrock and Spaeth, 1993). Notably, the $EL2^+$ defect is formed after electron irradiation at low temperature and after the sample was annealed to a temperature of 520 K and can be produced in concentrations higher than ever before found in semi-insulating as-grown GaAs. The antistructure pair is thermally destroyed just above room temperature (Koschnick et al., 1995). For a detailed discussion of the ODENDOR investigation and the center models, see also Spaeth and Krambrock (1993).

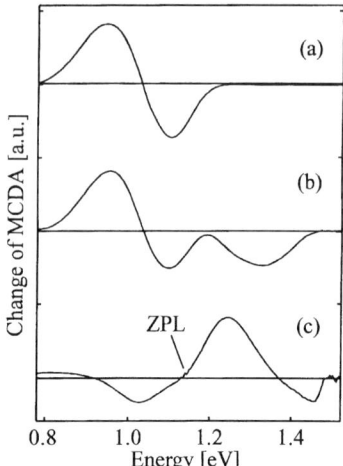

FIG. 14. MCDA excitation spectra (MCDA tagged by EPR) of the three different As antisite-related defects. (a) The isolated $As_{Ga}$ defect, (b) the EL2 defect, and (c) the next nearest antistructure pair. In (c), the ZPL of the antistructure pair is indicated. (From Spaeth and Krambrock, 1993, with permission.)

### 4. Spatially Resolved MCDA and ODEPR Spectra

In semiconductor physics, it is often necessary to have a nondestructive method for measuring the spatial distribution of defects. For example, one wants to know the spatial distribution of different charge states of defects across a wafer on which microelectronic devices are to be fabricated. Such measurements are not possible with conventional EPR. However, with optical detection, such experiments are possible within the limitations of the cross-section of the light beam. One integrates, of course, over the thickness of the sample.

The first example of such a measurement is shown in Fig. 15, where the spatial distribution of the two charge states of the EL2 defect in semi-insulating GaAs is investigated across half the diameter of a GaAs wafer of 0.3 mm thickness. The light beam of a halogen lamp was focused and limited in diameter by a pinhole such that the spatial resolution of about 300 μm was achieved. The charge state of the midgap EL2 (dots in Fig. 15) is monitored by its intracenter absorption band at 1.18 eV (see Fig. 2a), and the charge state of the paramagnetic EL2 by the MCDA at 1350 nm [1 T, 1.5 K (see Fig. 2b for the MCDA, crosses in Fig. 15)]. From Fig. 15 it is seen that the two charge states of the same defect are not equally distributed. In

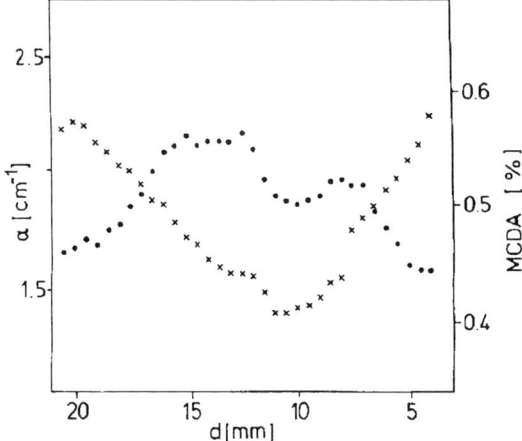

FIG. 15. Spatial distribution of midgap EL2 (crosses) and paramagnetic EL2 defects (dots) measured across half a diameter of a semi-insulating GaAs wafer. Position 4 mm is the center of the wafer. The spatial resolution was about 300 μm. The midgap EL2 was measured via the 1.18 eV intracenter absorption band, and the paramagnetic EL2 via the MCDA at 1350 nm ($B = 1$ T, $T = 1.5$ K). (From Spaeth et al. (1988). Int. Phys. Conf. Ser. **91**, ch. 4, 391, with permission of IOL Publishing Ltd.)

places where there is more midgap EL2, there is less paramagnetic EL2 and vice versa. This points to the role of acceptors ionizing the midgap EL2 and to their distribution in the wafer, which must be similar to that of the paramagnetic EL2 (Heinemann et al., 1987, Spaeth et al., 1988). Upon converting the optical absorption and the MCDA into defect concentrations, after performing a calibration of both the MCDA and the optical absorption (Spaeth et al., 1988), it is seen that there is, indeed, a significant anticorrelation between the two charge states of the EL2 defect and that the distribution of the defects in both charge states is not homogeneous.

This kind of mapping experiment can be improved in several ways. One is to use a focused laser beam to improve the resolution. The higher the spatial resolution, the less light is passed through the sample, and, therefore, the signal-to-noise ratio drops. With higher light intensity levels, one can improve the resolution, but this is feasible only if the defects are not light-sensitive, which was found to be the case for the EL2 defects. Also, normal lasers have a high output with a rather high noise (up to 1%). This is disadvantageous for the MCDA method.

One of the interesting possibilities of this mapping technique is that after identification of the MCDAs of several defects, one does not necessarily need to spatially resolve the ODEPR spectra but can simply measure their MCDAs at characteristic wavelengths, where there is no MCDA superposition of different defects. This mapping technique allows a simultaneous nondestructive investigation of the distribution of several defects. One can thus study their interactions upon external influences on the crystal, such as heat treatments.

This mapping can be done in two dimensions and one can establish topographic pictures of the distribution of defects. One must take care, however, that a large sample, when cooled to very low temperatures for the MCDA experiments, does not exhibit stress dichroisms, because it is not easy to correct reliably for this effect, especially if it has the same order of magnitude as the magnetic circular dichroism, and low temperatures are necessary for sensitivity reasons.

## 5. Determination of the Spin State with the MCDA Method

The knowledge of the charge of a defect is important. Connected with the charge state is the spin state. For many defects it is clear that $S = \frac{1}{2}$, but for many-electron systems, such as for transition metal ions ($3d^n$), the spin state can be derived from the EPR spectra only if a fine structure or crystal field splitting is observed. This splitting can be too small to be resolved. Then, the measurement of an EPR spectrum does not allow the spin and charge state of the defects to be determined. It may be determined from ENDOR if these measurements are possible.

However, there is an alternative way by using the MCDA method. The following is exact for orbital singlet states. According to Eq. (7), the MCDA is the sum of a diamagnetic part $\varepsilon_d(B)$ and a paramagnetic part $\varepsilon_p(P)$. The diamagnetic part is proportional to the magnetic field $B$ and usually small compared to the paramagnetic part $\varepsilon_p$, which, in turn, is proportional to the spin polarization $P$. For a general value of the spin, the polarization is given by the Brillouin function $B_s(x)$:

$$P = g\mu_B S B_s(x), \tag{14}$$

$$x = \frac{g\mu_B B_0}{kT}, \tag{15}$$

$$B_s(x) = \frac{1}{S}\left\{\left(S + \frac{1}{2}\right)\coth\left(S + \frac{x}{2}\right) - \frac{1}{2}\coth\frac{x}{2}\right\}, \tag{16}$$

where $g$ is the electronic $g$-factor of the defect.

According to Eq. (7), $P$, which contains the information on the spin state $S$, cannot be measured directly by an MCDA measurement due to the diamagnetic term $\varepsilon_d$. However, since $\varepsilon_d$ is linear in $B$, one can measure the total MCDA $\varepsilon$ for several magnetic fields and temperatures and form the following ratio, (e.g., for $B_1$ and $T_1$, $B_2$ and $T_2$):

$$R_{\exp} = \frac{\varepsilon(B_1, T_1) - \varepsilon(B_1, T_2)}{\varepsilon(B_2, T_1) - \varepsilon(B_2, T_2)} \tag{17}$$

$$= \frac{\varepsilon_p(B_1, T_1) - \varepsilon_p(B_1, T_2)}{\varepsilon_p(B_2, T_1) - \varepsilon_p(B_2, T_2)} \tag{18}$$

Eq. (18) follows from Eq. (17), since the diamagnetic term $\varepsilon_d$ cancels out. According to Eqs. (8, 13, and 14),

$$\varepsilon_p = CSB_s(x), \tag{19}$$

where $C$ is a proportionality constant, which contains $g_e \mu_B$ and the right and left circularly polarized absorption cross-sections and thus information on the nature of the excited states. As seen in Eq. (20), $C$ does not enter the spin determination. From Eqs. (18) and (19) it follows that

$$R(S) = \frac{B_s(S, B_1/T_1) - B_s(S, B_1/T_2)}{B_s(S, B_2/T_1) - B_s(S, B_2/T_2)}. \tag{20}$$

Therefore, a measurement according to Eq. (18) can be compared to the calculated $R(S)$ by varying $S$. For the correct value of $S$ both must be equal: $R_{\exp} = R(S)$ (Görger et al., 1988).

The application of the method is illustrated for $V^{2+}$ and $V^{3+}$ defects in GaAs. From the ODEPR spectrum (see Fig. 12a) the $g$ factor of $V^{3+}$ is known to be $g = 1.96 \pm 0.001$. A determination of $R_{\exp}$ from Eq. (17) at 1.1 eV (see Fig. 12b) and comparison with Eq. (20) for varying $S$ is given by Fig. 16 for three pairs of $B_1$ and $B_2$ at temperatures $T_1$ and $T_2$. It is seen that $S = 1$. This is in agreement with an earlier spin determination by ENDOR (Hage et al., 1984, 1989). Applying the same method for $V^{2+}$ in GaAs, it is found that $S = \frac{1}{2}$. The $3d^3$ configuration of $V^{2+}$ has a low spin configuration in agreement with theoretical predictions (Katayama-Yoshida and Zunger, 1986).

The foregoing method is strictly valid only for an orbital singlet state. In general, for an orbitally degenerate state, Eq. (20) cannot simply be used to determine the spin state. The spin polarization must be calculated according to the details of the system, and analogous measurements must then be performed as previously outlined.

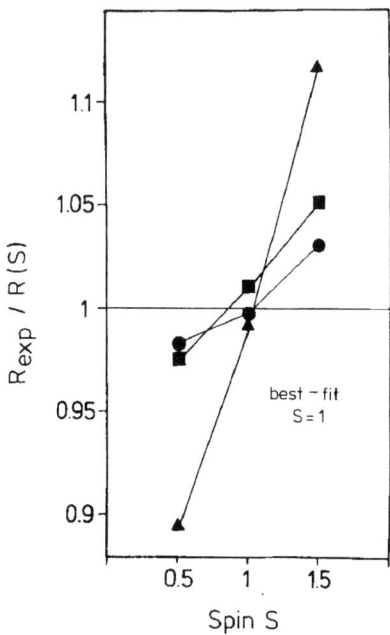

FIG. 16. Ratio of $R_{exp}$ and $R(S)$ as a function of $S$ for $V_{Ga}^{3+}$ centers measured at 1.1 eV in high-resistivity (p-type) GaAs for the pairs of magnetic field values $B_1$ and $B_2$ (in T): 1.3 and 0.5 (triangles), 1 and 0.8 (circles), 0.8 and 0.5 (squares); at temperatures of $T_1 = 1.65$ K and $T_2 = 4.2$ K. (From Görger et al. (1988). Transition elements in III-V-Semiconductors—A study with optically detected magnetic resonances. In: *Semi-insulating III-V Materials*, edited by G. Grossman and L. Lebedo. Adam Hilger, Bristol, with permission of IOL Publishing Ltd.)

## 6. ENDOR DETECTED WITH THE MCDA METHOD

The basis of an optically detected EPR experiment of the ground state is the measurement of its spin polarization by detecting the MCDA [see, e.g., Eq. (13)]. By a saturating or at least partially saturating microwave EPR transition, the spin polarization is diminished, which is measured as a decrease of the MCDA. In principle, the polarization can be driven to zero if the microwave transition rate is large enough compared to spin-lattice relaxations. This cannot be achieved if there is a hyperfine interaction, since the spin polarization can be only partly diminished due to the EPR selection rule. This is illustrated for the simple configuration of $S = \frac{1}{2}$ and $I = \frac{1}{2}$ in Fig. 17a. In this case, for a saturating EPR transition, the polarization can be diminished only by roughly 50%, and then the MCDA can be reduced only to half of its value in thermal equilibrium.

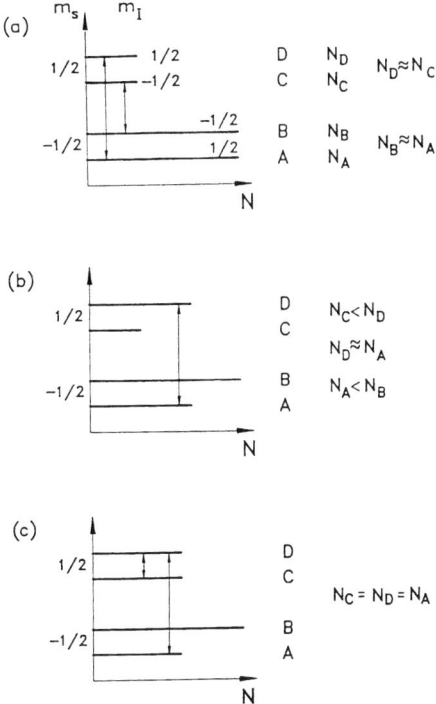

FIG. 17. Schematic explanation of optically detected ENDOR: (a) Energy level scheme for $S = \frac{1}{2}$ and one nucleus with $I = \frac{1}{2}$. The abscissa schematically indicates the relative occupation $N$ of the levels in thermal equilibrium; (b) occupation of the levels under a saturating EPR transition between A and D; (c) occupation of the levels upon the additional strong NMR transition induced between levels C and D. (From Spaeth, J.-M., Niklas, J. R., and Bartram, R. H. Structural Analysis of Point Defects in Solids. Springer-Verlag, Heidelberg, with permission. Copyright ©1992 by Springer-Verlag.)

The two possible EPR transitions are indicated by arrows between levels A and D and between B and C, respectively. The occupation of levels A and B is nearly identical in thermal equilibrium due to the small energy separation of these levels. The same is true for levels C and D, respectively. In Fig. 17 the abscissa gives the relative occupation of the levels. The signal observed in MCDA is proportional to $(N_B + N_A) - (N_C + N_D)$. Due to a saturating EPR transition between levels A and D, the occupation of these levels is equalized: $N_D = N_A$ (see Fig. 17b). In this case, the MCDA signal is only proportional to $(N_B - N_C)$, which amounts to roughly half the value considered in Fig. 17a. If an additional strong NMR transition is now induced between levels C and D, the occupation of these levels is also

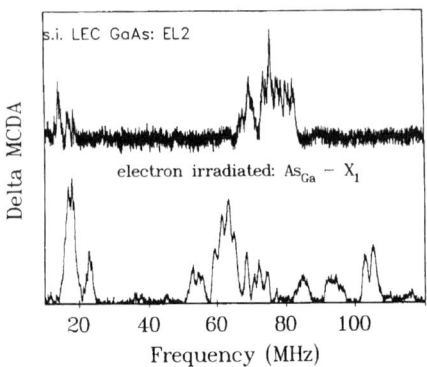

FIG. 18. Section of the MCDA-detected ENDOR spectra of two arsenic antisite-related defects in GaAs. $\mathbf{B} \parallel \langle 100 \rangle$; $T = 1.6$ K, $K$-band. Upper trace: EL2 defect in semi-insulating as-grown GaAs. Lower trace: $As_{Ga}$-related defect created by room-temperature electron irradiation of $n$-type GaAs (antistructure pair). (From Spaeth and Krambrock, 1993, with permission.)

equalized (see Fig. 17c). The saturating EPR transition is still present, with the result that the occupation of three levels, A, D, and C, is identical now, or at least nearly identical. However, the occupation of level C, $N_C$ is now bigger than in Fig. 17b, whereas the occupation of level B remains unchanged. As a result, the observed MCDA signal, still proportional to $N_B - N_C$, is additionally diminished compared to that of Fig. 17b by the NMR transition. Thus, the NMR transition is indirectly detected by the optical detection of the spin polarization as an additional decrease of the MCDA signal compared to the saturating EPR transition.

As an example, Fig. 18 shows the ODENDOR (Optically Detected Electron Nuclear Double Resonance) spectra of paramagnetic EL2 centers (upper trace) and another arsenic antisite-related defect created by room-temperature electron irradiation of $n$-type GaAs for the same field orientation (see also Section 3, Figs. 13 and 14). Their ODEPR spectra show the same hf splitting due to $^{75}$As; they differ only slightly in EPR linewidth. ODENDOR clearly shows that the microscopic structure of the two arsenic-antisite related defects is different. The lines above 40 MHz belong to As ligands (Spaeth et al., 1990; Spaeth and Krambrock, 1993).

In Fig. 8 the MCDA-detected EPR spectra of the shallow Al acceptors in 4H- and 6H-SiC were shown. MCDA-detected ENDOR was measured in line group II of Fig. 8b, (i.e., probably the hexagonal site Al acceptor in 6H-SiC for $B_0 \parallel \vec{c}$-axis). In the ODENDOR spectrum (Fig. 19), two groups of lines are seen symmetrically positioned about $m_s \cdot A_\parallel / h = \frac{1}{2} \cdot 19.82 =$

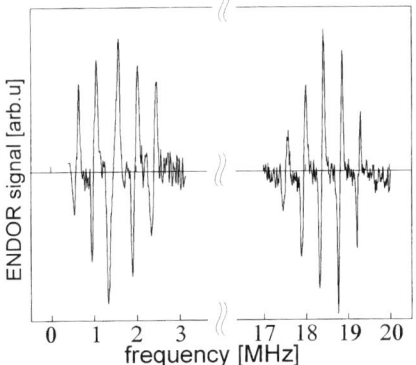

FIG. 19. MCDA-detected ENDOR spectrum measured in the MCDA-detected EPR line III of Fig. 8b in 6H-SiC containing the shallow Al-acceptors. $K$-band at $T = 1.5$ K, $B_0 \parallel \bar{c}$-axis. (From Spaeth et al., 1997, with permission.)

9.91 MHz and separated by twice the Larmor frequency of $^{27}$Al ($K$-band measurement). The fivefold splitting of each line group is due to a quadrupole interaction with $Q_\parallel/h = 0.44$ MHz (Spaeth et al., 1997).

## III. Electrical Detection of EPR (EDEPR)

1. EXPERIMENTAL OBSERVATION OF EDEPR

Various defects were investigated in donor- and acceptor-doped Czochralski (Cz)-grown silicon by EDEPR. To generate deep-level defects, the samples were irradiated with 2-MeV electrons at a fluence of $10^{17}$ cm$^{-2}$ at room temperature, producing oxygen–vacancy pair defects (so-called A centers) (Watkins and Corbett, 1960). Annealing of Cz-grown Si generated thermal double donors (NL8) (Muller et al., 1982). The photoconductivity was generated by above-bandgap light and was measured via electrical contacts. It was carefully checked that the contacts were ohmic. A constant current source was used, and the conductivity changes were measured as voltage changes across the sample. To induce microwave transitions, a standard homodyne X-band EPR spectrometer, operating between 10 and 300 K, was used.

In a silicon sample containing only shallow phosphorous donors (n-type, Fermi level near the conduction band), no EDEPR effect was detected, whereas, of course, in conventional EPR the P hf doublet was measured. However, upon compensation of the sample with boron, the ED-EPR signal

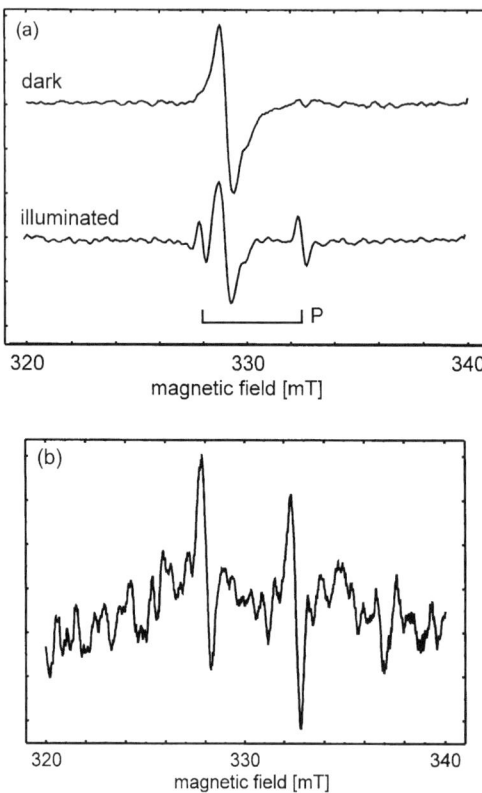

FIG. 20. (a) Conventional EPR spectrum of a compensated Si ample containing shallow $P$ donors and shallow $B$ acceptors, both in a concentration of about $10^{16}\,\text{cm}^{-3}$. $\nu_{\text{EPR}} = 9.23\,\text{GHz}$, $P_{\text{MW}} = 20\,\mu\text{W}$, $T = 10\,\text{K}$, $B_0 \parallel [100]$. Upper trace: measurement in the dark; lower trace: measurement with above-band gap light using a halogen lamp. (b) Electrical detection of the EPR of the sample described in Fig. 20a (Stich, 1996).

could be seen. This is shown in Fig. 20. Figure 20a, upper trace, shows the conventional EPR spectrum in the compensated P-doped sample in the dark. The signal of an unknown defect is detected. The phosphorous is not seen, since the Fermi level is below the $0/+$ level of the P donors. Upon illumination with above-bandgap light, the paramagnetic state of $P^0$ is occupied and the P hf doublet is seen. Figure 20b shows the EDEPR spectrum, in which the $^{31}$P hf doublet is clearly seen, not, however, that other EPR line seen in the dark with conventional EPR (Stich, 1996). Analogous experiments using, for example, thermal double donors (TDDs) as shallow donors generated by the annealing of the oxygen-containing silicon at 450°C gave the same result: if the sample contained B acceptors in such a concentration that in the dark the Fermi-level was below the $+/++$ level of $TDD^+$ (i.e., the number of B acceptors exceeded that of both TDDs and $TDD^+$s), the EDEPR signal was seen. Many experiments confirmed that the condition for the observation of an EDEPR signal is that a donor–acceptor pair recombination takes place, whereby the electrons and holes must recombine from separate paramagnetic donors and acceptors (Stich *et al.*, 1995; Stich, 1996).

In a phosphorous- and TDDs-containing sample, irradiated with 2 MeV electrons at room temperature, the spectrum of Fig. 21 was measured. The lower trace represents the conventional EPR spectrum measured under above-bandgap illumination, and the upper trace represents the EDEPR spectrum. The signal-to-noise ratio of the EDEPR spectrum is about the same as that for the conventional EPR spectrum for the P donors but is greatly enhanced for the so-called SL1 center, an excited triplet state of the oxygen–vacancy pair defect (A-center) (Brower, 1971) for various orientations of the defect. In the EDEPR spectrum, the linewidth is the same as that in the conventional EPR spectrum, and the structural information resolved is also identical: hf structure and shf structure of $^{29}$Si ligands are

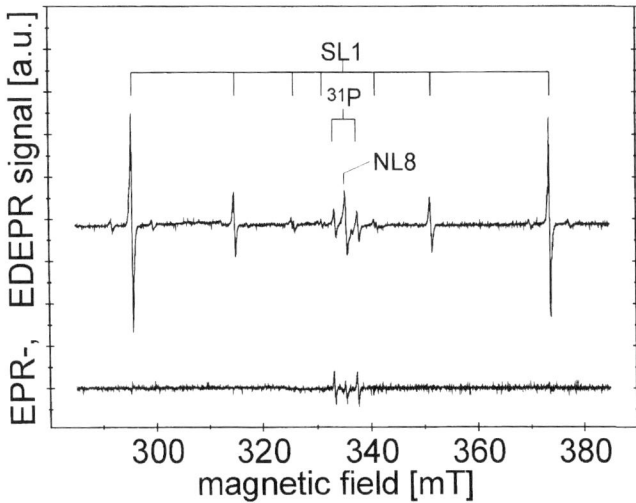

FIG. 21. Conventional EPR (lower trace) and EDEPR (upper trace) of an Si sample containing shallow P donors, thermal double donors (NL8), and SL1 centers under above-band gap illumination. $T = 20$ K and $B \parallel [011] + 2°$, $v_{EPR} = 9.3$ GHz.

resolved. A clear difference between conventional EPR and EDEPR is the dependence of the signal intensity on the microwave power measured at the same temperature (Stich et al., 1995). The conventional EPR spectrum saturates and has a maximum signal intensity at relatively low microwave power, following the law

$$S_{EPR} \propto (1 + S)^{-3/2}(P_{MW})^{1/2}. \qquad (21)$$

where $S$ is the saturation factor

$$S = \left(\frac{1}{4}\right)\gamma^2 B_1^2 T_1 T_2 \qquad (22)$$

and $P_{MW}$ is the microwave power. At high microwave power, an inverse relation, $S_{EPR} \propto (P_{MW})^{-1}$, is observed. Equation (21) describes the saturation of the ground-state polarization and takes into account the response function of the detector (Mendz et al., 1979). Examples are shown in Fig. 22a for the P donors, the thermal donors TDD$^+$, and the SL1 defect. In contrast to this behavior, Fig. 22b shows the microwave power dependence of the corresponding EDEPR signals. The saturation curves are of a completely different character: at high microwave power a high maximum signal is reached but not like in conventional EPR a saturated low signal intensity. The solid lines in Fig. 22b are experimental fits using the relation

$$S_{EDEPR} = C \frac{\beta P_{MW}}{1 + \beta P_{MW}}. \qquad (23)$$

Where $\beta$ and $C$ are parameters. For low microwave power, there is a linear relationship between $S_{EDEPR}$ and $P_{MW}$, and for high $P_{MW}$ values, there is constant signal C.

Interestingly, a very similar power dependence has been measured for a luminescence-detected EPR signal of the residual donor in GaN (Fig. 23) (Koschnick and Spaeth, 1996; Koschnick, 1997). The basis for the optical detection of EPR via luminescence is a radiative donor-acceptor pair recombination (see, e.g., Spaeth et al., 1992). Thus, this observation together with the foregoing experimental results point to a donor–acceptor pair recombination mechanism to explain the occurrence of EDEPR effects, at least for low and moderate defect concentrations (discussion follows). It should be noted that the B acceptors in silicon are difficult to observe with EPR. Their lines are very broad, and they consequently escape detection (Neubrand, 1978).

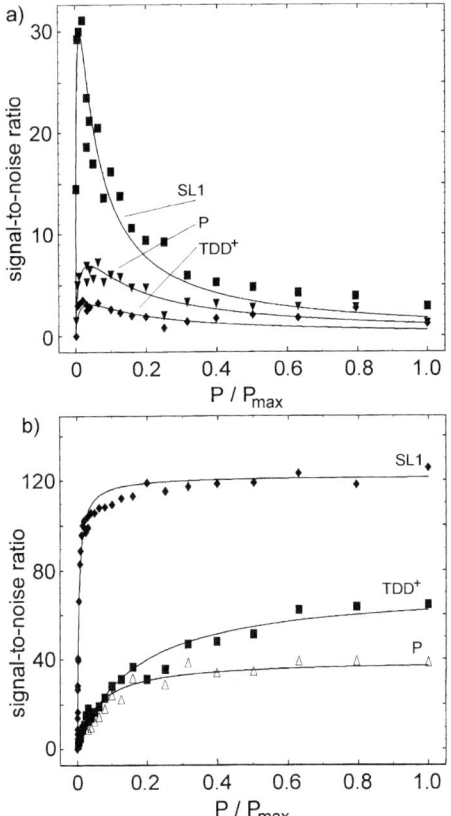

FIG. 22. EPR microwave power dependence of the lines of SL1, P, and TDD$^+$ (NL8) defects shown in Fig. 21. The various symbols are the experimental signal-to-noise ratios of the EPR lines. The solid curves are fitted according to theory. $P_{max} = 200$ mW ($v_{EPR} = 9.3$ GHz, $B \parallel [011]$, $T = 20$ K). (a) Microwave power dependence of the conventionally detected EPR lines, fits are according to Eq. (21). (b) Microwave power dependence of the EDEPR lines, fits are according to Eq. (23). (Reprinted with permission from Stich, B., Greulich-Weber, S., and Spaeth, J.-M. (1995). *Journal of Applied Physics* **77**, 1546. Copyright ©1995 American Institute of Physics.)

The recombination of weakly exchange-coupled neutral donors and acceptors, be it radiative or nonradiative, is a spin-dependent process and can thus be influenced by magnetic dipole transitions, as is well known from ODEPR (Spaeth *et al.*, 1992). This was proposed to be the mechanism that can quantitatively explain the observation of EDEPR (Stich *et al.*, 1995). At low concentration, single defects, such as P donors alone, cannot be detected

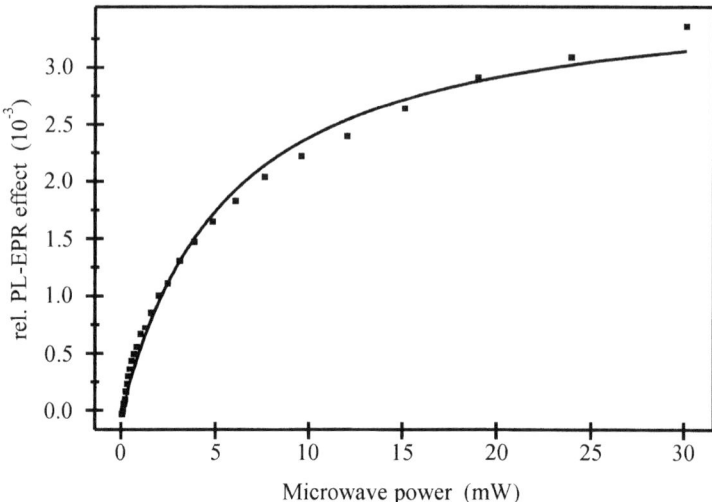

FIG. 23. Dependence of the PL-detected EPR signal of the residual donor in GaN (on sapphire substrate) on microwave power. $B_0 \parallel \bar{c}$-axis, photon energy of the excitation light was $E = 3.76\,\text{eV}$; the $D^0 - A^0$ pair recombination luminescence peaked at 2.2 eV ("Yellow Luminescence"). The maximum microwave power was approximately 30 mW. The solid line was calculated according to Eq. (28) (Koschnick, 1997).

by EDEPR in contrast to earlier models (Rong et al., 1992; Honig, 1966; Lannoo et al., 1993). The concept that electron-hole recombination is responsible for EDEPR has been suggested previously (Kaplan et al., 1978; Kaplan and Pepper, 1980), however, with the difference that the electron-hole recombination was thought to occur at one defect and not between separated donors and acceptors. At least, this mechanism of separated but weakly exchange-coupled pairs of $D^0$ and $A^0$ defects was not thought to be the decisive mechanism. If the electron-hole recombination could occur at one paramagnetic defect alone, P donors alone would have been seen, or, for example, deep chalcogen paramagnetic centers independent of the compensating B concentration. This was not the case (Stich et al., 1995). The situation is different when the concentration of defects increases. Then, for instance, when hopping conductivity between donors can occur, other mechanisms seem to operate (discussion follows).

2. THE DONOR-ACCEPTOR RECOMBINATION MODEL

A general description of the EDEPR signal with the donor-acceptor pair recombination model leads to a complicated set of rate equations, by which the temporal change of the excess charge carriers is calculated as a function of their generation (e.g., by above-bandgap light) and recombination via the spin-dependent, exchange-coupled, donor-acceptor pair states. For the concentration $n$ of conduction-band electrons, one obtains, for example (Stich, 1996; de le Broise et al., 1994)

$$\frac{d}{dt}n = g_0 - c_n n(N - n_D - n_{DA}) + e_n(n_D + n_{DA}), \qquad (24)$$

where $g_0$ is the generation rate of the charge carriers, $N$ is the number of possible $D^0-A^0$ pairs, $c_n$ is the capture cross-section of the donors, $e_n$ is the emission rate, $n_D$ is the concentration of $D^0-A^0$ pairs having captured an electron, and $n_{DA}$ is the concentration of $D^0-A^0$ pairs having captured an electron and a hole. An analogous expression holds for the concentration of holes in the valence band. For $n_A$, $n_D$, $n_{DA}$ also rate equations hold containing recombination rates, microwave transition probability, spin-lattice relaxation, and so on. Such a complicated system of rate equations cannot be solved analytically. Furthermore, non-spin-dependent recombination processes (shunt processes) are not yet taken into account in Eq. (24), and many quantities are not well enough known, such as the capture cross-sections and emission rates. For a certain set of parameters, a numerical solution of the rate equation systems can be calculated. Some examples and a detailed discussion are found in Stich (1996).

However, many of the essential features of the experimental observations can be explained within a simplified donor–acceptor pair recombination model, in which the capture and emission of electrons and holes by the donors and acceptors (which may have different concentrations) are neglected; that is, the details of the connection between the charge carrier concentration in the bands and the single and coupled donors and acceptors are not taken into account explicitly, as in Eq. (24). The temporal development of the charge carrier concentration is assumed to be

$$\frac{dn}{dt} = g_0 - Rn, \qquad (25)$$

where $R$ is the total recombination rate of the charge carriers determined by the $D^0-A^0$ pair recombination process. Since only the stationary solution is needed ($dn/dt = 0$) and one is only interested in the difference effect with and without microwave transitions, one obtains

$$S_{\text{EDEPR}} \propto \Delta n_{\text{EPR}} = \frac{\partial n}{\partial R}\Delta R = -\frac{g_0}{R}\frac{\Delta R}{R} = -n\frac{\Delta R}{R} \qquad (26)$$

The calculation of $\Delta R$ requires the solution of a system of rate equations describing the temporal evolution of the occupancy of the exchange-coupled

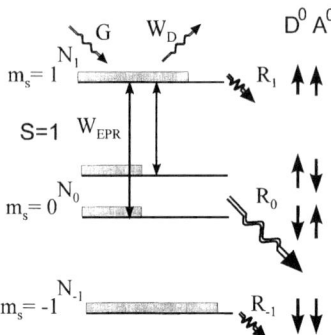

FIG. 24. Schematic representation of donor–acceptor $D^0$–$A^0$ pair recombination. Spin-lattice relaxation is neglected. The levels with $m_s = 1, 0, -1$ are occupied with a pair generation probability $G$. $N_{1,0,-1}$ are the occupation numbers of the $m_s$ states. $W_{EPR}$ is the EPR transition probability, $W_D$ denotes a dissociation probability, and $R_{1,0,-1}$ are the (nonradiative) recombination probabilities of the $m_s$ states.

$D^0$–$A^0$ system according to Fig. 24, which is a schematic representation of their states and of the recombination to the singlet ground state. Spin-lattice relaxation is neglected. The levels with $m_s = 1, 0, -1$ are occupied with a pair generation probability $G$. $N_1$, $N_0$, and $N_{-1}$ are the occupancies of the $m_s$ states. Exchange coupling is assumed to be very weak; since the ground state is a singlet state, the recombination probability $R_0$ is expected to be much longer than those of the other two states, $R_1$ and $R_{-1}$, so that the occupancy of the $m_s = \pm 1$ levels under stationary above-bandgap illumination conditions is much higher than that with $m_s = 0$. $W_{EPR}$ is the transition probability between the $m_s = \pm 1$ levels and the $m_s = 0$ levels. $W_D$ denotes the dissociation probability describing a thermally induced dissociation of electron-hole pairs. $N$ is the number of states that can be occupied. For this scheme, one obtains the following rate equations for the occupancies $N_i (i = 1, 0, -1)$ (Stich et al., 1995; Stich, 1996).

$$\frac{dN_i}{dt} = G\left(N - \sum_{j=-1}^{+1} N_j\right) - N_i\left(W_{EPR} + W_D + R_i\right) + \sum_{j \neq i} N_j W_{EPR}. \quad (27)$$

The simplest case would be to assume $R_0 \neq 0$, $R_1 = R_{-1} = 0$, and $W_D = 0$. The EDEPR signal is proportional to the difference of the occupancies with and without inducing EPR transitions. In this approximation, with a stationary solution of Eq. (27), one obtains

$$S_{EDEPR} \propto \sum_i (N_i(EPR) - N_i(0)) = C \frac{W_{EPR} T_R}{1 + W_{EPR} T_R} \quad (28)$$

with

$$C = \frac{2GN}{4G + R_0} \text{ and } T_R = \frac{4G + R_0}{GR_0} \qquad (29)$$

The result of Eq. (28) qualitatively describes the experimental saturation curves of Fig. 22b. The parameter $\beta$ in Eq. (23) is the parameter $T_R$ of Eq. (28). It was fitted to the experimental curves in Fig. 22b and reflects the different recombination probabilities $R_0$. $G$ is also probably not exactly the same in all experiments. For low microwave power, the signal is proportional to $W_{EPR}$, which in turn is proportional to the microwave power $P_{MW}$:

$$W_{EPR} = \frac{1}{4} g\mu_B B_1^2 \propto P_{MW}, \qquad (30)$$

where $g$ is the $g$-factor of either the donor or the acceptor, $\mu_B$ is the Bohr magneton, and $B_1$ is the amplitude of the microwaves. The maximum effect depends on the pair generation probability $G$ as well as on the number of defects. The former is given by the electron-hole capture probability into the $D^+ - A^-$ states as well as by the above-bandgap light intensity producing electron-hole pairs. The results are in qualitative agreement with the experimental observation within the approximation of Eq. (25) (i.e., regarding only the recombination of the coupled pairs). Inclusion of spin-lattice relaxation or $W_D$ does not change the form of Eq. (28). Only the expressions for $C$ and $T_R$ becomes more complicated (Stich, 1996).

An increase of the exchange coupling between donors and acceptors $D^0$ and $A^0$ would result in a relative shift of the level position in Fig. 24 and probably results in a line broadening compared to the conventional EPR lines because of a statistical distribution of couplings due to the statistical distribution of the separations between donors and acceptors. Such a broadening has not yet been observed. The probable reason is that, due to the increase in the recombination rates upon increasing the exchange coupling, the EPR transitions cannot be induced anymore within the lifetimes of the spin states because of the limited available microwave power of 100 to 200 mW. The observation of the line broadening or line shift probably requires higher microwave powers. In the limiting case of strong coupling, one would deal with a triplet system. The SL1 center provides such a triplet system for which EDEPR could be observed very well (see Fig. 21). Its recombination mechanism is discussed in detail by Stich (1996).

A discussion of limiting cases, such as $W_D \to \infty$, $N \to 0$, $G \to \infty$, 0 showed that Eqs. (26) and (28) do not provide an adequate description of the EDEPR effects. The reason is not that the $D^0 - A^0$ pair mechanism is not correct, but that the model discussed so far does not adequately consider the connection between the charge carrier concentrations and the valence and conduction bands, respectively, and the generation, dissociation, and recombination of the $D^0-A^0$ pairs. For details, the reader is referred to Stich (1996).

Xiong and Miller (1993) presented rate equations for the occupation and recombination of weakly exchange-coupled electron-hole pairs, in which they also included spin-lattice relaxation and the electron-hole dissociation probability $W_D$. Their expression for $S_{EDEPR}$ generally has the same form as Eq. (28); however, in their work, $C$ is proportional to $W_D$. For vanishing $W_D$, the whole EDEPR effect vanishes, which does not seem compatible with the $D^0-A^0$ recombinations observed (Stich et al., 1995) in EDEPR as well as in ODEPR at very low temperature (6 K). At very low temperature, $W_D$ is certainly small, considering the ionization energies of the defects involved. The temperature dependence of EDEPR-discussion follows shows that the EDEPR effect increases still below 6 K (Stich, 1996). The explanation for the discrepancy probably is that in the rate equations of Xiong and Miller (1993), the occupation rate is constant for all four levels of Fig. 24, which is not true if there is only a finite number of electron-hole pairs available of the same order of magnitude as levels $N_i$, which can be distributed over the four weakly coupled electron-hole pair states.

### 3. Concentration and Temperature Dependence of EDEPR Signals

The concentration dependence of EDEPR is not well understood and needs further study. High-defect concentrations seem unfavorable. When producing $10^{14}$ cm$^{-3}$ SL1 centers in Si by electron irradiation, no EDEPR signals are observed. It was qualitatively observed that signals appear after annealing out the SL1 centers. The signals have a rather broad maximum for concentrations where the conventional EPR signal is hardly detectable (Fig. 25) (Greulich-Weber et al., 1995) until it decreases for very low concentrations again. In a series of experiments, the concentration of thermal donors TDD$^+$ was varied by variation of the annealing time of Cz-Si at 450°C. Surprisingly, it was observed that EDEPR signals were still measured, with a very good signal-to-noise ratio for concentrations well below $10^{12}$ cm$^{-1}$, at which IR control measurements could no longer detect TDDs. It sufficed to quench the Cz samples from high temperature (770–1200°C) in order to observe the EDEPR signals of TDDs, a new experience (Greulich-Weber et al., 1995; Stich, 1996). Experimentally it seems that there is a broad range of concentrations of "distant" donor and acceptor pairs, in

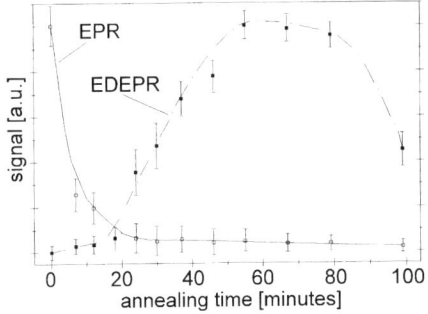

FIG. 25. SL1 EPR signal and EDEPR signal, depending on the SL1 concentration due to successive annealing. (From Greulich-Weber et al., 1995, with permission.)

which EDEPR can be observed with very good signal-to-noise values. As mentioned, if the recombination times become too fast compared to the inverse of the EPR transition probability, as they probably do for closer pairs, EDEPR is no longer observable. The limit for $R_0^{-1}$ seems to be of the order of 100 ns, for a microwave power of 100 to 200 mW, usually available in conventional EPR spectrometers. However, a quantitative relation between defect concentration and EDEPR signal intensity must yet be established. It also will probably depend on the energy level of the defects in the gap (i.e., whether the recombination involves only shallow or also deep level defects).

First experiments with a high concentration of $N$ donors in 6H-SiC revealed interestingly different features compared to what has been discussed so far. For a concentration of about $10^{18}$ cm$^{-3}$ $N$ donors, a single line of about $g = 2$ was observed as an increase of conductivity and not as a decrease upon enhanced recombinations, as in the case of the $D^0-A^0$ pair recombination mechanism. When investigating the EDEPR of $N$ donors in a $B$-compensated 6H-SiC sample with lower concentration, the $N$ hf triplet $(I(^{14}N) = 1)$ was observed (Grasa-Molina et al., 1998). At high concentration, the mechanism seems to be different from the one discussed so far. The effect is probably connected with hopping conductivity occurring at higher donor concentrations. The loss of hf-splitting points to a rapid motion of the donors (Spaeth et al., 1992). From the observed temperature dependence, according to our experiments, it does not seem to be clear whether the mechanism proposed earlier by Morigaki and Onda (1972) on the basis of observations in germanium is correct, according to which electrons are excited to an impurity conduction band by a thermal effect due to microwaves. Further investigations are underway (Grasa-Molina et al., 1998).

It is very possible that many observations made in diodes, in which usually only a structureless line of about $g = 2$ is observed without hf-splittings, are also the result of donor concentrations that were too high, such that hopping effects dominated. It will be important to measure whether the conductivity increases or decreases upon microwave transitions and which is the size of the effect, in order to learn more about the underlying processes. (The relative change of the conductivity typically is of the order of $10^{-3}$ to $10^{-4}$ for $D^0$–$A^0$ pair recombinations in silicon.)

Little is known about the temperature dependence of the EDEPR effects. It was reported that signals measured in diodes at room temperature disappeared upon cooling (Stich et al., 1993). The general observation for the $D^0$–$A^0$ pair recombination, as studied in Si (Stich et al., 1995; Stich, 1996), was that, upon cooling, the signals increased. Most experiments were done between 6 and 20 K. For thermal donors in Si, an investigation between 6 and 16 K revealed a temperature dependence following a $T^{-2}$ law: $S_{EDEPR} \propto T^{-2}$ (Stich, 1996). This can be explained by assuming that the dominant influence is the temperature dependence of the spin-lattice relaxation time $T_1$, for which a direct process, including the photon bottleneck effect, was assumed. For the latter process $T_1 \propto T^{-2}$ is expected (Abragam and Bleaney, 1986). Surprisingly, it was found that above 16 K, the EDEPR effect increased again, possibly due to increasing $D^0$–$A^0$ pair dissociation effects, restoring the concentration of carriers in the conduction band (Stich, 1996). However, more systematic studies of temperature dependence are needed.

## IV. Electrical Detection of ENDOR (EDENDOR)

In a Cz-grown P-doped sample, which contained, apart from shallow P donors, the SL1 center after electron irradiation, ENDOR of the P donors could be detected in the photoconductivity (Stich et al., 1996). Figure 26 shows the ENDOR spectrum for $B_0 \parallel [100]$ measured in the low-field EDEPR line as an rf and microwave-induced change of the photoconductivity. The best signals were obtained with frequency modulation of 1 kHz and at 4.5 K. The spectrum of Fig. 26a contains two ENDOR lines at about 52.60 and 64.93 MHz for $m_s = \pm\frac{1}{2}$ due to the well-known $^{31}P$ hf interaction separated by $2v_n(^{31}P) = 12.33$ HHz (see, e.g., Spaeth et al., 1992), $v_n$ being the Larmor frequency of the free $^{31}P$ nucleus. At $v_n(^{29}Si) = 2.84$ MHz, the $^{29}Si$ "distant" ENDOR line is observed. Up to 6 MHz, well-known $^{29}Si$ shf lines could be measured (Feher, 1959; Hale and Mieher, 1969), the frequency positions of which are indicated by a stick diagram in Fig. 26c. The P

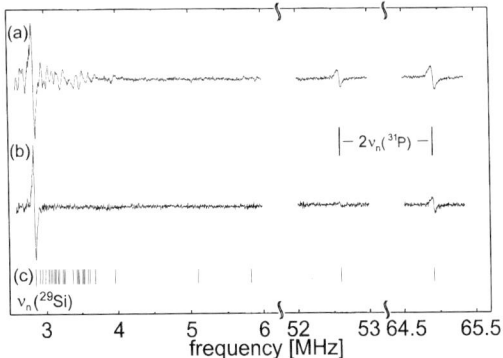

FIG. 26. (a) EDENDOR spectrum of the shallow P donors in Si measured with frequency modulation and with above-band gap illumination in the low-field EDEPR line (see Fig. 21) at $B = 333.3$ mT and at $T = 4.5$ K ($P_{MW} = 1$ mW). (b) Conventional stationary ENDOR spectrum of the same sample as in (a) measured with above-bandgap illumination. (c) Stick spectrum of the $^{31}$P and $^{29}$Si ENDOR lines calculated from the known hf and shf interactions after Hale and Mieher (1969) (Reprinted with permission from Stich, B., Greulich-Weber, S., and Spaeth, J.-M. (1996). *Applied Physics Letters* **68**, 1102. Copyright ©1996 American Physics Institute.)

concentration was chosen to be small enough that EDEPR and EDENDOR could be measured but high enough for conventional ENDOR measurements. Figure 26b shows the conventional stationary ENDOR spectrum measured with the same sample as in Fig. 26a at the same orientation, using the same illumination and frequency modulation. The $^{31}$P hf lines and the $^{29}$Si distant ENDOR line are detected, but the concentration was not sufficient to detect the $^{29}$Si ligands. Their signals seem to be particularly intense in electrical detection.

A systematic study of the dependence of the EDENDOR signals on the microwave power $P_{MW}$ and on the rf power $P_{HF}$ qualitatively gave the result that the signals increased with both $P_{MW}$ and $P_{HF}$ (Fig. 27) (Stich et al., 1996). The experimental results could be described with a quantitative relation for the EDENDOR signals, which was also derived from rate equations similar to Eq. (27), however, expanded by terms comprising the rf NMR transition probabilities and the nuclear Zeeman levels (Stich, 1996):

$$S_{EDENDOR} \propto \frac{W_{EPR} T_R}{1 + W_{EPR} T_R} \frac{W_{NMR} T_{HF}}{1 + W_{NMR} T_{HF}} \quad (31)$$

$W_{EPR}$ and $W_{NMR}$ are the transition probabilities of the EPR and NMR

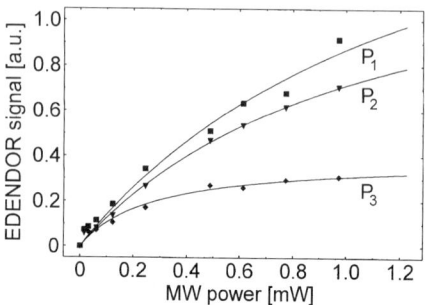

FIG. 27. Microwave power dependence of the $^{31}$P EDENDOR signal at $v = 64.93$ MHz at $T = 4.5$ K for three different rf power levels ($P_1 > P_2 > P_3$). The solid lines were fitted according to Eq. (31). The rf circuit was fed by a 100-W rf on 50 $\Omega$ (Reprinted by permission from Stich, B., Greulich-Weber, S., and Spaeth, J.-M. (1996). *Applied Physics Letters* **68**, 1102. Copyright ©American Physics Institute.)

transitions, respectively, and $T_R$ and $T_{HF}$ are parameters, depending on the details of the $D^0-A^0$ pair recombination model assumed. In the simplest case, $T_R$ is given by the expression of Eq. (29). $T_{HF}$ also depends on $W_{EPR}$:

$$T_{HF} \propto \frac{(W_{EPR} + 1/T_R)T_{R'}}{1 + W_{EPR} T_{R'}} \quad (32)$$

which, for large values of $W_{EPR}$, becomes an expression independent of $W_{EPR}$, depending only on the recombination rates $W_D$, $G$, and so on (for details, see Stich, 1996).

Thus, the EDENDOR signal depends on the microwave power, similar to the EDEPR signals, and on the rf power in an analogous way. This latter result is very different from conventional ENDOR, where, upon high rf power, the ENDOR signals saturate and decrease (see, e.g., Spaeth *et al.*, 1992).

## V. New Possibilities

To see an EDEPR effects, it is not necessary to illuminate the whole sample. Even when illuminating only a small spot—for example, with an HeNe laser (632 nm, 50 mW)—the EDEPR spectrum can be detected (Figs. 28 and 29) (Stich *et al.*, 1995). Thus, with electrical detection via photoconductivity, it is possible to measure EPR with spatial resolution. In the experiment of Fig. 29 (Stich *et al.*, 1995), the light spot had a diameter of

## 2 MAGNETO-OPTICAL, ELECTRICAL DETECTION OF PR

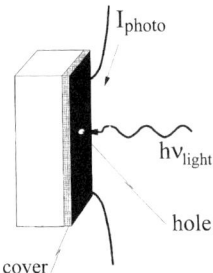

FIG. 28. Experimental setup for measuring spatially resolved EDEPR. The sample ($3 \times 4 \times 5\,\text{mm}^3$) was covered except for a small hole of 0.6-mm diameter, which was illuminated with an HeNe laser (632 nm, 50 mW). EDEPR was measured, detecting directly the change in photoconductivity via electrical contacts. (Reprinted with permission from Stich, B., Greulich-Weber, S., and Spaeth, J.-M. (1995). *Journal of Applied Physics* **77,** 1546. Copyright ©American Institute of Physics.)

0.6 mm. The light beam can be scanned throughout the sample surface, and a two-dimensional distribution can, in principle, be measured, whereby, in this kind of experiment, the sample is integrated over the penetration depth of the light (in Si, about 5 $\mu$m). One must be certain, however, that no other diamagnetic recombination processes take place simultaneously. Further work is needed to study in detail such kinds of mapping experiments.

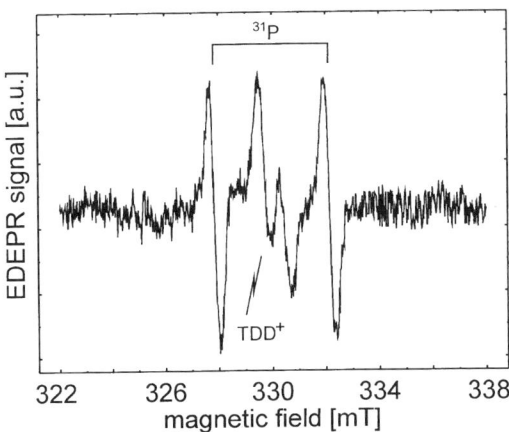

FIG. 29. Part of the EDEPR spectrum of Fig. 21, measured when illuminating only a small spot on the sample (see Fig. 28) with above-bandgap light (B ∥ [111] $T = 20\,\text{K}$, $\nu_{\text{EDEPR}} = 9.3\,\text{GHz}$).

The sensitivity of EPR and EDENDOR can be several orders of magnitude higher compared to conventional detection. In Fig. 21, for example, for the SL1 center, the signal-to-noise ratio was higher than in conventional EPR, while that for the P donors was about equal. Yet, one has to consider that in conventional EPR, the whole sample is measured, while in EDEPR only a layer of the thickness of the above-bandgap light penetration depth is measured. Thus, even for the P donors in Si in Fig. 21, the sensitivity gain is about a factor of 600 (sample thickness, 3 mm; light penetration, 5 μm). Since it is possible to illuminate only a spot, defects in small volumes can be measured. For the Si sample containing P and the SL1 center (Fig. 29) (Stich et al., 1995), it was estimated that $6 \times 10^7$ P atoms were detectable. Similarly, in the EDENDOR experiment, about $6 \times 10^9$ P nuclei were detectable, which is a sensitivity gain of $10^4$ to $10^5$, compared to conventional ENDOR. As seen from Fig. 26, the sensitivity gain is particularly high for the ligand ENDOR lines.

A further interesting feature is that the EDEPR signal intensities do not depend on the size of the magnetic field. This is a direct consequence of the $D^0-A^0$ pair recombination model and in contrast to earlier models for the effect by Lépine et al. (1972), who claimed that the effect of EDEPR is due to spin-dependent carrier capture and should be proportional to the spin polarization of the conduction electrons and proportional to the polarization of the Zeeman ground state of the defect. According to this model, the decrease of the photoconductivity (i.e., the EDEPR effect) should vary with

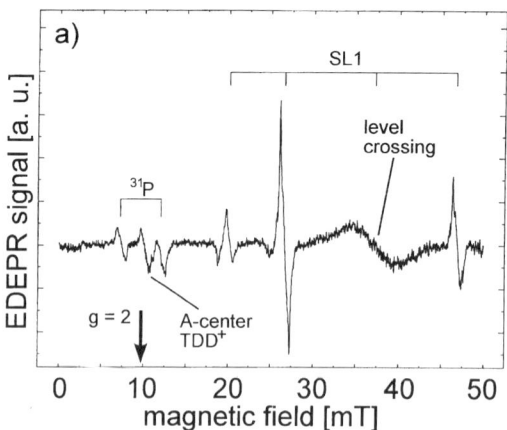

FIG. 30. EDEPR spectrum of an electron-irradiated Cz-Si:P crystal also containing thermal double donors NL8 measured with $v_{EPR} = 280$ MHz at 11 K ($B \parallel$ [110]) using a low-frequency EDEPR spectrometer of special design (see Fig. 32) (Pinkawa, 1996).

$(B/T)^2$. However, it was observed that, given the $B_1$ amplitude at the sample is the same, the signal-to-noise ratio of the EDEPR signal does not depend on the resonance frequency (or field), respectively. Figure 30 shows the EDEPR spectrum of a sample containing P shallow donors, thermal donors, and the SL1 center measured at about 280 MHz. The signal-to-noise ratio is excellent and only depends on $P_{MW} \propto B_1^2$ and the temperature (see also Stich et al., 1995). This failure of the Lépine model was also observed previously (Mima et al., 1980). In Fig. 30 there is also a level crossing of the SL1 center seen at about 40 mT, also previously observed by Vlasenko and Khramtsov (1985). The angular dependence measured at 280 MHz shows the typical features of a low field spectrum with loops rather than open branches (Fig. 31) (Stich et al., 1995). The possibility to measure at low field or low frequency opens up the possibility to construct inexpensive and small spectrometers to measure EDEPR. Figure 32 shows the cross-section of such a low-frequency spectrometer, built at the University of Paderborn. The spectrum of Fig. 30 was measured with it (Pinkawa, 1996). At low frequency, one is not restricted to a small microwave cavity. For frequencies such as 300 MHz, a system of solenoids can be used instead of a cavity, and

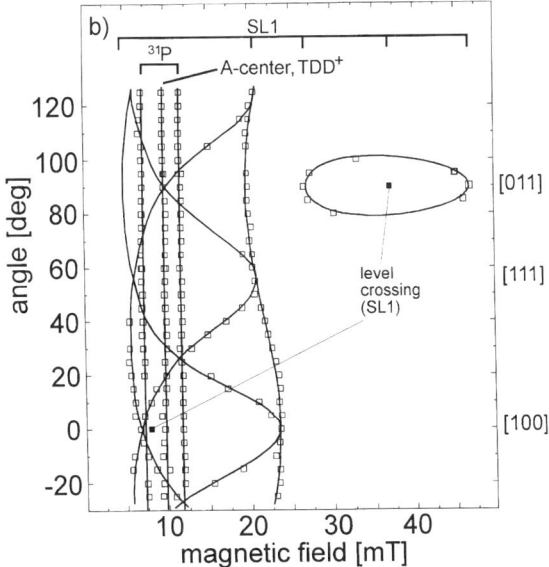

FIG. 31. Angular dependence of the EDEPR spectrum of Fig. 30. The open squares are the experimental line positions; the solid lines of the SL1 center were calculated with the EPR parameters after Brower (1971).

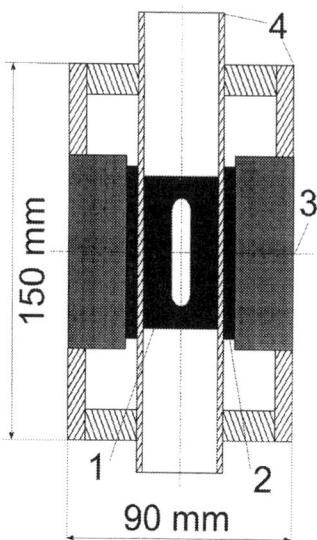

FIG. 32. Cross-section of a low-frequency EDEPR spectrometer (Pinkawa, 1996): 1, rf solenoids; 2, modulation coils; 3, magnetic field coils; 4, holder for cryosystem.

thereby, larger samples can be investigated (e.g., for mapping purposes using the feature of spatial resolution). All of this new potential has not yet been exploited, except for a few first steps.

REFERENCES

Abragam, A. and Bleany, B. (1986). *Electron Paramagnetic Resonance of Transition Ions.* Dover, New York.
Baeumler, M., Meyer, B. K., Kaufmann, U., and Schneider, J. (1989). *Mat. Sci. Forum* **38–41,** 797.
Broise, de le, X., Lannoo, M. M., and Stiévénard, M. D. (1994). Rapport de project, Univ. des Sciences et Techniques de Lille, unpublished.
Brower, K. L. (1971). *Phys. Rev. B* **4,** 1968.
Christmann, P., Wetzel, C., Meyer, B. K., Asenov, A., and Endrös, A. (1992). *Appl. Phys. Lett.* **60,** 1857.
Clerjaud, B. C., Nand, B., Deveaud, B., Lambert, B., Plot, B., Bremond, G., Benjeddon, C., Guillot, G., and Nouailhat, A. (1985). *J. Appl. Phys.* **59,** 4207.
Feher, G. (1959). *Phys. Rev.* **114,** 1219.
Fockele, M., Spaeth, J.-M., Gibart, P. (1990). *Mat. Sci. Forum* **65–66,** 443.
Geschwind, S. (1972). Optical techniques in EPR in Solids. In: *Electron Paramagnetic Resonance,* edited by S. Geschwind. Plenum, New York.

Görger, A., Meyer, B. K., Spaeth, J.-M., and Hennel, A. (1988). Transition elements in III-V-semiconductors—A study with optically detected magnetic resonance. In: *Semi-insulating III-V Materials*, edited by G. Grossmann and L. Lebedo. Adam Hilger, Bristol, 331.
Grasa-Molina, I., Greulich-Weber, S., and Spaeth, J.-M. (1998). To be published.
Greulich-Weber, S., Stich, B., and Spaeth, J.-M. (1995). *Mat. Sci. Forum* **196-201**, 1509.
Hage, J., Niklas, J. R., and Spaeth, J.-M. (1984). *J. Electron. Mat.* **A14**, 1051.
Hage, J., Niklas, J. R., and Spaeth, J.-M. (1989). *J. Phys. C.* **4**, 773.
Hale, E. B. and Mieher, R. L. (1969). *Phys. Rev.* **184**, 739.
Heinemann, M., Meyer, B. K., Spaeth, J.-M., and Löhnert, K. (1987). The occupation of the two charge states of EL2 in LEC-grown GaAs-Wafers—a mapping investigation. In: *Defect Recognition and Image Processing in III-V Compounds II*, edited by E. R. Weber. Elsevier, New York.
Honig, A. (1966). *Phys. Rev. Lett.* **17**, 186.
Kaplan, D. and Pepper, M. (1980). *Solid State Commun.* **34**, 803.
Kaplan, D., Solomon, I., and Mott, N. F. (1978). *Le Journal de Physique Lettres* **39**, L51.
Katayama-Yoshida, H., and Zunger, A. (1986). *Phys. Rev. B* **33**, 2961.
Koschnick, F. K. and Spaeth, J.-M. (1996). *Phys. Rev. B* **54**, 11042.
Koschnick, F. K. (1997). Private communication.
Koschnick, F. K., Krambrock, K., Hesse, M., and Spaeth, J.-M. (1995). *Appl. Phys.* **A60**, 551.
Krambrock, K. and Spaeth, J.-M. (1993). *Phys. Rev. B* **47**, 3187.
Krambrock, K., Linde, M., Spaeth, J.-M., Look, D. C., Bliss, D., and Walukiewicz, W. (1992a). *Semicond. Sci. Technol.* **7**, 1037.
Krambrock, K., Spaeth, J.-M., Delerue, C., Allan, G., and Lannoo, M. (1992b). *Phys. Rev. B* **45**, 1481.
Kurylev, V. V. and Karyagin, S. N. (1974). *Phys. Stat. Solidi (A)* **21**, K127.
Lannoo, M., Vuillaume, D., Deresmes, D., and Stiévenard, D. (1993). *Microelectron. Eng.* **22**, 143.
Lépine, D. J. (1972). *Phys. Rev. B* **6**, 436.
Lépine, D. J. and Prejean, J. J. (1970). *Proceedings of the 9th International Conference on the Physics of Semiconductors*, Boston.
Maier, K., Schneider, J., Wilkening, W., Leibenzeder, S., and Stein, R. (1990). Proceedings of the Symposium C on Properties and Applications of SiC, Natural and Synthetic Diamond and Related Materials of the 1990 E-MRS Fall Conference. *Mat. Sci. Eng.* **B11**, 27.
Mendz, G. and Haneman, D. (1978). *Phys. C: Solid State Phys.* **11**, L197.
Mendz, G., Miller, D. J., and Haneman, D. (1979). *Phys. Rev. B* **20**, 5246.
Meyer, B. K. (1988). *Rev. Phys. Appl.* **23**, 809.
Meyer, B. K., Hofman, D. M., Niklas, J. R., and Spaeth, J.-M. (1987). *Phys. Rev. B* **36**, 1332.
Meyer, B. K., Spaeth, J.-M., and Scheffler, M. (1984). *Phys. Rev. Lett.* **52**, 851.
Mima, L. S., Strikha, V. I., and Tretyak, O. V. (1980). *Sov. Phys. Semicond.* **14**, 1328.
Mollenauer, L. F. and Pan, S. (1972). *Phys. Rev. B* **6**, 772.
Morigaki, K. and Onda, M. (1972). *J. Phys. Sci. Jpn.* **33**, 1031.
Muller, S. H., Tuynman, G. M., Sieverts, E. G., and Ammerlaan, C. A. J. (1982). *Phys. Rev. B* **25**, 25.
Neubrand, A. (1978). *Phys. Stat. Solidi (B)* **86**, 269.
Paus, H. (1980). *Farbzentren und zweiwertige Fremdkationen in Alkalihalogenid-Kristallen*. Habilitationsschrift, Stuttgart.
Pinkawa, C. (1996). Diplomarbeit, University of Paderborn.
Reinke, J., Greulich-Weber, S., and Spaeth, J.-M. (1993a). *Solid State Commun.* **85**, 1017.
Reinke, J., Greulich-Weber, S., and Spaeth, J.-M. (1996). *Solid State Commun.* **96**, 835.
Reinke, J., Weihrich, H., Greulich-Weber, S., and Spaeth, J.-M. (1993b). *Semicond. Sci. Technol.* **8**, 1862.

Rong, F. C., Buchwald, W. R., Poindexter, E. H., Warren, W. L., and Keeble, D. J. (1991). *Solid State Electron.* **34**, 835.
Rong, F. C., Gerardi, G. J., Buchwald, W. R., Poindexter, E. H., Umlor, M. T., Keeble, D. J., and Warren, W. L. (1992). *Appl. Phys. Lett.* **60**, 610.
Schiff, L. I. (1949). *Quantum Mechanics.* McGraw-Hill, New York.
Schmidt, J. and Solomon, I. (1966). *C. R. Acad. Sci. Paris* **263**, 169.
Schneider, J., Müller, H. D., Maier, K., Wilkening, W., Fuchs, F., Dörnen, A., Leibenzeder, S., and Stein, R. (1990). *Appl. Phys. Lett.* **56**, 1184.
Solomon, I. (1976). *Solid State Commun.* **20**, 215.
Spaeth, J.-M. and Krambrock, K. (1993). *Festkörperprobleme (Advances in Solid State Physics)* **33**, 111.
Spaeth, J.-M. and Linde, M. (1994). *Defect and Diffusion Forum* **108**, 35. Ch. 2 in "DX centers, donors in AlGaAs", edited by E. Muñoz (1994).
Spaeth, J.-M., Hofmann, D. M., Heinemann, M., and Meyer, B. K. (1988). *J. Int. Phys. Conf. Ser.* **91**, ch. 4, 391.
Spaeth, J.-M., Krambrock, K., and Hofmann, D. M. (1990). *Proc. of the ICPS, Thessaloniki,* edited by E. M. Anastassakis and J. D. Joannopoulus. World Scientific, Singapore, vol. **1**, 441.
Spaeth, J.-M., Niklas, J. R., and Bartram, R. H. (1992). *Structural Analysis of Point Defects in Solids: An Introduction to Multiple Magnetic Resonance Spectroscopy.* Springer-Verlag, Heidelberg, New York, Vol. **43**.
Spaeth, J.-M., Greulich-Weber, S., März, M., Reinke, J., Feege, M., Kalabukhova, E. N., and Lukih, S. N. (1997) *Mat. Sci. Forum* **239–241**, 149.
Stich, B. (1996). Ph.D. Thesis, University of Paderborn.
Stich, B., Greulich-Weber, S., and Spaeth, J.-M. (1993). *Semicond. Sci. Technol.* **8**, 1385.
Stich, B., Greulich-Weber, S., and Spaeth, J.-M. (1995). *J. Appl. Phys.* **77**, 1546.
Stich, B., Greulich-Weber, S., and Spaeth, J.-M. (1996). *Appl. Phys. Lett.* **68**, 1102.
Vlasenko, L. S. and Khramtsov, V. A. (1985). *Sov. Phys. JETP Lett.* **42**, 38.
Vlasenko, L. S., Vlasenko, M. P., Lomasov, V. N., and Khramtsov, V. A. (1986). *Sov. Phys. JETP* **64**, 612.
Watkins, C. D. and Corbett, J. W. (1960). *Phys. Rev.* **121**, 1001.
Xiong, Z. and Miller, D. J. (1993). *Appl. Phys. Lett.* **63**, 352.

CHAPTER 3

# Magnetic Resonance of Epitaxial Layers Detected by Photoluminescence

*T. A. Kennedy and E. R. Glaser*

NAVAL RESEARCH LABORATORY
WASHINGTON, DC

| | |
|---|---|
| I. INTRODUCTION | 93 |
| II. FUNDAMENTALS OF ODMR AND EPITAXY | 94 |
|    1. *ODMR Detected on Photoluminescence* | 94 |
|    2. *Epitaxy* | 101 |
| III. ILLUSTRATIVE EXAMPLE: BULK InP:Zn | 104 |
|    1. *Introduction* | 104 |
|    2. *ODMR of the Effective-Mass Donor* | 105 |
|    3. *ODMR of $P_{In}$* | 106 |
|    4. *ODMR of the Effective-Mass Acceptor* | 110 |
|    5. *Conclusion* | 112 |
| IV. EXAMPLES IN EPILAYERS | 113 |
|    1. *Effective-Mass Donors* | 113 |
|    2. *Deep Centers* | 122 |
|    3. *Acceptors* | 129 |
| V. SUMMARY AND FUTURE DIRECTIONS | 133 |
| References | 134 |

## I. Introduction

The enormous varieties of layers and heterostructures grown by epitaxial methods in recent years contain point defects, which often determine their electrical and optical properties. Optically detected magnetic resonance (ODMR) using photoluminescence (PL) offers a method to identify residual defects and elucidate the properties of dopants in these layers. The method relies on the spin-dependence of recombination processes and combines PL with electron paramagnetic resonance (EPR). The PL provides very high sensitivity to allow the study of defects at residual levels in state-of-the-art samples. The EPR provides very high resolution to reveal the local symmetry and hyperfine structure of the defect. In favorable cases, ODMR provides a full picture of both the atomic structure and the electronic structure of the defect.

This chapter describes the method and some recent examples of its application. ODMR was first applied to defects in bulk semiconductors, and this work serves as a background to the work in epitaxial structures. Excellent reviews of the earlier work are available (Cavenett, 1981; Davies, 1985, 1988). In the studies of defects in epilayers, two new elements appear. First, the growth conditions are quite different from those for bulk semiconductors. The temperatures are much lower, and kinetic effects are important. Second, most epilayers are grown on substrates of a different chemical composition. In the resulting heterostructures, strain plays a key role.

This chapter is organized as follows. Section I is this introduction. Section II describes the fundamentals of ODMR and gives pertinent information about epitaxy. Section III gives a comprehensive example of the application of ODMR to the defects in bulk-grown InP:Zn. Section IV provides the examples of donors, deep centers, and acceptors in epitaxial layers. Section V gives a summary and some possible future directions.

## II. Fundamentals of ODMR and Epitaxy

1. ODMR Detected on Photoluminescence

*a. Donor–Acceptor Pair Recombination*

While a great variety of optical processes are spin-dependent and hence can exhibit ODMR, the process that has proved to be most useful for the study of defects in epitaxial layers is radiative recombination by "distant" donor–acceptor pairs (DAPs). Hence we will start by neglecting the spin and considering what happens in a PL experiment—both in k-space and in real space.

For most of the work to be described herein, the PL cycle begins as the sample is excited by a laser with photon energy above the bandgap of the semiconductor (Fig. 1a). The excited electrons and holes quickly thermalize to their respective band edges. While some electrons and holes may recombine directly or form excitons, many will be captured at donors and acceptors. The electron at a donor and the hole at an acceptor will recombine at a rate that depends on their separation in real space (Fig. 1b). Those pairs that produce clear, well-resolved ODMR are distant (i.e., separated by many Bohr radii). For a III–V semiconductor with a moderate

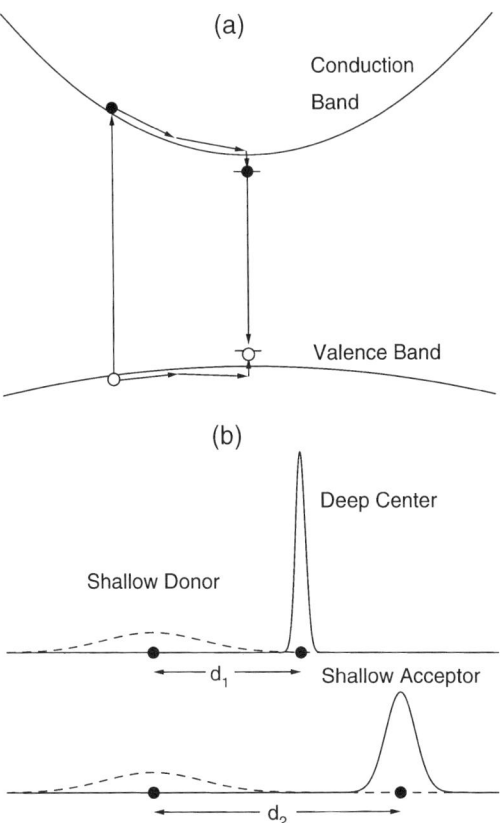

FIG. 1. (a) The optical cycle for a donor–acceptor pair: photoexcitation, thermalization, capture, and recombination. (b) Wave functions in real space for a shallow donor–deep center pair and a shallow donor–shallow acceptor pair.

bandgap, such as GaAs or InP, the Bohr radius ($a_0$) of a donor electron is about 200 Å. Isolating donors by placing them in boxes $5a_0$ on a side would give cubes with a volume of $10^{-15}$ cm$^3$. Hence a concentration of occupied donors of $10^{15}$ cm$^{-3}$ would be more than enough for distant DAP recombination. More importantly, note that concentrations of the order of $10^{15}$ cm$^{-3}$ (or about 0.1 ppm) are readily detectable by PL and ODMR. Of course, the donors and acceptors will have a Poisson distribution of separations. Since the recombination rate is inversely proportional to the separation, the more distant pairs will have the longest lifetimes. For a direct-gap semiconductor such as InP or GaAs, the lifetimes for DAP recombination range from about 1 ns for very close pairs to 100 ms for very distant pairs. For more details on DAP recombination, see the review by Dean (1982).

### b. ODMR Model

ODMR is possible because both the electron and hole have spin. Conservation of angular momentum leads to selection rules, which render the recombination of an electron and a hole spin-dependent. Magnetic resonance of either particle alters the populations of the spin states and, hence, alters the recombination. Thus, magnetic resonance can be detected by looking for changes in the intensity of the PL.

To illustrate the underlying physics, we take a slightly simplified example. If the electron and hole have $S = \frac{1}{2}$ and isotropic g-values, the spin Hamiltonian for the pair is

$$H = g_e \beta \mathbf{S}_e \cdot \mathbf{H} + g_h \beta \mathbf{S}_h \cdot \mathbf{H} + J\mathbf{S}_e \cdot \mathbf{S}_h, \tag{1}$$

where the last term describes the exchange interaction with coupling strength J between the electron and hole. We further assume that the excited states associated with the electron and hole are well separated from any other excited states and from the ground state — the unoccupied donor and acceptor — so that we may take the product states of an isolated donor and acceptor as a basis set:

$$|m_e, m_h\rangle = \begin{cases} |\tfrac{1}{2}, \tfrac{1}{2}\rangle \\ |\tfrac{1}{2}, -\tfrac{1}{2}\rangle \\ |-\tfrac{1}{2}, \tfrac{1}{2}\rangle \\ |-\tfrac{1}{2}, -\tfrac{1}{2}\rangle \end{cases} \tag{2}$$

The matrix elements for the spin-Hamiltonian taken with these states are easily calculated or found (Orton, 1968). Using the following definitions,

$$\Sigma = (g_e + g_h)\beta H, \quad \Delta = (g_e - g_h)\beta H, \tag{3}$$

the energy matrix for this problem, entered in MATHCAD, is shown in Fig. 2. Inspection of the matrix indicates the general character of the energies and states. The $|\tfrac{1}{2}, \tfrac{1}{2}\rangle$ and $|-\tfrac{1}{2}, -\tfrac{1}{2}\rangle$ states are unaffected by exchange. These two states are purely triplet in character (i.e., have $S = 1$) and hence, following the selection rule for spin, cannot recombine, since the ground state has $S = 0$. The exchange mixes the $|\tfrac{1}{2}, -\tfrac{1}{2}\rangle$ and $|-\tfrac{1}{2}, \tfrac{1}{2}\rangle$ states. These two are partly triplet and partly singlet, and hence can recombine.

**Energy Matrix for e h Pair** $\left(\frac{1}{2},\frac{1}{2}\quad \frac{1}{2},-\frac{1}{2}\quad -\frac{1}{2},\frac{1}{2}\quad -\frac{1}{2},-\frac{1}{2}\right)$

$$\begin{bmatrix} \frac{1}{2},\frac{1}{2} \\ \frac{1}{2},-\frac{1}{2} \\ -\frac{1}{2},\frac{1}{2} \\ -\frac{1}{2},-\frac{1}{2} \end{bmatrix} \begin{bmatrix} \frac{\Sigma}{2}+\frac{J}{4} & 0 & 0 & 0 \\ 0 & \frac{\Delta}{2}-\frac{J}{4} & \frac{J}{2} & 0 \\ 0 & \frac{J}{2} & -\frac{\Delta}{2}-\frac{J}{4} & 0 \\ 0 & 0 & 0 & -\frac{\Sigma}{2}+\frac{J}{4} \end{bmatrix}$$

**Solve for Eigenvalues**

$$\begin{bmatrix} \left(\frac{\Sigma}{2}+\frac{J}{4}\right)-E & 0 & 0 & 0 \\ 0 & \left(\frac{\Delta}{2}-\frac{J}{4}\right)-E & \frac{J}{2} & 0 \\ 0 & \frac{J}{2} & \left(-\frac{\Delta}{2}-\frac{J}{4}\right)-E & 0 \\ 0 & 0 & 0 & \left(-\frac{\Sigma}{2}+\frac{J}{4}\right)-E \end{bmatrix}$$

**Find Determinant**

$$\left(\frac{1}{2}\cdot\Sigma+\frac{1}{4}\cdot J - E\right)\cdot\left(\frac{1}{8}\cdot\Delta^2\cdot\Sigma - \frac{1}{16}\cdot\Delta^2\cdot J + \frac{1}{4}\cdot\Delta^2\cdot E + \frac{3}{32}\cdot J^2\cdot\Sigma - \frac{3}{64}\cdot J^3 + \frac{5}{16}\cdot J^2\cdot E - \frac{1}{4}\cdot J\cdot E\cdot\Sigma - \frac{1}{4}\cdot J\cdot E^2 - \frac{1}{2}\cdot E^2\cdot\Sigma - E^3\right)$$

**Factor and Set Equal to Zero**

$$\left(\frac{1}{2}\cdot\Sigma+\frac{1}{4}\cdot J - E\right)\cdot\left[\frac{1}{64}\cdot\left(4\cdot\Delta^2 + 3\cdot J^2 - 8\cdot J\cdot E - 16\cdot E^2\right)\cdot(4\cdot E + 2\cdot\Sigma - J)\right] = 0$$

**Solve for E**

$$\begin{bmatrix} \frac{1}{2}\cdot\Sigma+\frac{1}{4}\cdot J \\ -\frac{1}{4}\cdot J - \frac{1}{2}\sqrt{J^2+\Delta^2} \\ -\frac{1}{4}\cdot J + \frac{1}{2}\sqrt{J^2+\Delta^2} \\ -\frac{1}{2}\cdot\Sigma+\frac{1}{4}\cdot J \end{bmatrix} = \begin{bmatrix} E1 \\ E2 \\ E3 \\ E4 \end{bmatrix}$$

FIG. 2. The eigenvalue problem for an electron and a hole coupled by exchange.

The full solutions to this eigenvalue problem are given as equations in the lower part of Fig. 2 and are plotted in Fig. 3 for $g_e = 2$ and $g_h = 3$. For the experiments to be described in this chapter, we focus on the high-field range where the Zeeman terms ($\Delta$ and $\Sigma$) are large compared to J. At high fields, the states are close to those for an uncoupled electron and hole. Hence the transitions can be assigned to either the electron or hole. However, each

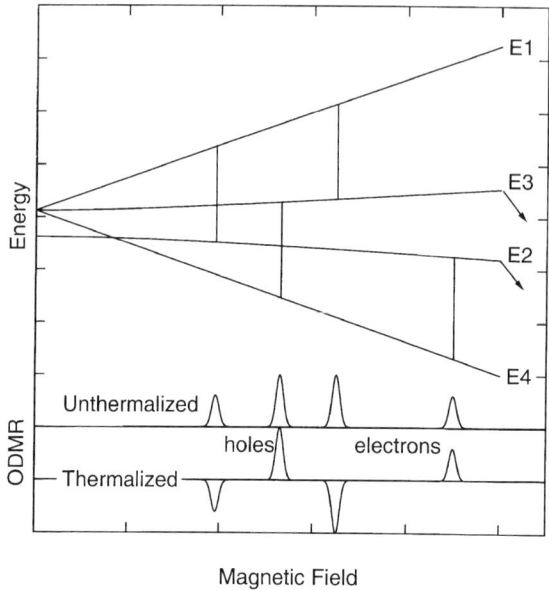

FIG. 3. The energy levels and ODMR transitions for the electron and hole. Levels E3 and E4 are radiative. Net population transfer at resonance leads to increasing signals for unthermalized states and both increasing and decreasing signals for thermalized states.

transition is split by the exchange interactions by approximately $J/g_e\beta$ for the electron and $J/g_h\beta$ for the hole. These splittings arise from the diagonal parts of the exchange interaction, as shown in the energy matrix. In practice, the exchange varies with the separation between the donor and acceptor. This variation produces a Lorentzian lineshape, which can be very broad if the concentration of defects is high.

When microwaves are applied to this system, they tend to equalize the populations of the states in resonance. The character of the ODMR observed then depends on the original populations of the manifold of excited states (see lower part of Fig. 3). Before resonance, the populations depend on whether the recombination rate ($1/\tau$) from the emitting states (E2 and E3) is faster than the thermalization rate within the manifold, which is governed by spin-lattice relaxation ($1/T_1$). In the unthermalized case, the recombination rate is faster, and resonance produces a net transfer of population to emitting states. Hence, the ODMR is an increase in emission. In the thermalized case, the spin-lattice relaxation is faster, and resonance produces both increases and decreases in emission, depending on whether the net transfer of population is to an emitting or to a dark state. In practice, most of the thermalized ODMR involves shallow acceptors in cubic symmetry since they have very fast spin-lattice relaxation.

c. *Resonance Parameters and Their Meaning*

In the foregoing example, we chose a spin-Hamiltonian for the electron and hole that included their Zeeman interactions with the magnetic field and their interaction with each other by exchange. Further interactions are possible, with the hyperfine interaction with nuclei central to or near the defect of primary importance. Since it is through these interactions that ODMR provides information about defects, we give an overview of these parameters here. More detailed treatments are available (Orton, 1968; Watkins, 1975; Watkins, this volume).

The g-tensor reflects the symmetry of the defect. For effective-mass defects, the g-tensors can depart strongly from the free-electron value of 2.0023. The departures arise from strong contributions of orbital angular momentum. Measured g-values provide important input for k·p theory. For deep states, the orbital angular momentum is quenched and the g-values stay very close to the free-electron value. However, they still reveal the symmetry of the deep defect, which may be lower than that of the lattice due to pairing or a Jahn-Teller effect.

Hyperfine interactions are extremely valuable for studies of defects. When resolved, they reveal, at a minimum, the spin of the nucleus involved and the nuclear abundance. For many elements of the periodic table, the nuclear spin and abundance are sufficient to chemically identify the atom in question. In the best cases, the central atom can be identified, as well as a finite number of neighboring shells. From these data, the atomic structure of the defect and the wave function of the defect can be deduced. These cases illustrate the full power of ODMR for defect studies.

Returning to consideration of the DAP, it is the overlap of the two wave functions that makes recombination possible and, through exchange, makes the ODMR possible. When both donor and acceptor have been identified, a study of the spectral properties of the emission and excitation provides the energy levels and lattice relaxation associated with the defects. Often, one of the defects of the pair is an effective-mass (EM) defect and the other is deep. For this case, the strength of the exchange is usually small enough that the hyperfine parameters can be detected for the deep state. When both the donor and acceptor are EM-like, exchange is usually the dominant contribution to the linewidth. Deep-donor to deep-acceptor PL is fairly rare, due to the small overlap of wave functions. Very close DAPs may bind excitons rather than emit radiation that is characteristic of separated carriers.

*d. ODMR Spectrometers*

The equipment required for ODMR consists of a suitable merger of an EPR spectrometer with a PL spectrometer. Typically, the EPR part is equipped for power modulation and the PL part is run CW. There are three examples of such mergers that have proved to be valuable for studies of defects in semiconductors. Many details can be found in the book by Spaeth *et al.* (1992).

At X-band, most EPR spectrometers are equipped with gas-flow cryostats. The cavities have optical access and hence the spectrometers can readily be used for ODMR (Godlewski *et al.*, 1988). These spectrometers have the advantages of ease of use and the ability to apply large amounts of microwave power, since the power is not dissipated in the low-temperature region. This allows the study of faster ODMR processes, since the excited states must be affected by the microwaves within their lifetime.

At K-band, the cavities are smaller, and many spectrometers operate with the microwave cavity inside a simple optical cryostat. The magnetic field is provided by an electromagnet, as in the gas-flow spectrometer (Fig. 4). This naturally leads to the Voight geometry where B is perpendicular to the wave vector of the exciting and emitted light. Higher frequency spectrometers have an advantage for ODMR, since the higher field enhances the field-dependent terms in the spin-Hamiltonian with respect to the exchange

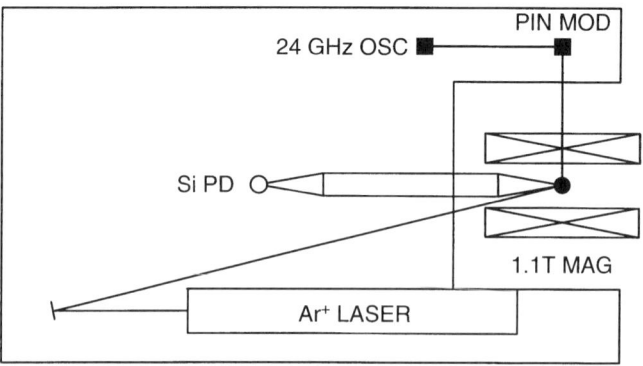

FIG. 4. An ODMR spectrometer utilizing an optical table surrounding a 9-inch electromagnet.

term. Hence, the conditions for ODMR of distant DAPs are more readily achieved.

Many spectrometers at K- and Q-bands are based on cryostats that include split-pair superconducting magnets. In addition to the advantages from the higher microwave frequency, these spectrometers can operate in either Faraday ($\mathbf{B} \parallel \mathbf{k}$) or Voight ($\mathbf{B} \perp \mathbf{k}$) geometry. When a defect, crystal, or heterostructure has a natural axis, it is advantageous to be able to place the field and k-vector along that axis. This often simplifies the analysis and makes studies of the polarization of the light feasible.

2. EPITAXY

*a. Perfect Epitaxy*

The development of molecular beam epitaxy (MBE) and organometallic vapor-phase epitaxy (OMVPE) in the 1970s led to a spectacular increase in the growth of semiconducting materials and structures. This growth can be attributed to two driving forces. First, the techniques produce highly pure materials. Second, the methods open up the possibility for a variety of physically and technologically interesting heterostructures when different materials are grown sequentially. These structured materials allow tailoring of both the optical and electronic properties.

Homoepitaxy is the growth of a particular semiconductor on a chemically identical substrate. Because it makes a larger range of properties possible, heteroepitaxy, the growth of layers of different chemical composition, is more interesting and more challenging. Different chemical composition implies a difference in lattice constant and in thermal expansion coefficient. Assume first that this difference does not introduce defects into the epitaxial growth; the new layer grows in perfect registry, or "pseudomorphically." In this case, the thicker substrate will define the in-plane lattice constant and the epilayer will adjust (Fig. 5). Through the Poisson effect, the epilayer will elongate along the growth direction if $a_{epi} > a_{sub}$ and will contract along the growth direction if $a_{epi} < a_{sub}$. This effect will depend on temperature when the thermal expansion coefficients of the two materials differ. Hence, it is possible for two materials to be closely matched at the growth temperature but experience a much larger mismatch at room temperature. The AlAs/GaAs system has a mismatch at room temperature of about 0.14%. The extensive work and fine results in this system make it the prototype for heteroepitaxy and quantum structures.

Under the condition of pseudomorphic growth, the resulting homogeneous strain in the epilayer will split degenerate electronic states. Hence, we

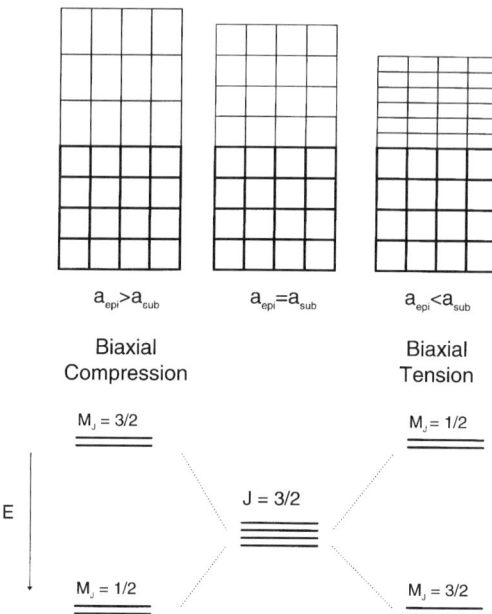

FIG. 5. Strain in epilayers and the response of the valence band to strain. The upper part of the illustration shows the response of the unit cells of the epilayer for different lattice constants. The lower part of the illustration shows the splitting of the $J = \frac{3}{2}$ valence band for the two cases of biaxial compression and biaxial tension.

consider the nature of the strain further. The biaxial compression (tension) induced by the mismatch can be decomposed into a hydrostatic compression (expansion) and a uniaxial tension (compression) along the growth axis. While the hydrostatic part only changes the bandgap of the semiconductor, the uniaxial part will induce splittings not present in unstrained (cubic) materials. As an example, the typical valence band structure of a cubic semiconductor is split with the heavy hole ($M_J = \frac{3}{2}$) lower for uniaxial tension and the light hole ($M_J = \frac{1}{2}$) lower for uniaxial compression (Fig. 5). The case of uniaxial tension is similar to that for Wurtzite crystals, where the axial crystal field arising from the crystal structure induces a splitting in the valence band.

### b. *Mismatch-Induced Defects*

Ths mismatch of 0.14% for the prototypical AlAs/GaAs system is one of the smallest for a pair of semiconductors. For example, the mismatch is

0.27% for ZnSe/GaAs, 3.4% for GaN/SiC, and 16% for GaN/Al$_2$O$_3$ (sapphire). Increasing mismatch implies greater strain, and a large energy develops associated with the volume of strained material. At a certain critical thickness, the crystal will lower the strain energy by accommodating the mismatch through misfit dislocations (van der Merwe, 1963; Matthews and Blakeslee, 1976). This critical thickness depends on the mismatch and is about 1 μm for ZnSe/GaAs. The misfit dislocation consists of an edge part lying at or near the interface between the different materials and a threading part, which extends through the epilayer (Fig. 6a). The perpendicular dislocations lying in the plane of growth relieve the biaxial strain.

For very large mismatches, such as the GaN/Al$_2$O$_3$ system, it becomes difficult to nucleate planar growth and the material grows by forming islands, which subsequently join to form a solid. The result has a kind of mosaic structure (Fig. 6b). The material contains low-angle grain boundaries, which themselves may be made up of dislocations. The boundaries may also have antiphase character. Materials with large mismatch may also contain stacking faults or regions with different crystal phases. This type of defect is more common for GaN and ZnS, which can occur in both cubic and Wurtzitic phases.

Mismatched epitaxy leads to dislocation densities as high as $10^8$ to $10^{10}$ cm$^{-3}$ for the GaN/Al$_2$O$_3$ system. The existence of such high densities of extended defects has important effects on the distribution and character of point defects. Two are important to the present work. First, the point defects will have a preference for locating at or near dislocations and surfaces. Second, the strain field associated with the dislocation will affect the point defects. The strain field of a dislocation is complex, with tensile and compressive components that depend on the angle around the dislocation and fall off from the dislocation as $1/r$. Nondegenerate donors will be

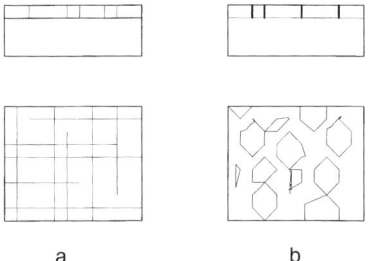

a  b

FIG. 6. Epitaxial layers with large mismatch. In (a), stress is relieved by misfit dislocations in the plane, which may have threading segments through the epilayer. In (b), three-dimensional (island) growth leads to grains that intersect with boundaries formed from dislocations. In GaN grown on sapphire, the hexagonal grains range from 1 to 40 μm.

least affected by this strain, although the wave function will be altered by the variation of the conduction band with strain. Deep centers may be oriented or distorted by the strain, depending on their degeneracy. Shallow acceptors will be very sensitive to strain because of their degenerate, or nearly degenerate, states.

### III. Illustrative Example: Bulk InP:Zn

1. INTRODUCTION

Before addressing the results in epitaxial layers, we present results on bulk InP that was lightly doped with Zn as a comprehensive example. When grown by the liquid encapsulated Czochralski (LEC) method, the crystal has a dislocation density of $10^4 \, \text{cm}^{-2}$, corresponding to a relatively low degree

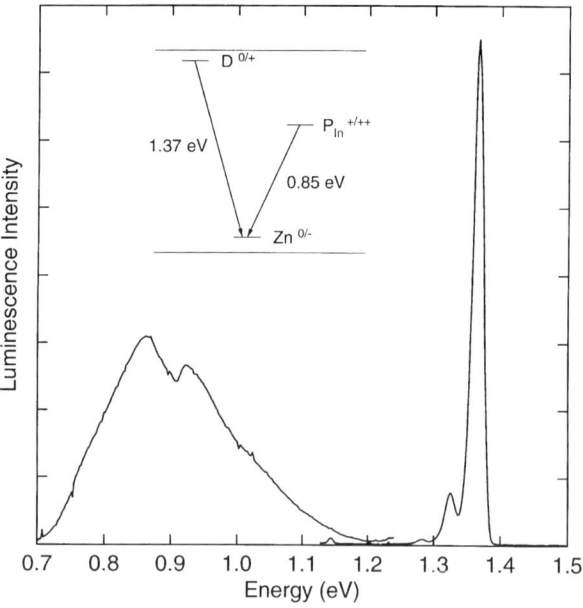

FIG. 7. Photoluminescence for Zn-doped InP. The inset shows, schematically, the major processes: Shallow donor–shallow acceptor recombination with the zero phonon line at 1.37 eV and antisite–shallow acceptor recombination peaking around 0.85 eV.

of native strain. Furthermore, the sample contains native, residual shallow donors; native, deep P-antisite ($P_{In}$) defects, which were identified by ODMR; and doped, shallow (Zn) acceptors. Results on this sample and others like it serve to illustrate the power of ODMR to identify and study defects.

Analysis of transport data for this sample shows that it has a shallow donor concentration of $2 \times 10^{15}$ cm$^{-3}$, a $P_{In}$ concentration of $2 \times 10^{15}$ cm$^{-3}$, and a Zn concentration of $10^{16}$ cm$^{-3}$. These concentrations are ideal for ODMR detected on PL. The emission spectrum at 1.6 K for this sample is dominated by shallow DAP emission at 1.37 eV and broad deep emission bands (Fig. 7). The most prominent of these peaks is at 0.85 eV and has been assigned (from ODMR) to DAP recombination involving the singly occupied antisite ($P_{In}^+$) as the donor and the shallow Zn acceptor. These two processes are illustrated in the inset to Fig. 7.

The ODMR detected on these emission bands will be described next. The discussion is organized by the particular defect under consideration.

## 2. ODMR OF THE EFFECTIVE-MASS DONOR

ODMR of the shallow DAP band taken at two modulation frequencies illustrates the richness of spin-dependent recombination (Fig. 8). At 40 kHz, a positive signal with a g-value of 1.20 is detected. This g-value identifies the signal as arising from effective-mass donors in InP (Lawaetz, 1971) and is consistent with the energy and character of the emission band itself. No acceptor is detected in free-standing material due to broadening that arises from the interaction of the degenerate valence band with random strains due to dislocations. Finally, this spin-dependence is dominant for high-modulation frequencies consistent with the fact that the lifetimes for processes involving a shallow donor and a shallow acceptor are faster due to the stronger overlap of wave functions.

In contrast, the dominant spin-dependent process for slower modulation frequencies lies outside the radiative 1.37-eV channel itself (see Fig. 8b). Negative signals indicate that a competing process has the stronger spin-dependence. Two defects are involved: the shallow donor seen at high-modulation frequencies and a two-line spectrum due to $P_{In}$ antisites. The spin-dependent process is capture from the occupied shallow donor to the singly occupied $P_{In}$.

Shallow donor states associated with a nondegenerate ($\Gamma$-point) conduction band edge are easy to observe with ODMR. Because they are singlets, they are relatively insensitive to random or inhomogeneous strain.

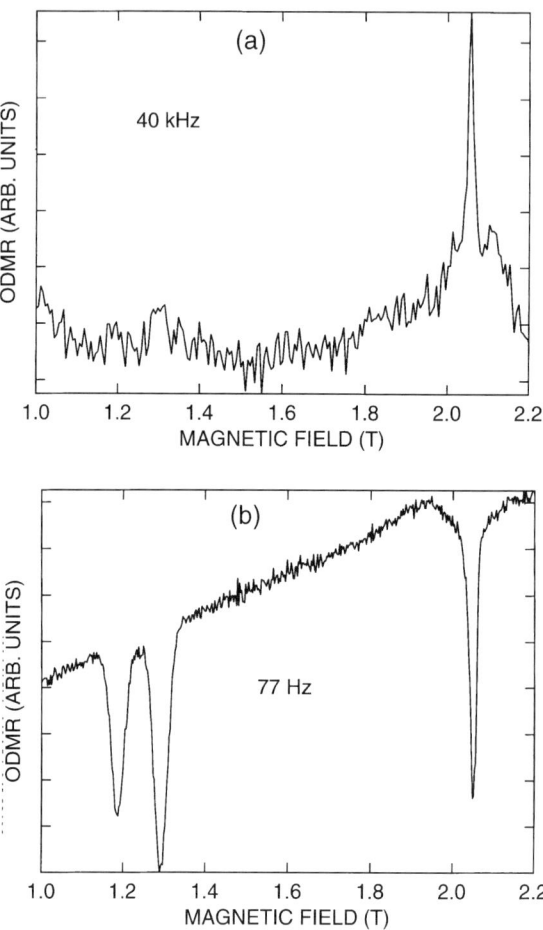

FIG. 8. ODMR for 35-GHz microwaves detected on the 1.37-eV emission for two different modulation frequencies. At 40 kHz, the dominant spin-dependent response involves the shallow donor at 2.05 T with the unseen shallow acceptor. A linear baseline was subtracted from this spectrum. At 77 Hz, the dominant spin-dependent response is a capture process from the shallow donor to the P antisite.

3. ODMR OF $P_{In}$

Although deep centers can be detected as negative signals on band-edge emission, it is natural to study them on their own deep emissions. Hence, the $P_{In}$ antisite is observed as a positive signal on the 0.85-eV band (Fig. 9).

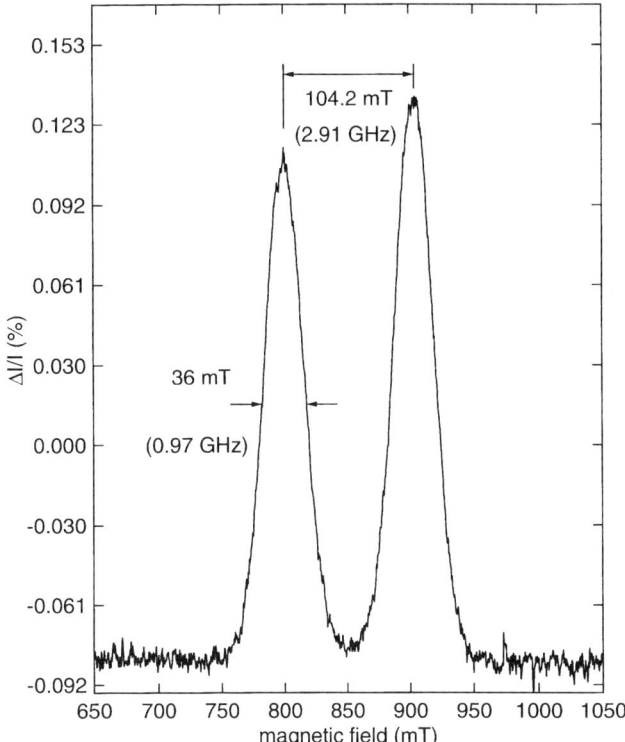

FIG. 9. ODMR of the P antisite for 24-GHz microwaves detected on the 0.85-eV emission. The splitting arises from hyperfine interaction with the central $^{31}$P atom, and the linewidth arises from unresolved hyperfine interactions, primarily with the four $^{31}$P nearest neighbors.

The antisite was identified from its strong, isotropic central hyperfine splitting into two lines by the 100% abundant $^{31}$P nucleus (Kennedy and Wilsey, 1984). For native antisites, this splitting is 104.2 mT or 2.91 GHz. Its g-value, 2.004(4), is close to the free-electron value of 2.0023 since the orbital angular momentum of deep centers is quenched.

Unlike for the antisite in GaP, further information on the atomic structure of the defect is not available from the ODMR spectrum since the component lines show no resolved structure. However, this information can be gained by performing electron-nuclear double-resonance (ENDOR). A detailed description of ENDOR and its advantages can be found in the book by Spaeth *et al.* (1992). The chief advantage is that hyperfine interactions not resolved in the EPR spectrum are easily resolved in the ENDOR, in which the nuclear magnetic resonance (NMR) of each nucleus coupled to the paramagnetic defect can be studied. The transition energy for each nucleus is

$$h\nu = |M_S A(\theta) \pm g_n \beta_n B_0|, \quad (4)$$

where $\nu$ is the frequency for ENDOR, $M_S$ is the azimuthal quantum number for the electron spin (usually $\frac{1}{2}$), $A(\theta)$ is the hyperfine coupling of the

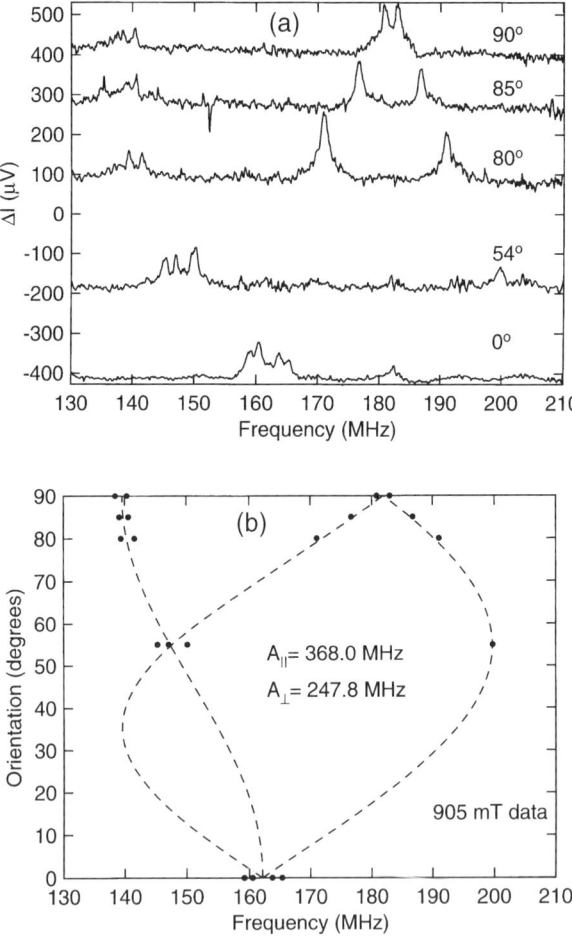

FIG. 10. ODENDOR at 905 mT of the four $^{31}$P nearest neighbors to the antisite. The upper part is the data for five directions of B with respect to the [001]. The lower part gives a fit to the data and illustrates the trigonal symmetry of each neighbor.

particular nucleus and the electron, $g_n$ is the nuclear g-factor and $\beta_n$ the nuclear magneton, and $B_o$ is the magnetic field. In ENDOR, both the hyperfine interactions and the nuclear moment can be measured by performing experiments at different fields.

The first results for isolated antisites in InP were obtained by optically detecting ENDOR on magnetic circular dichroic (MCD) absorption (Jeon et al., 1987). These data showed that the four $^{31}$P nearest neighbors to the antisite are equivalent and the symmetry of the antisite is precisely tetrahedral. Similar results can be obtained from ENDOR detected on PL (Crookham et al., 1992). Lowering the modulation frequency to 17 Hz to select very isolated pairs and setting the field on the upper resonance at 905 mT, one looks for changes in the ODMR as an rf field is swept from 130 to 210 MHz (Fig. 10a). A study of the angular dependence reveals a pattern characteristic of the symmetry of the four neighbors (see Figs. 10a and 10b). Changing the field at which the ENDOR is performed measures the nuclear moment and confirms that these signals arise from four $^{31}$P nuclei. The hyperfine constants, $A_{\parallel} = 368.0$ MHz and $A_{\perp} = 247.8$ MHz, can be used to construct a detailed model of the wave function of the $P_{In}$. The ENDOR of the next shell, consisting of 12 In atoms, has also been detected (Jeon et al., 1987; Crookham et al., 1992), but a detailed analysis has not been possible so far.

With this detailed picture of the atomic structure of the defect, the optical properties can be probed in detail to reveal the electronic structure. The spectral dependencies of both the emission and absorption associated with the $P_{In}$ ODMR signature give an equally detailed picture of the electronic structure (Sun et al., 1993). The emission peaking at 0.85 eV is due to $P_{In}^+$-$Zn^0$ recombination with the $P_{In}^{+/++}$ level 1.1 eV above the valence band. Furthermore, there is a band peaking at 1.1 eV that is attributed to $P_{In}^0$-$Zn^0$

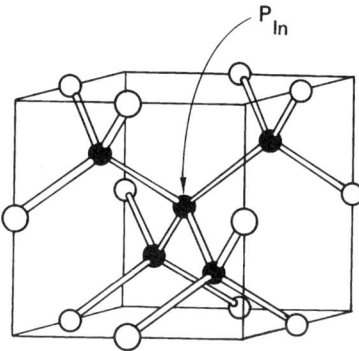

FIG. 11. The atomic structure of the P antisite.

TABLE I

PROPERTIES OF THE ISOLATED P ANTISITE

| Atomic Structures | Central Ion | Nearest Neighbors | Second Neighbors |
|---|---|---|---|
| | P | P4 | In 12 |
| Symmetry | $T_d$ | $C_{3v}$ | |
| Hyperfine interactions | 2.998 GHz | $A_\parallel = 368.0$ MHz | ? |
| | | $A_\perp = 247.8$ MHz | |
| Electronic levels | $0/+$ | $+/++$ | |
| (from valence band) | 1.39 eV | 1.1 eV | |

recombination with the $P_{In}^{0/+}$ level 1.39 eV above the valence band. Hence, the antisite is a double donor with the normal ordering of energy levels with a Coulomb energy of 0.3 eV.

These studies provide a very detailed picture of the deep defect, which is summarized in Fig. 11 and Table I. The atomic structure consists of a P atom on an In site surrounded by a tetrahedron of P atoms. The hyperfine constants for these atoms provide information about the wave function of the defect. The antisite is a double donor with levels 30 meV and 300 meV from the conduction band. Further studies have probed the nuclear polarization of the $P_{In}$ antisite (Mao et al., 1993).

## 4. ODMR OF THE EFFECTIVE-MASS ACCEPTOR

As the work of the previous sections shows, the ODMR of a shallow acceptor in a cubic semiconductor is generally not observed. In the effective-mass approximation, a hole bound to a shallow acceptor has the character of the valence-band maximum. Thus the hole state is degenerate in the absence of strain. This degeneracy makes the g-factor very sensitive to random strain and dynamic Jahn-Teller distortion, which broaden the resonance so much that it cannot be detected (Mehran et al., 1972). However, applying an external uniaxial stress lowers the symmetry at the acceptor, thereby removing the degeneracy. The splitting produced by the applied stress alters the magnetic resonance condition while at the same time reducing its sensitivity to the broadening factors mentioned previously. As a result, an acceptor resonance, whose position and width depend on the magnitude and direction of the applied stress, can be observed.

Applying uniaxial stress to the Zn-doped InP sample reveals the ODMR of the shallow acceptor (Fig. 12 and Crookham et al., 1993). The sample

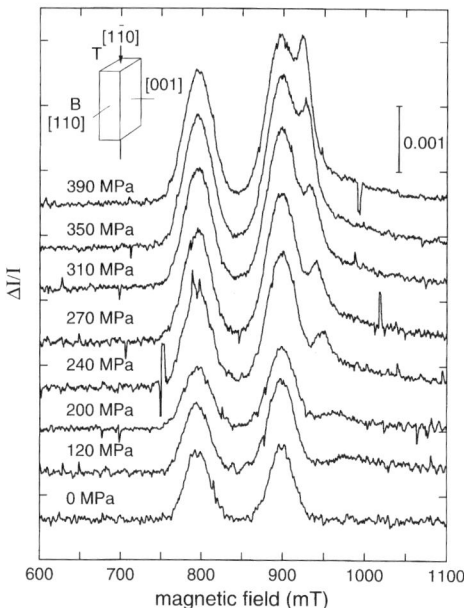

FIG. 12. ODMR in the region around g = 2 for different values of uniaxial stress. The inset illustrates the geometry. The acceptor line appears at around 950 mT with increasing stress and moves to lower fields. (From Crookham et al., *Physical Review B*, 1993, with permission.)

dimensions were $0.5 \times 0.75 \times 2$ mm$^3$ and the force was applied along the $[1\bar{1}0]$. Pressures up to 400 MPa were obtained and ODMR on the 0.85-eV emission was studied. For B parallel to the [110] direction, a new line appears with stress and becomes stronger and sharper as stress increases. At 390 MPa, the spectrum shows increasing signals from both the $P_{In}$ donor and the Zn acceptor, confirming that these defects are responsible for this band. Similar data were obtained for B parallel to the [001], with the acceptor line growing at the exact position of the upper $P_{In}$ line.

The data were analyzed within the following approximation (Mehran et al., 1972). The interaction between the magnetic field and the $J = \frac{3}{2}$ state is described by the spin-Hamiltonian,

$$H = \mu_0[g_1 \mathbf{J} \cdot \mathbf{H} + g_2(J_x^3 H_x + J_y^3 H_y + J_z^3 H_z)]. \quad (5)$$

Compressive stress applied along the $[1\bar{1}0]$ removes the degeneracy of the $J = \frac{3}{2}$ hole state, leaving a nearly pure $M_J = \pm\frac{1}{2}$ doublet lower in energy (see

Section II.2). Effective g-values for transitions between these $M_J = \pm\frac{1}{2}$ states have been calculated (Mehran et al., 1972). An anisotropy occurs when the magnetic field is in the plane perpendicular to the stress because of the $J^3$ terms. $g_1$ and $g_2$ can be deduced from the resonances observed for $B \parallel [110]$ and $B \parallel [001]$ from the following equations:

$$g_1 = (10/9)g_{[110]} - (11/18)g_{[001]}, \qquad (6a)$$

$$g_2 = (4/9)\{g_{[001]} - g_{[110]}\}. \qquad (6b)$$

Using these equations, the data for a stress of 390 MPa give $g_1 = 0.869(3)$ and $g_2 = 0.031(2)$.

Since the behavior of this spectrum with stress is exactly what is expected for a shallow acceptor and since the sample was Zn-doped, the spectrum is assigned to the Zn acceptor. Further implications of the results are described in Crookham et al. (1993). We emphasize here that these results confirm the expected sensitivity of EM acceptors to the condition of strain. Generally the random strains in cubic materials render the acceptor resonances unobservable. The broadening arises from the spread in g-values, however, and hence is proportional to the magnetic field for resonance. The acceptor may have been observed in a free-standing sample at 3 GHz (Viohl et al., 1991). Finally, the observation of the acceptor resonance confirms its participation as the acceptor in a DAP process with $P_{In}$ for the 0.85-eV band. This gives confidence to the analysis of the electronic level for $P_{In}^{+/++}$ described in the previous section.

## 5. CONCLUSION

The results obtained in lightly Zn-doped InP illustrate the power of ODMR to reveal the identities of the electronically active defects and associate them with specific recombination processes. The shallow donors are revealed from their g-value and insensitivity to strain corresponding to the nondegenerate state. The identification of the native $P_{In}$ defects shows the power of magnetic resonance techniques for defect studies. The detail achieved for the structural and electronic properties of the antisite sets a standard for the work in epitaxial materials. The EM acceptor is only clearly identified when a state of homogeneous uniaxial strain is achieved. Hence, the acceptors, with their degenerate states, are clearly the most sensitive to the condition of strain in a sample.

## IV. Examples in Epilayers

### 1. EFFECTIVE-MASS DONORS

*a. Γ Donors and Hydrostatic Strain*

Effective-mass donors, particularly those in singlet states, are readily observed in the ODMR of epitaxial layers since they are relatively insensitive to strain. Donors associated with the Γ-minimum are always in singlet states. They provide spin-dependence to the recombination either directly, as the donor in a DAP process, or indirectly, through electron capture or excitation transfer to deeper centers. ODMR of the Γ donors may respond to the hydrostatic part of the mismatch-induced strain.

Since hyperfine-interactions for the EM donors in III–V and II–VI hosts are rarely resolved, the Γ states produce single lines in ODMR. The g-values reported for various epitaxial systems and one ion-implanted layer are collected in Table II. For cubic materials, the g-value is isotropic, but for wurtzitic GaN, the g-tensor is axial reflecting the crystal structure. In every case, the g-values reflect the band structure of the host and provide important input for obtaining, or verifying, band parameters.

Strain affects the energies of the host bands and, hence, also affects the g-values. Shifts from the free-electron g-value are inversely proportional to the separations between energy bands. Hydrostatic pressure (or tension)

TABLE II

g-VALUES FOR EFFECTIVE-MASS DONORS IN THIN LAYERS

| System | Donor | g or $g_\parallel$ | $g_\perp$ | g in Bulk | Reference |
|---|---|---|---|---|---|
| Γ-Donors | | | | | |
| ZnSe/GaAs | (Residual) | 1.11 | — | 1.14[a] | Murdin et al. (1993) |
| Ar-implanted ZnSe | (Residual) | 1.19(1) | — | 1.14[a] | Verity et al. (1981) |
| ZnS/GaAs | (Residual) | 1.82–1.86 | — | 1.882[b] | Poolton et al. (1987) |
| GaN/Al$_2$O$_3$ | (Residual) | 1.9515(2)[c] | 1.9485(2) | | Glaser et al. (1995) |
| X-Donors | | | | | |
| AlAs/GaAs | Si | 1.917[d] | 1.976 | | Glaser et al. (1991) |
| Al$_{0.41}$Ga$_{0.59}$As/GaAs | Si | 1.938[d] | 1.946 | | Ibid. |

[a]Schneider et al. (1968).
[b]Müller and Schneider (1963).
[c]Axes with respect to Wurtzite c-axis.
[d]Axes with respect to ellipsoidal energy minima.

increases (or decreases) the separations between the bands. The biaxial part of the strain splits the degenerate valence-band states.

The effect of the hydrostatic strain can be seen for the EM donors in ZnSe and ZnS. First, the g-values in heteroepitaxial and implanted layers differ from those in the bulk. Second, it may be possible to relate the signs of the shifts observed in thin layers to their states of strain. For ZnSe/GaAs, the mismatch and thermal expansion difference have opposite effects (Mayer et al., 1995). Experimentally, the donor shows a larger shift from $g = 2.00$ than that observed in bulk material. Ion implantation adds atoms to the lattice, creating a hydrostatic compression. In ZnSe, this results in greater energy splittings and a smaller g-shift (Verity et al., 1981). For ZnS/GaAs, we have the same effect as for ZnSe/GaAs. The smaller ZnS is under tension, resulting in smaller energy splittings and a larger g-shift.

### b. X-Donors and Axial Strain

The band diagram for $Al_xGa_{1-x}As$ (Fig. 13) illustrates the symmetries that occur for conduction-band minima in cubic crystals. For $x \leqslant 0.4$, the $\Gamma$-minimum is lowest. The g-values for electrons in this range were measured by conduction-electron spin resonance (CESR) (Weisbuch and Hermann,

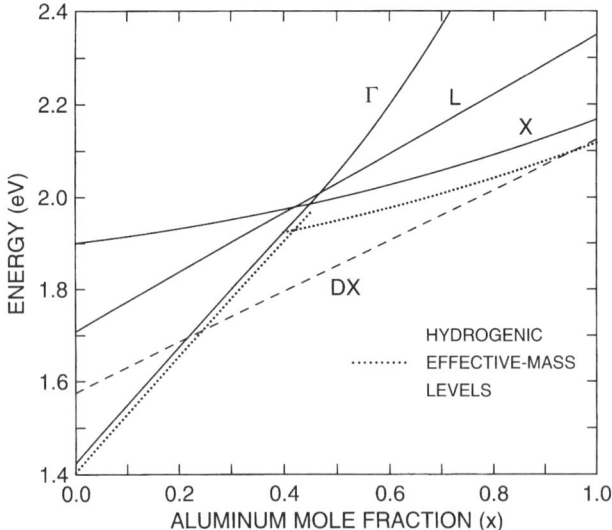

FIG. 13. Conduction-band states and donor states in $Al_xGa_{1-x}As$ as a function of Al mole fraction (x). (From Glaser et al., Physical Review B, 1991, with permission.)

1977). Near x = 0.4, there is a near degeneracy of the Γ-, L-, and X-bands. For x ⩾ 0.4, the X-band is lowest.

For 0.3 ⩽ x ⩽ 1.0, there is also a deep state associated with donors, which is referred to as DX. This state is readily detected in transport measurements through persistent photoconductivity or by its deep-level transient spectroscopy signature (Mooney, 1990). For low-temperature PL experiments, this deep state ($DX^0$) first captures a hole to become $d^+$ (Brunthaler et al., 1989). At low temperatures, electrons cannot overcome the barrier to repopulate $DX^0$, and hence $d^+$ participates in the "normal" recombination process. This $d^+$ state is detected by ODMR.

Donors bound to the triply degenerate X valleys respond to a variety of potentials to produce a rich story. Results on these donors were first obtained using EPR (Böttcher et al., 1973; Wartewig et al., 1975) on samples grown by liquid-phase epitaxy (LPE). Major studies have been performed using ODMR (Montie et al., 1990; Glaser et al., 1991) and EPR (Mooney et al., 1989) for samples grown by OMCVD and MBE.

Some background on the theory of EM donors in a binary semiconductor like AlAs is necessary (Morgan, 1968; Glaser et al., 1991; Glaser and Kennedy, 1992). The constant-energy surfaces about the X-point are ellipsoids in k-space, with their long axes along the ⟨001⟩ directions (Fig. 14a). In the hydrogenic effective-mass (HEM) approximation, the wave function of the ground state is the product of (1) a Bloch function for the $X_x$, $X_y$, or $X_z$ valley, and (2) a 1-s envelope function. The higher electronegativity of the group V host atom compared to the group III host atom lowers the

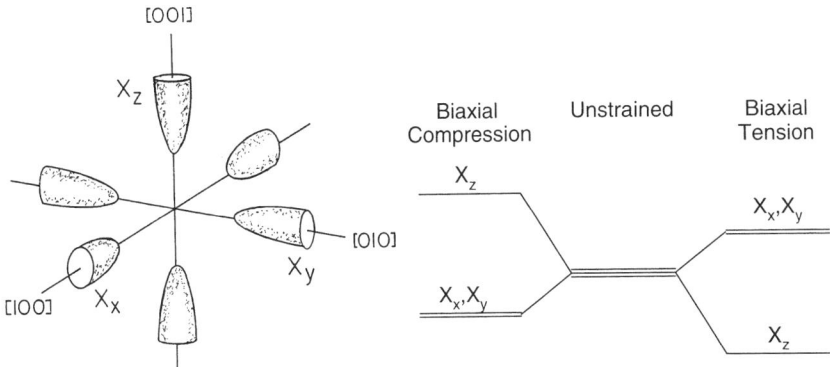

FIG. 14. X valleys in k-space and their response to strain. The three valleys are degenerate in unstrained material but split into a doublet and singlet whose ordering depends on the sign of the strain. (From Glaser et al., Physical Review B, 1991, with permission.)

band state having a node at the Al site and an antinode at the As site. Thus, the location of the substitutional donor atom on either the group III or group V site determines the degree to which the states from the three valleys interact. For group IV donors on the group III site in AlAs, the donor potential will not mix the three HEM states, since the Bloch function has a node at the donor. Thus, the ground state of an electron bound to a group IV donor is an orbital (valley) triplet ($T_2$) in full tetrahedral symmetry ($T_d$) (see Fig. 14b). For group VI donors on the group V site, the central-cell potential strongly admixes the states of the three valleys, since the Bloch function has an antinode at the donor. As a result, an orbital singlet ($A_1$) ground state with increased binding energy is formed from a symmetric combination of the valley states. A twofold degenerate (E) excited state remains near the original effective-mass level. The difference in energy between the singlet ground state and the doublet excited state is commonly referred to as the valley-orbit splitting ($E_{12}$), or chemical shift.

With these basic ideas in mind, we present some selected results from the work on donors, which illustrate the interaction of these donor states with the uniform strain that arises in pseudomorphic $Al_xGa_{1-x}As$ on GaAs and with the alloy disorder present for $x < 1$.

The magnetic resonance spectra of an Si-doped AlAs/GaAs heterostructure are shown in Fig. 15a. With the magnetic field applied in the ($1\bar{1}0$) plane, a single line is observed that shifts from g = 1.945 for B ∥ [110] to g = 1.978 for B ∥ [001]. In the (001) plane, two resonances with approximately equal amplitudes are found with g = 1.917 and g = 1.976 for B ∥ [100]. A single line is observed again with g = 1.947 with the magnetic field in the [110] direction. No differences are observed in the linewidths or g-values of the two well-resolved resonances with B along the third cube-edge direction ([010]). The angular rotation pattern for this sample is given in Fig. 15b. An overall tetragonal symmetry of the resonance about the [001] axis is evident.

These results can be understood using an independent valley model in the presence of strain (Glaser et al., 1991). The mismatch of the AlAs and GaAs lattice constants ($1.4 \times 10^{-3}$) leads to a biaxial compression in the plane of the epilayer and an elongation along the [001] due to the Poisson effect. The heteroepitaxial stress raises the $X_z$ valley relative to the $X_x$ and $X_y$ valleys by approximately 14 meV (van Kesteran et al., 1989). Thus, the $X_z$ valley at 1.6 K does not participate in the recombination. Two well-resolved resonances of about equal intensities are observed with B along the [100] direction because the field is simultaneously oriented parallel to the long axis of the $X_x$ valley and parallel to the short axis of the $X_y$ valley. A single anisotropic resonance is observed as the field is rotated in the ($1\bar{1}0$) plane since this plane mirrors the $X_x$ and $X_y$ valleys.

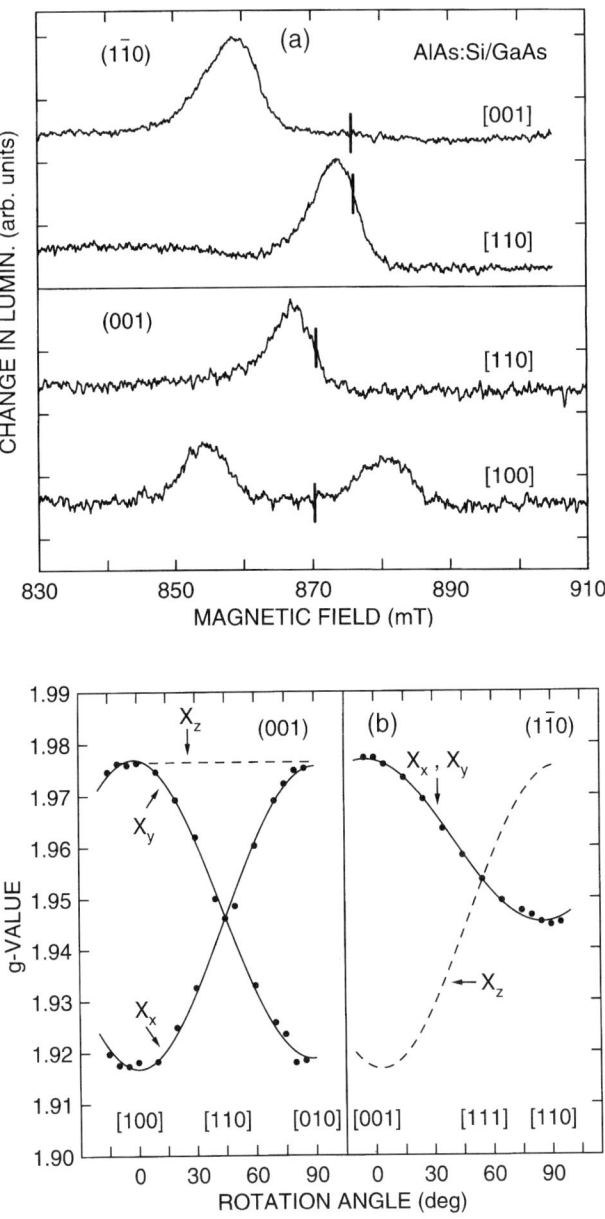

FIG. 15. Angular dependence of the ODMR of a Si-doped AlAs/GaAs heterostructure. (a) The data for two field directions in two crystal planes. Markers denote the field for g = 2. (b) The full rotation patterns for these planes. (From Glaser *et al., Physical Review B*, 1991, with permission.)

Fits were made to the data with the usual expression for the g-values in the case of axial symmetry

$$g = (g_\parallel'^2 \cos^2\theta + g_\perp'^2 \sin^2\theta)^{1/2},$$

where $\theta$ refers to the angle between the applied magnetic field and the [100] or [001] axes and $g_\parallel'$ and $g_\perp'$ are the g-values associated with B oriented parallel and perpendicular to the [100] or [001] direction. The fits yield the single-valley g-values associated with the X-point conduction band minima in AlAs: $g_\perp = 1.976(1)$ and $g_\parallel = 1.917(1)$. These values are in good agreement with those obtained from band-structure parameters for AlAs. Thus, the shallow donor state in AlAs can be described as an orbital doublet consisting of $X_x$ and $X_y$.

The independent-valley picture employed to describe the Si donor ground state in AlAs is consistent with the weak valley–orbit interaction for a group IV donor on a group III site. However, due to the multiplicity of valleys associated with the X-point and the finite spin ($\frac{1}{2}$) of the donor electron, a spin-valley coupling must also be considered since this interaction ($\lambda L \cdot S$) can mix the $X_x$, $X_y$, and $X_z$ valleys. It has been suggested that $\lambda$ scales with the difference between the atomic spin–orbit coupling constants of the impurity and that of the host atom it replaces (Dean et al., 1974). The ODMR results on Si-doped AlAs reflect the fact that the difference between the atomic spin–orbit constants of $^{27}$Al and $^{28}$Si is quite small.

Si donors in AlAs were studied under different conditions of strain from the biaxial compression present in AlAs/GaAs. In a free-standing sample, the two in-plane valleys and single [001] valley produce resonances that reflect their number (Fig. 16). A free-standing sample was also glued to a silica substrate and cooled to produce a biaxial tension. This leaves the $X_z$ valley lower in energy (Fig. 14b), whose resonance is observed at the lower field for H ∥ [110] (see Fig. 16). The splitting in energy is smaller than for AlAs/GaAs and the resonance associated with the partially populated $X_x$, $X_y$ valleys is also observed. These results confirm the independent-valley model and show that the Si donor acts as a local probe of the strain on the scale of the Bohr radius.

Moving from AlAs/GaAs to $Al_xGa_{1-x}As$/GaAs, three effects occur. First, the mismatch-induced strain decreases linearly with the AlAs mole fraction x. Second, disorder enters through the distribution, random or otherwise, of Al and Ga on the cation sites. Third, the band-structure changes at x = 0.4 (see Fig. 13). Not surprisingly, there are changes in the ODMR spectra as a function of AlAs mole fraction. Figure 17 shows the spectra for Si donors for several values of x with B ∥ [100]. The data show decreasing splitting between the two resonances with decreasing x. Only one resonance with a g-value of around 1.94 is observed for the sample with x = 0.41.

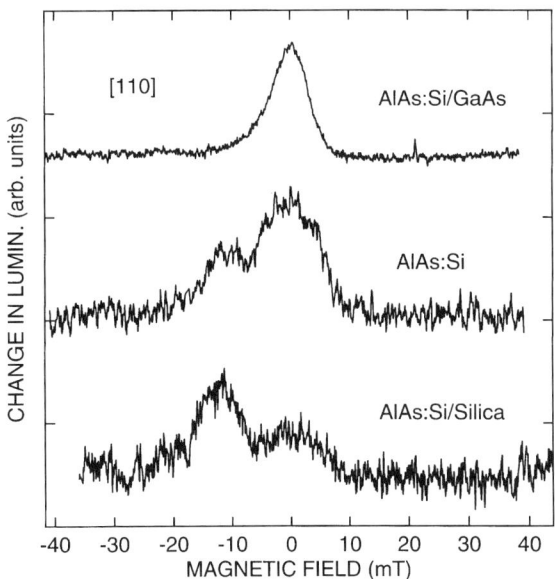

Fig. 16. ODMR spectra for AlAs:Si under different states of strain. For AlAs/GaAs, the layer is under biaxial compression. For free-standing AlAs, the layer is unstrained. For AlAs on silica, the layer is under biaxial tension. (From Glaser et al., *Physical Review B*, 1991, with permission.)

To explain the alloy dependence, we first apply the independent-valley model, which was successful for AlAs/GaAs. For B ∥ [100], the two resonances correspond to the $g_\parallel$ and $g_\perp$ for a single valley. Within the virtual crystal approximation (VCA), changes of the g-values with composition can be obtained from the formulae that relate the g-values to band parameters (Glaser et al., 1991). These equations predict $g_\parallel = 1.954$ and $g_\perp = 2.013$ for $x = 0.41$. The observation of a single resonance at this concentration clearly indicates a breakdown of the independent-valley model. Hence, the behavior of the g-values for Si-doped $Al_xGa_{1-x}As$ samples with $x < 1$ strongly suggests the presence of a coupling among the individual X-valleys, or a coupling of the X valleys with the Γ- or L-bands. Various mechanisms that can produce this coupling include a finite spin–valley interaction, L-X or Γ-X interband mixing, and alloy disorder.

Uniaxial stress experiments indicate that alloy disorder plays the crucial role in altering the nature of the Si donor ground state for $x < 1$ (Glaser and Kennedy, 1991). The Bohr radius for the Si shallow donor state is $\cong 15$ Å. Thus the donor wave-function senses the alloy disorder over three lattice constants. The potential associated with the disorder can mix states

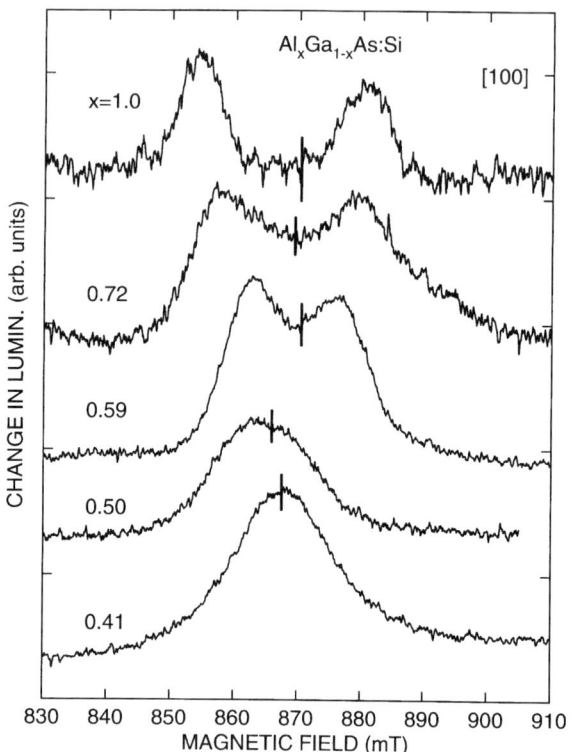

FIG. 17. ODMR spectra for Si-doped $Al_xGa_{1-x}As$ with different Al mole fractions. Vertical bars indicate the position corresponding to g = 1.94. (From Glaser et al., *Physical Review B*, 1991, with permission.)

of different k-vectors. This leads to a coupling betwen states derived from the $X_x$ and $X_y$ valleys or between states derived from the $\Gamma$- and X-point conduction band minima. These intervalley and interband scattering interactions can explain the decrease in g-anisotropy and the increasing linewidth of the Si resonances as x varies from 1 to 0.4. Additional evidence for the importance of alloy disorder in this system comes from the increase of the linewidths observed with increasing microwave frequency (Glaser et al., 1991).

The group VI donors S, Se, and Te, have also been studied in $Al_xGa_{1-x}As/GaAs$ heterostructures with x = 0.4 and 0.6 (Glaser et al., 1991). The strikingly different ODMR spectra for samples with x = 0.6 (Fig. 18) demonstrate the difference between the Si donors and the group VI donors. Two resonances are resolved in the Si-doped sample, while the lines

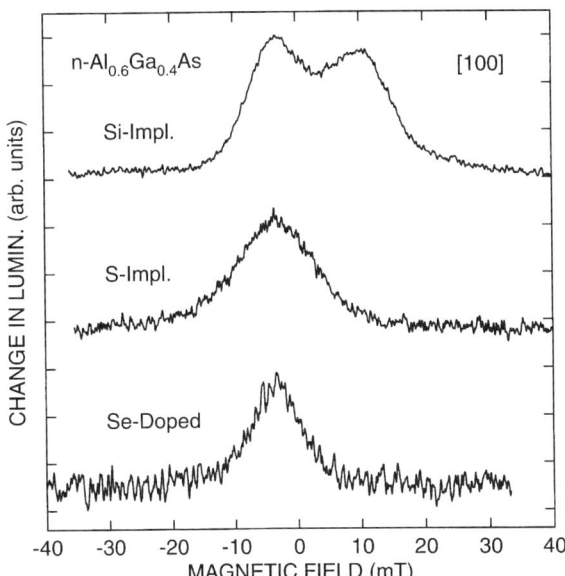

FIG. 18. ODMR spectra for Si-, Se-, and S-doped $Al_xGa_{1-x}As$ with x = 0.6. The spectra are displayed about g = 1.95 (B = 0) to account for slightly different microwave frequencies. (From Glaser et al., Physical Review B, 1991, with permission.)

in the S- and Se-doped samples are unsplit. Similar behavior was observed for the Te-doped sample.

The results obtained for S, Se, and Te can be described by hydrogenic effective-mass theory modified by the heteroepitaxial strain and a finite valley–orbit interaction that mixes the $X_x$, $X_y$, and $X_z$ states to form an $A_1$ ground state with an isotropic g-value. However, a small g-anisotropy about the [001] axis is observed in the $(1\bar{1}0)$ plane because of the slight mixing between the ground ($A_1$) and excited (E) states. This interaction is due to the tensile strain from the lattice mismatch. The observed g-anisotropy reflects the anisotropy of the doublet state.

The results show that the valley–orbit interaction for group VI atoms is much larger than the scattering by alloy disorder for x = 0.6. Similar admixing effects on the behavior of donor g-values associated with an $A_1$ singlet ground state have been observed in EPR investigations of shallow donors in Si under applied uniaxial stress by Wilson and Feher (1961). Their analysis can be used for $Al_xGa_{1-x}As$ with x = 0.6 if the virtual crystal approximation is assumed (Glaser et al., 1991). The valley–orbit splitting energies (chemical shifts) are 16.5 to 20 meV for the S-, Se-, and Te-doped $Al_{0.6}Ga_{0.4}As$ samples.

In summary, ODMR provides a detailed picture of donors in high-x $Al_xGa_{1-x}As$. The optically active states can be described using HEM theory with weak perturbations. The association with the X-minima is clearest for Si donors in AlAs where the only perturbation is from the mismatch of AlAs on GaAs. The mild alloy disorder perturbs the states for Al mole fractions less than 1. The group VI donors, S, Se, and Te, exhibit a central-cell interaction consistent with the symmetry of the wave function relative to the lattice site. The sizes of the perturbations vary with Al mole fraction and chemical species but in all the cases studied are smaller than the HEMT binding energy of 45 meV. The overriding importance of an X-derived state is evident through the average g-value of 1.945.

## 2. Deep Centers

### a. The $Ga_i$ in AlGaAs

In contrast to shallow donors, deep centers have highly compact wave functions that lead to strong central and ligand hyperfine interactions. These can be used to chemically identify the defect and reveal its symmetry and wave function. A good example is the Ga interstitial ($Ga_i$) in $Al_xGa_{1-x}As$/GaAs heterostructures.

The first observations were made on MBE-grown layers with varying Al mole fraction, doping, and substrate temperature (Kennedy and Spencer, 1986; Kennedy et al., 1988). Layer thicknesses varied from 1 to 5 μm. Optimal growth temperature for high-quality $Al_xGa_{1-x}As$ is around 680°C. Lower than optimal growth temperatures favor the formation of $Ga_i$ defects. Samples grown around 620°C exhibit much weaker band-edge PL and stronger deep emission in the range from 0.8 to 1.2 eV.

The ODMR for a sample with x = 0.26 consists of a partially resolved four-line spectrum (Fig. 19). The resonance is negative (luminescence quenching) in the 0.8- to 1.2-eV range and may be weakly anisotropic. The weaker and broader outer lines provide the key to the analysis. The splittings are due to a hyperfine interaction with an element with two isotopes of spin $\frac{3}{2}$ of nearly equal natural abundance. A simulation for $^{69}Ga$ and $^{71}Ga$ that provides the best fit is shown. The most likely Ga point defects are the Ga interstitial and the Ga antisite. However, only $Ga_i$ has the $a_1$ (s-like) character that leads to a strong hyperfine splitting (Baraff and Schlüter, 1985), and thus the spectrum is uniquely assigned.

The weak anisotropy observed for the spectrum was previously attributed to alloy disorder (Kennedy et al., 1988). This attribution came from considering the local symmetry for interstitial sites in $Al_xGa_{1-x}As$. Both theory

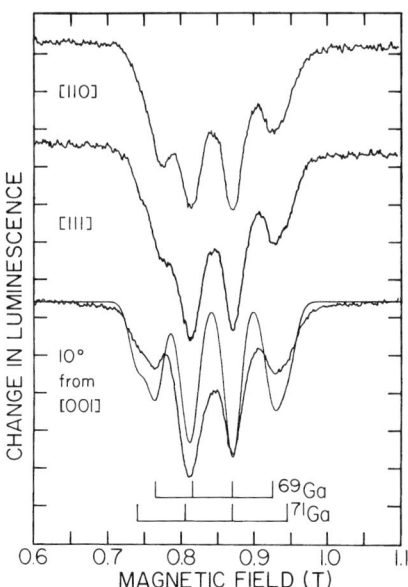

FIG. 19. ODMR of the Ga interstitial in $Al_xGa_{1-x}As$ for three field directions. A simulation for the $^{69}Ga$ and $^{71}Ga$ nuclei in natural abundance is shown with the lowest data. (From Kennedy and Spencer, *Physical Review Letters*, 1986, with permission.)

(Baraff and Schlüter, 1985) and simple ionic-bonding considerations indicate that the $Ga_i$ occupies the tetrahedral interstitial site surrounded by four As atoms (Fig. 20). Although the Ga in this site only senses the alloy disorder from its second-nearest neighbors, the second (cation) shell is only 15% farther from the $Ga_i$ site than the first (As) shell. However, it is not clear how alloy disorder in this shell would affect the central hyperfine interaction. It seems more likely that the changes in the low-field side of the spectrum

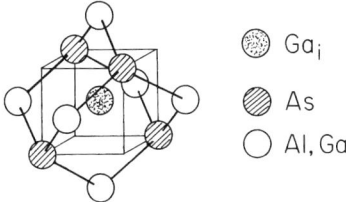

FIG. 20. The atomic structure of the Ga interstitial. (From Kennedy et al., *Physical Review B*, 1988, with permission.)

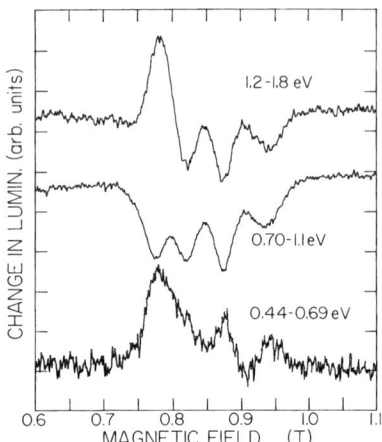

FIG. 21. ODMR of Be-doped $Al_xGa_{1-x}As$ for three spectral regions. The response at 0.77 T is from Be. Both the signals from the Ga-interstitial and the Be are positive for the energy range from 0.44 to 0.69 eV. (From Kennedy et al., *Physical Review B*, 1988, with permission.)

with angle are due to a second spectrum. This second spectrum is likely to be from the Be acceptor in the Be-doped sample (Fig. 21 and Section IV.3) and from Si on the group-V site acting as an acceptor in Si-doped samples (see Fig. 19). In this analysis, the $Ga_i$ is an isolated defect.

Study of the spectral dependence of the ODMR leads to a determination of the energy level for the interstitial. Since the Si-doped samples grown at 620°C showed high resistivity, it is likely that these samples are not n-type. The best evidence for assigning an energy level comes from the ODMR of a Be-doped sample (Fig. 21). As in the Si-doped cases, the $Ga_i$ signals are negative down to 0.7 eV. However, below this energy there is positive ODMR from the $Ga_i$ and the Be acceptor (which will be discussed in Section IV.3). This result can be understood as arising from the process

$$Ga_i^{++} + A^0 \rightarrow Ga_i^{3+} + A^- + \text{photon}.$$

This implies that the $+2/+3$ energy level is found 0.56 eV above the acceptor level, in rough agreement with the theoretical prediction (Baraff and Schlüter, 1985).

The lack of any observation of Al-interstitials deserves comment. The MBE growth process can be thought of as occurring in three steps. First, atomic and molecular species are adsorbed on the surface. Second, the species may diffuse and react at the surface. Third, atoms are incorporated into the growing layer. The ODMR reveals a difference between Al and Ga since Ga interstitials are observed while Al interstitials are not. $^{27}$Al has nuclear spin $\frac{5}{2}$, which would produce a six-line spectrum. Ga-atoms are more mobile, while Al atoms are more reactive. Thus the Al might initially take an interstitial site but then exchange with substitutional Ga because of the greater bond strength. The reaction is written as

$$Al_i + Ga_{Ga} \to Ga_i + Al_{Ga}.$$

The incorporation becomes less likely at higher temperatures where the surface mobilities are greater.

The $Ga_i$ has been observed also in bulk GaP:O (Lee, 1988), in GaAs/AlAs superlattices grown by MBE (Trombetta *et al.*, 1991), and in a 50-nm GaAs quantum well grown by MBE (Mochizuki *et al.*, 1995). From the doping levels of these samples and from the known sensitivity of ODMR, the concentration of interstitials is inferred to be on the order of $10^{16}$ cm$^{-3}$. Interstitial concentrations at elevated temperatures are much higher, where they play an important role in diffusion processes (Tan, 1995).

*b. Donors in N-Doped ZnSe/GaAs*

While defects are best identified by a nuclear signature, such as a hyperfine interaction, g-tensors can also provide enough information for a nearly complete identification. We review the case of defects in ZnSe grown by MBE on GaAs. This heterostructure contains a mismatch of 0.27% at room temperature.

The wider bandgap of ZnSe makes it potentially useful as a light emitter in the blue range. Optoelectronic devices in this material became possible when p-doping was achieved in MBE growth by doping with N from a plasma source. However, efforts to achieve high p-type conductivity have been thwarted by the occurrence of compensating donors that limit the concentration of holes at room temperature to $10^{18}$ cm$^{-3}$. PL and ODMR have been applied to try to identify the compensating donors.

Although the PL at low doping levels shows emission from excitons bound to neutral N acceptors, at higher doping levels DAP recombination dominates (Hauksson *et al.*, 1992). For a sample with $N_A - N_D = 9 \times 10^{16}$ cm$^{-3}$, the N acceptor recombines with a donor of energy 26 meV. At the higher concentration of $4 \times 10^{17}$ cm$^{-3}$, the recombination is dominated by pairs made of the N acceptor with a deeper donor at 44 meV. These

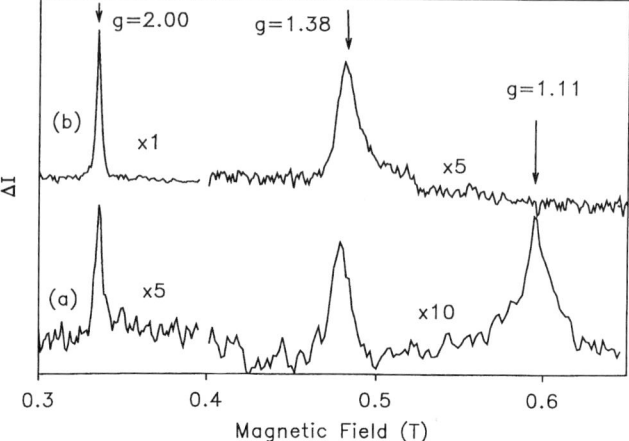

FIG. 22. ODMR at 9 GHz for N-doped ZnSe grown on GaAs for (a) a lower doping level and (b) a higher doping level. Three resonances with distinctive g-factors are observed. (Reprinted with permission from Murdin et al., *Applied Physics Letters* **63**, 2411 (1993). Copyright © 1993 American Institute of Physics.)

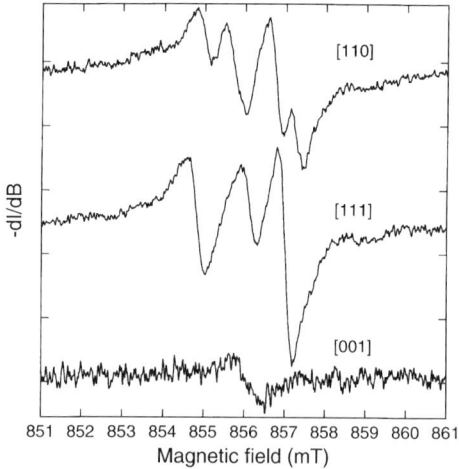

FIG. 23. The angular dependence of the ODMR for N-doped ZnSe/GaAs near g = 2.00 taken at 24 GHz. Field modulation produced derivative spectra, enhancing the resolution of the lines. Most of these can be described by a trigonal spin-Hamiltonian. (Reprinted with permission from Kennedy et al., *Applied Physics Letters* **65**, 1112 (1994). Copyright © 1994 American Institute of Physics.)

bands exhibit ODMR (Murdin et al., 1993) at widely different g-values (Fig. 22).

The g-value of 1.11 readily identifies one of the resonances as from shallow donors in ZnSe (see Section IV.1.a). These are probably impurities such as Cl or Ga, which may have diffused in from the substrate.

The resonance with g = 2.00 was originally thought to be due to a deep acceptor. However, further study at 35 GHz with microwave power modulation and at 24 GHz with magnetic field modulation (Kennedy et al., 1994) revealed splittings and anisotropy not evident in the original 9-GHz data (Fig. 23) and led to a reassignment to a complex acting as a deep donor. A plot of the data for different crystallographic directions in the ($1\bar{1}0$) plane reveals a pattern with trigonal symmetry, although some lines are unaccounted for. The g-anisotropy is very small, with $g_{\parallel} = 2.0072(2)$ and $g_{\perp} = 2.0013(2)$. Since the database of EPR and ODMR in II–VI compounds is so extensive, it was possible from this g-tensor to assign the spectrum to a $V_{Se}$-X pair, where X could be Cu or Ag. The small shifts from the free-electron g-value are characteristic of a Se vacancy and the trigonal

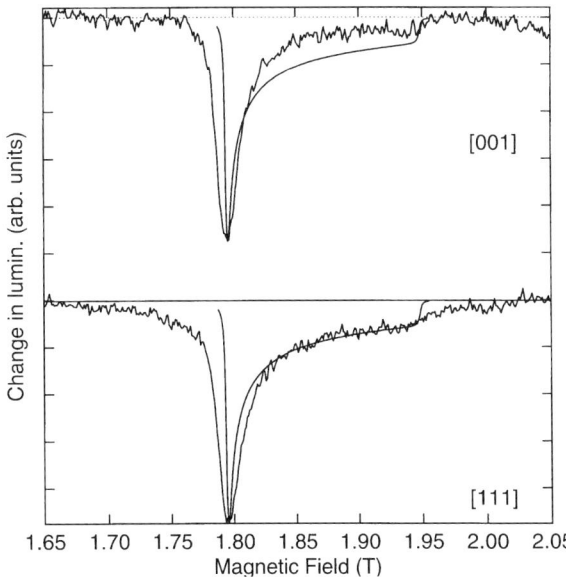

FIG. 24. ODMR at 35 GHz for two field directions in the region near g = 1.38. The lines are simulated by a powder pattern with $g_{\parallel} = 1.27$ and $g_{\perp} = 1.38$. (Reprinted from *Journal of Crystal Growth*, **159**, Kennedy et al., "ODMR of MBE-Grown ZnSe:N," 325–328, 1996 with kind permission of Elsevier Science-NL, Sara Burgerhartstraat 25, 1055 KV Amsterdam, The Netherlands.)

symmetry denotes a pair. The g-tensor is particularly close to those reported for $V_S$-Ag and $V_S$-Cu in ZnS (Dieleman et al., 1964).

The resonance with g = 1.38 is intermediate between the shallow donor (g = 1.11) and the deep, vacancy-pair defect (g = 2.00). Furthermore, it has a tail that extends to higher fields and has the lineshape of a powder pattern. Data taken at 35 GHz (Fig. 24) confirms the shape seen at 9 GHz with a scaling that indicates that the linewidth arises from g-anisotropy (Kennedy et al., 1996). Powder patterns arise from centers whose axes occur in random directions in space, and the lineshapes are readily calculated (Weil et al., 1994). The simulations were done with a Lorenztian single line convoluted with an axial g-tensor and yield $g_{\parallel}$ = 1.27 and $g_{\perp}$ = 1.38. Hence, the defects giving rise to this spectrum are axial, with directions highly randomized. This contrasts sharply with the spectrum for the $V_{Se}$-X center, which has the pattern for defects with equivalent orientations in an oriented single crystal. Hence, the two types of defects must arise from different regions of the heterostructure. Putting the evidence together, the g-value indicates a great deal of conduction-band character, and the powder pattern dictates randomization. Together, these facts suggest delocalized electrons associated with the dislocations in the heterostructures, which are grown beyond critical thickness. Independent studies have found a correlation between threading dislocations and compensation of the N acceptors in ZnSe (Kuo et al., 1994).

c. *Others*

In a similar fashion to the work in ZnSe, the Zn vacancy in OMCVD-grown ZnS on GaAs was identified from its g-tensor and optical-emission band and comparison to previous results (Poolton et al., 1987). The samples were grown with a high VI/II ratio. The ODMR detected on emission around 500 to 600 nm consisted of the donor line (see Section IV.1.a) and an anisotropic center with $g_{\parallel}$ = 2.005(2) and $g_{\perp}$ = 2.055(2) with the principal axis along $\langle 111 \rangle$ directions. Since this g-tensor is consistent with that for both A centers and zinc vacancies, the defect was identified from the spectral dependence, which peaked around 550 nm. The observation of $V_{Zn}^-$ centers in epitaxial ZnS was unexpected, since in bulk crystalline ZnS, these vacancies are seldom observed except after electron irradiation at low temperature.

In some cases, the information provided by ODMR so far is insufficient to provide a convincing identification. In Cu-doped GaP grown by LPE, there is a deep emission band that shows a strong resonance with g = 2.006(3) (Greidanus and Roosmalen, 1987). Although a convincing analysis is presented to show that the emission is due to DAPs, the line cannot be assigned. Similarly, the deep center observed on the 2.2-eV emission band in CVD-grown GaN on sapphire has so far evaded a convincing identification (Glaser *et al.*, 1995). Hope for greater information on the defect in GaN springs from the first observation of optically detected ENDOR (ODENDOR) (Koschnick *et al.*, 1996).

3. ACCEPTORS

*a. Effective-Mass Acceptors and Strain*

By definition, effective-mass acceptors have the character of the valence-band edge. Since this edge is fourfold degenerate, the states interact strongly with any strain in the system. To address the results on acceptor impurities in epitaxial systems, we need to develop a model to predict what the effective g-tensors will be for cases of interest.

The model follows that of Mehran *et al.* (1972) and starts by taking the case where the spin–orbit splitting is larger than any strain splittings. This leaves the fourfold degenerate $J = \frac{3}{2}$ state for which the Zeeman Hamiltonian is

$$H_z = \mu_0[g_1 \mathbf{J} \cdot \mathbf{H} + g_2(J_x^3 H_x + J_y^3 H_y + J_z^3 H_z)]. \tag{7}$$

For cases where uniaxial stress is externally applied, such as in the InP:Zn example in Section III.4, the coefficient of $J^3$ is important, even though it is small, since it produces an anisotropy in the experimentally accessible plane perpendicular to the applied stress. However, for epitaxial materials and Wurtizitic crystal structures, the magnetic field can be aligned both parallel and perpendicular to the axis of strain. Hence we will neglect $g_2$ and consider what happens for strain splittings that are large compared to the Zeeman interaction. Furthermore, we take $g_1$ to be $\frac{4}{3}$, the Lande g-factor for a pure $p_{3/2}$ state. Two cases arise.

Compressive uniaxial stress leaves the $M_J = \pm\frac{1}{2}$ doublet lower in energy. The effective g-factors along and perpendicular to the strain axis are $g'_{\parallel} = g_1 = 1.333$ and $g'_{\perp} = 2g_1 = 2.667$. A good experimental example of this case comes from the EPR study of B in Si, for which $g'_{\parallel} = 1.21$ and $g'_{\perp} = 2.43$ (Feher *et al.*, 1960).

Tensile uniaxial stress (from biaxial compression) and the crystal field in hexagonal crystal structures leave the $M_J = \pm\frac{3}{2}$ doublet lower in energy. The effective g-factors (corresponding to $S' = \frac{1}{2}$) are $g'_{\parallel} = 3g_1 = 4$ and $g'_{\perp} = 0$. A good experimental example comes from the ODMR study of

TABLE III

g'-VALUES FOR ACCEPTORS

| Host | Impurity | $g'_\parallel$ | $g'_\perp$ | Reference |
|---|---|---|---|---|
| $M_J = \pm\frac{1}{2}$ Manifold | | | | |
| Atomic $p_{3/2}$ derived | | 1.33 | 2.66 | |
| Si | B | 1.21 | 2.43 | Feher et al. (1960) |
| InP | Zn | 0.869 | | Crookham et al. (1993) |
| $M_J = \pm\frac{3}{2}$ Manifold | | | | |
| Atomic $p_{3/2}$ derived | | 4.0 | 0 | |
| CdS | Li | 2.829 | 0.3 | Patel et al. (1981) |
| $Al_xGa_{1-x}As/GaAs$ | Be | 2.15 | 2.20 | Cf. Kennedy et al. (1988) |
| GaN/sapphire | Mg | 2.08 | 1.99 | Kunzer et al. (1994) |

residual Li and Na acceptors in (Wurtzite) CdS where, for Li, $g'_\parallel = 2.829(7)$ and $g'_\perp = 0.3(1)$ (Patel et al., 1981).

These g-tensors provide dramatic evidence of effective-mass behavior for suitable impurities in suitably strained hosts (see Table III). However, there are a number of interactions that lead to the loss of pure effective-mass behavior. As the central-cell potential deepens the energy level for other impurities (such as Al, Ga, and In in Si), the orbital contribution to the g-tensor gets smaller. Furthermore, the degeneracy of the states leaves them subject to Jahn-Teller interactions, mixing due to the strain caused by the size mismatch of the impurity, and strong interaction with random strains from dislocations or nearby impurities (see Mehran et al., 1972, and references therein). Hence, a variety of behaviors has been observed for acceptors in bulk materials and can be expected for acceptors in epitaxial heterostructures.

b. *Be in AlGaAs/GaAs*

AlGaAs grown pseudomorphically on GaAs is under biaxial compression, which leads to the expectation that the $M_J = \pm\frac{3}{2}$ states are lower in energy. The ODMR for a Be-doped sample (Fig. 21) shows a strong broad resonance with $g_\parallel = 2.15$ and $g_\perp = 2.20$. This resonance seems to be the acceptor that is the partner to the $Ga_i$ donor in pair recombination. While this g-tensor has a strong positive g-shift and moderate anisotropy, it does not have the proper sense ($g_\parallel > g_\perp$) expected for $M_J = \pm\frac{3}{2}$ (see Table III). Since Be-implanted samples show a similar, though not identical, ODMR, this g-tensor is characteristic of Be in AlGaAs.

The measured g-tensor indicates that the Be acceptors observed are not in purely EM states. Yet the large g-shifts and anisotropy indicate strong contribution of orbital angular momentum. The reversed anisotropy is similar to the behavior observed for holes bound to tightly bound electrons at Ti impurities in SiC (Lee et al., 1985). In AlGaAs, the first-neighbor shell is sometimes aligned around C acceptors (Zheng et al., 1997). A similar effect for the second neighbors of a Be acceptor might produce the reversal of the local ordering of valence-band states.

c. *Acceptors in GaN*

Undoped GaN often shows PL with well-resolved phonon sidebands and a zero phonon line at 3.27 eV. Since the energy of the phonon sidebands corresponds with the longitudinal optical (LO) lattice modes (92 meV), this emission can be confidently attributed to pair recombination by effective-mass donors *and* acceptors (Dingle and Ilegems, 1971). Hence, ODMR of this band should exhibit both a shallow donor and a shallow acceptor signal. However, experiments performed on this emission from undoped

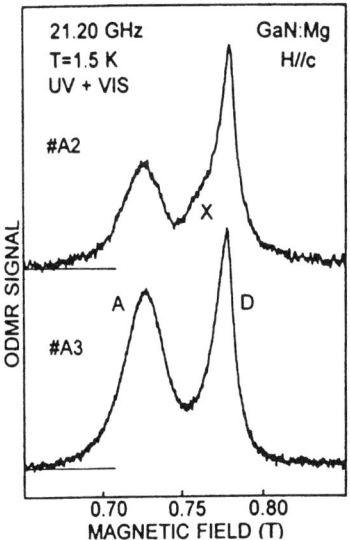

FIG. 25. ODMR at 21 GHz for GaN grown on sapphire with two levels of Mg-doping. The line labeled A is only observed in Mg-doped samples. (From Kunzer et al., *Materials Science Forum* **143**–**147**, 87 (1994), with permission.)

GaN/SiC samples showed only the signal from the shallow donor (Glaser et al., 1996). Thus, the residual strain due to dislocations in these films is sufficient to wash out the acceptor resonance, even though the PL still shows a clear EM signature.

Samples doped with Mg do show an acceptor-related ODMR. The resonance is broader than that from the donor with $g_{\parallel} = 2.08$ and $g_{\perp} = 1.99$ (Fig. 25) (Kunzer et al., 1994). Although this anisotropy has the proper sense, it is much smaller than what is expected for an EM acceptor (see Table III). However, the g-shift and anisotropy do have similar magnitudes to those observed for Be in AlGaAs. In heavily Mg-doped samples, the PL is broad and shifted to the low-energy side of 3.27 eV (Fig. 26). The spectral dependence of the Mg-related acceptor has been measured and compared with the PL (Glaser et al., 1995). The ODMR-active luminescence comes from the lower-energy portion of the luminescence. This shift is larger than what is expected from the Coulomb shift due to close pairs. Thus, some of the Mg acceptors in heavily doped samples are in deepened states, which do produce an optically detected resonance.

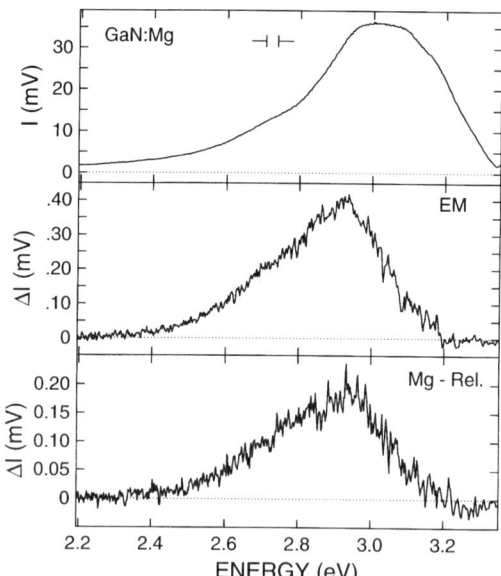

FIG. 26. The spectral dependence of the luminescence and ODMR for Mg-doped GaN. The ODMR response peaks at an energy slightly lower than that for the PL. (From Glaser et al., Physical Review B, 1995, with permission.)

Summarizing the results for acceptors in GaN gives the following picture. At low concentrations, the residual acceptors are fully EM-like but undetectable by ODMR due to random strain. These strains must produce crystal fields comparable to the Wurtzite-structure crystal field. At higher concentrations, some Mg atoms in more highly strained environments are deeper and produce a distinct resonance. These atoms may either sense the strain from nearby compensating donors or be nearer to dislocations because of their statistical distribution. This deepening with concentration is similar to the behavior of P in ZnSe (Yao and Okada, 1986).

## V. Summary and Future Directions

We presented the ODMR of Czochralski-grown InP:Zn as an illustration and a standard. In that case, three point defects were detected. The EM donor is distinguished by its unique g-factor. A range of experiments, including ODENDOR, provide a complete picture of the native deep center, the $P_{In}$ antisite. This picture includes the complete atomic structure, information about the wave function of the defect, and the electronic energy levels. The dopant, Zn, is detected and identified from its g-tensor and response to uniaxial stress.

The examples presented of defects in epilayers can be compared and contrasted to the results in InP:Zn. AlGaAs/GaAs samples contain uniform strain and alloy disorder. The ODMR of donors in this system reflects these conditions and provides rich information through the g-tensors. The work gives the most detailed picture for the optically active states of donors in AlGaAs with high Al mole fraction achieved so far. For deep states in epilayers, significant information comes from the hyperfine interaction for $Ga_i$ in AlGaAs and from the g-tensors for the vacancy-related defects in ZnSe and ZnS. Inhomogeneous strain is directly evident in the $g = 1.38$ donor in ZnSe. For these defects, more work is necessary before a full picture is achieved. The acceptor impurities are most sensitive to inhomogeneous strain and alloy disorder. Neither Be in AlGaAs nor Mg in GaN shows classic EM behavior.

While homogeneous strain from mismatch preserves order and translational symmetry, inhomogeneous strain and randomness in the alloys do not. The disorder potentials interact with the point defects to produce inhomogeneous broadening in their spectra. These effects are not new. Changes for the hyperfine interaction for P in Si were reported (using EPR) in going from pure Si to SiGe alloys (Feher, 1959) and in going from crystalline Si to amorphous Si (Stutzmann and Street, 1985).

In addition to the current interest in epitaxial systems with large mismatch, there is growing interest in defects in semiconductors with structure on the nanometer scale. Such structured systems include quantum wells, wires and dots, and nanocrystals. Defects in these structures are near interfaces and experience a variety of strained environments. The spectroscopy in semiconductors needs to evolve to address these defect problems.

Exiting developments in the spectroscopy of other systems offer solutions to this challenge. The observation of ODMR from single molecules is, perhaps, the best example. Using resonant excitation from a single-frequency laser, the magnetic resonance of a single molecule of pentacene in a p-terphenyl crystal was detected (Köhler et al., 1995; Wrachtrup et al., 1993). This spectroscopy represents the ultimate in both sensitivity and resolution. Spin coherence for the electron and hyperfine splitting from a single nucleus have also been demonstrated (Wrachtrup et al., 1995; Köhler et al., 1995). Since hyperfine interactions are so important to defect studies, optical NMR may also start to play a role (Marohn et al., 1995). Greater application of these and other advances in spectroscopy offers a bright future for the study of defects in epitaxial and structured semiconductors.

ACKNOWLEDGMENTS

We thank J. M. Trombetta, H. C. Crookham, and S. W. Brown for their contributions to the ODMR work at NRL and W. E. Carlos for a careful reading of the manuscript. The work at NRL was supported in part by the Office of Naval Research.

REFERENCES

Baraff, G. A. and Schlüter, M. (1985). *Phys. Rev. Lett.* **55**, 1327.
Böttcher, R., Wartewig, S., Bindemann, R., Kuhn, G., and Fischer, P. (1973). *Phys. Stat. Solidi B* **58**, K23.
Brunthaler, G., Ploog, K., and Jantsch, W. (1989). *Phys. Rev. Lett.* **63**, 2276.
Cavenett, B. C. (1981). *Adv. Physics* **30**, 475.
Crookham, H. C., Kennedy, T. A., and Treacy, D. J. (1992). *Phys. Rev. B* **46**, 1377.
Crookham, H. C., Glaser, E. R., Henry, R. L., and Kennedy, T. A. (1993). *Phys. Rev. B* **48**, 14, 157.
Davies, J. J. (1985). *J. Cryst. Growth* **72**, 317.
Davies, J. J. (1988). *J. Cryst. Growth* **86**, 599.
Dean, P. J., Schaierer, W., Lorenz, M., and Morgan, T. N. (1974). *J. Lumin.* **9**, 343.
Dean, P. J. (1982). *Prog. Crystal Growth Charact.* **5**, 89.
Dieleman, J., de Bruin, S. H., van Doorn, C. Z., and Haanstra, J. H. (1964). *Philips Res. Rep.* **19**, 311.

Dingle, R. and Ilegems, M. (1971). *Solid State Commun.* **9**, 175.
Feher, G. (1959). *Phys. Rev.* **114**, 1219.
Feher, G., Hensel, J. C., and Gere, E. A. (1960). *Phys. Rev. Lett.* **5**, 309.
Glaser, E. R. and Kennedy, T. A. (1991). *Semicond. Sci. Technol.* **6**, B97.
Glaser, E. R., Kennedy, T. A., Molnar, B., Sillmon, R. S., Spencer, M. G., Mizuta, M., and Kuech, T. F. (1991). *Phys. Rev. B* **43**, 14,540.
Glaser, E. R. and Kennedy, T. A. (1992). *Mat. Sci. Forum* **83-87**, 775.
Glaser, E. R., Kennedy, T. A., Doverspike, K., Rowland, L. B., Gaskill, D. K., Freitas, J. A., Jr., Khan, M. A., Olson, D. T., Kuznia, J. N., and Wickenden, D. K. (1995). *Phys. Rev. B* **51**. 13,326.
Glaser, E. R., Kennedy, T. A., Brown, S. W., Freitas, J. A., Jr., Perry, W. G., Bremser, M. D., Weeks, T. W., and Davis, R. F. (1996). *Mat. Res. Soc. Symp. Proc.* **395**, 667.
Godlewski, M., Chen, W. M., and Monemar, B. (1988). *Phys. Rev. B* **37**, 2570.
Greidanus, F. J. A. M. and van Roosmalen, J. T. C. (1987). *Solid State Commun.* **61**, 653.
Hauksson, I. S., Simpson, J., Wang, S. Y., Prior, K. A., and Cavenett, B. C. (1992). *Appl. Phys. Lett.* **61**, 2208.
Jeon, D. Y., Gislason, H. P., Donegan, J. F., and Watkins, G. D. (1987). *Phys. Rev. B* **36**, 1324.
Kennedy, T. A. and Wilsey, N. D. (1984). *Appl. Phys. Lett.* **44**, 1089.
Kennedy, T. A. and Spencer, M. G. (1986). *Phys. Rev. Lett.* **57**, 2690.
Kennedy, T. A., Magno, R., and Spencer, M. G. (1988). *Phys. Rev. B* **37**, 6325.
Kennedy, T. A., Glaser, E. R., Murdin, B. N., Pidgeon, C. R., Prior, K. A., and Cavenett, B. C. (1994). *Appl. Phys. Lett.* **65**, 1112.
Kennedy, T. A., Glaser, E. R., Brown, S. W., Cavenett, B. C., and Prior, K. A. (1996). *J. Cryst. Growth* **159**, 325.
Köhler, J., Brouwer, A. C. J., Groenen, E. J. J., and Schmidt, J. (1995). *Science* **268**, 1457.
Köhler, J., Disselhorst, J. A. J M., Donckers, M. C. J. M., Groenen, E. J. J., Schmidt, J., and Moerner, W. E. (1993). *Nature* **363**, 242.
Koschnick, F. K., Michael, K., Spaeth, J.-M., Beaumont, B., and Gibart, P. (1996). *Phys. Rev. B* **54**, R11,042.
Kunzer, M., Kaufmann, U., Maier, K., Schneider, J., Herres, N., Akasaki, I., and Amano, H. (1994). *Mat. Sci. Forum* **143-147**, 87.
Kuo, L. H., Salamanca-Riba, L., DePuydt, J. M., Cheng, H., and Qiu, J. (1994). *J. Electron. Mat.* **23**, 275.
Lawaetz, P. (1971). *Phys. Rev. B* **4**, 3460.
Lee, K. M., Dang, L. S., Watkins, G. D., and Choyke, W. J. (1985). *Phys. Rev. B* **4**, 2273.
Lee, K. M. (1988). *Mat. Res. Soc. Symp. Proc.* **104**, 449.
Mao, D., Taylor, P. C., and Ohlsen, W. D. (1993). *Phys. Rev. B* **49**, 7952.
Marohn, J. A., Carson, P. J., Hwang, J. Y., Miller, M. A., Shykind, D. N., and Weitekamp, D. P. (1995). *Phys. Rev. Lett.* **75**, 1364.
Matthews, J. W. and Blakeslee, A. E. (1976). *J. Cryst. Growth* **32**, 265.
Mayer, H., Rössler, U., Wolf, K., Elstner, A., Stanzl, H., Reisinger, T., and Gebhardt, W. (1995). *Phys. Rev. B* **52**, 4956.
Mehran, F., Morgan, T. N., Title, R. S., and Blum, S. E. (1972). *J. Magn. Res.* **6**, 620-627.
Mochizuki, Y., Mizuta, M., and Mochizuki, A. (1995). *Mat. Sci. Forum* **196-201**, 1927.
Montie, E. A., Henning, J. C. M., and Cosman, E. C. (1990). *Phys. Rev. B* **42**, 11,898.
Mooney, P. M., Wilkening, W., Kaufmann, U., and Kuech, T. F. (1989). *Phys. Rev. B* **39**, 5554.
Mooney, P. M. (1990). *J. Appl. Phys.* **67**, R1.
Morgan, T. N. (1968). *Phys. Rev. Lett.* **21**, 819.
Müller, K. A. and Schneider, J. (1963). *Phys. Lett.* **4**, 288.
Murdin, B. N., Cavenett, B. C., Pidgeon, C. R., Simpson, J., Hauksson, I., and Prior, K. A.

(1993). *Appl. Phys. Lett.* **63**, 2411.
Orton, J. W. (1968). *Electron Paramagnetic Resonance.* Gordon and Breach, New York.
Patel, J. L., Nicholls, J. E., and Davies, J. J. (1981). *J. Phys. C.* **14**, 1339.
Poolton, N. R. J., Nicholls, J. E., Davies, J. J., Cockayne, B., and Wright, P. J. (1987). *Semicond. Sci. Technol.* **2**, 448.
Schneider, J., Dischler, B., and Raeuber, A. (1968). *J. Phys. Chem. Solids* **29**, 451.
Spaeth, J.-M., Niklas, J. R., and Bartram, R. H. (1992). *Structural Analysis of Point Defects in Solids.* Springer-Verlag, Berlin and Heidelberg.
Stutzmann, M. and Street, R. A. (1985). *Phys. Rev. Lett.* **54**, 1836.
Sun, H. J., Gislason, H. P., Rong, C. F., and Watkins, G. D. (1993). *Phys. Rev. B* **48**, 17,092.
Tan, T. Y. (1995). *Mat. Chem. Phys.* **40**, 245.
Trombetta, J. M., Kennedy, T. A., Tseng, W., and Gammon, D. (1991). *Phys. Rev. B* **43**, 2458.
van der Merwe, J. H. (1963). *J. Appl. Phys.* **34**, 117.
van Kesteran, H. W., Cosman, E. C., Dawson, P., Moore, K. J., and Foxon, C. T. (1989). *Phys. Rev. B* **39**, 13,426.
Verity, D., Davies, J. J., Nicholls, J. E., and Bryant, F. J. (1981). *J. Appl. Phys.* **52**, 737.
Viohl, I., Ohlsen, W. D., and Taylor, P. C. (1991). *Phys. Rev. B* **44**, 7975.
Wartewig, S., Böttcher, R., and Kuhn, G. (1975). *Phys. Stat. Solidi B* **70**, K23.
Watkins, G. D. (1975). In: *Point Defects in Solids,* Vol. 2 (J. H. Crawford and L. M. Slifkin, eds.). Plenum Press, New York and London, Chap. 4.
Watkins, G. D. (This volume).
Weil, J. A., Bolton, J. R., and Wertz, J. E. (1994). *Electron Paramagnetic Resonance.* Wiley, New York, p. 94.
Weisbuch, C. and Hermann, C. (1977). *Phys. Rev. B* **15**, 816.
Wilson, D. K. and Feher, G. (1961). *Phys. Rev.* **124**, 1068.
Wrachtrup, J., von Borczyskowski, C., Bernard, J., Orrit, M., and Brown, R. (1993). *Nature* **363**, 244.
Wrachtrup, J., von Borczyskowski, C., Bernard, J., Orrit, M., and Brown, R. (1995). *Phys. Rev. Lett.* **71**, 3565.
Yao, T. and Okada, Y. (1986). *Jpn. J. Appl. Phys.* **25**, 821.
Zheng, J.-F., Stavola, M., Abernathy, C. R., and Pearton, S. J. (1997). *Mat. Res. Soc. Symp. Proc.* **442**, 387.

CHAPTER 4

# $\mu$SR on Muonium in Semiconductors and Its Relation to Hydrogen

## K. H. Chow

DEPARTMENT OF PHYSICS
LEHIGH UNIVERSITY
BETHELEHEM, PA

## B. Hitti

TRIUMF
VANCOUVER, B.C., CANADA

## R. F. Kiefl

PHYSICS DEPARTMENT
UNIVERSITY OF BRITISH COLUMBIA
VANCOUVER, B.C., CANADA

|    |                                                                       |     |
|----|-----------------------------------------------------------------------|-----|
| I. | INTRODUCTION                                                          | 138 |
| II.| FUNDAMENTALS OF $\mu$SR IN SEMICONDUCTORS                             | 143 |
|    | 1. Production and Decay of Spin-Polarized Muons                       | 144 |
|    | 2. Calculation of the Muon Spin Polarization Function $\vec{P}(t)$    | 145 |
|    | 3. Effective-Field Approximation                                      | 152 |
|    | 4. Influence of Nuclear Spins                                         | 153 |
| III.| EXPERIMENTAL TECHNIQUES AND EXAMPLES                                 | 155 |
|    | 1. Muon Spin Rotation in a Transverse Magnetic Field (TF-$\mu$SR)     | 157 |
|    | 2. Muon Level-Crossing Resonance ($\mu$LCR)                           | 167 |
|    | 3. Zero-Field Muon Spin Relaxation/Rotation (ZF-$\mu$SR)              | 176 |
|    | 4. Muon Spin Relaxation in a Longitudinal Magnetic Field (LF-$\mu$SR) | 180 |
|    | 5. Muon Spin Resonance in an RF Magnetic Field (RF-$\mu$SR)           | 193 |
|    | 6. Comparison of Techniques and Facilities                            | 201 |
| IV.| SUMMARY                                                               | 202 |
|    | References                                                            | 204 |

## I. Introduction

The structure and properties of isolated atomic hydrogen (H) in semiconductors have been under intense investigation in recent years. Much of this work was motivated by the discovery in the early 1980s that atomic hydrogen can passivate the electrical activity of shallow donors and acceptors in Si [see Pankove and Johnson (1991) for a concise introduction]. Subsequent work showed that similar passivation occurs in compound semiconductors such as GaAs (Chevallier *et al.*, 1985; Johnson *et al.*, 1986). It is reasonable to expect that isolated H can exist in three distinct charge states in a semiconductor ($H^0$, $H^+$, and $H^-$) and that its interstitial location, diffusivity, and reactivity with other defects are strongly dependent on the charge state. Unfortunately, direct information on isolated hydrogen is extremely limited due to its high diffusivity and reactivity with other defects (Pearton *et al.*, 1992). For example, only a single isolated H center has been characterized with electron paramagnetic resonance (EPR), the so called AA9 center in silicon [Gorelkinskii and Nevinnyi (1987, 1991)]. This has since been confirmed by an independent measurement [Bech Nielsen *et al.* (1994)]. Most of the experimental information on isolated hydrogen comes indirectly from studies of muonium (Mu = $\mu^+ e^-$), which is, in essence, a light hydrogen-like atom. Muonium in semiconductors is studied using a set of complementary methods that are collectively referred to as $\mu$SR, an acronym for muon spin *relaxation, rotation*, and *resonance*. In this chapter, we describe how these methods work and then give specific examples in each case which illustrate the type of information one obtains.

Muonium centers are formed when positively charged muons are implanted into a semiconductor. As with H, we expect that muonium can exist in all three charge states: $Mu^0$, $Mu^+$, and $Mu^-$. The center(s) that is observed in an experiment depends primarily on external factors such as temperature and dopant level, although under certain circumstances the details of the thermalization process of the muon that occurs within about 1 ns after implantation can be important. Over 20 distinct muonium centers have been identified in tetrahedrally coordinated semiconductors. Until a few years ago, the most significant contribution that $\mu$SR had made to the understanding of hydrogen had come from structural determinations of the neutral paramagnetic centers, particularly in silicon, diamond, GaAs, and GaP. As in the case of EPR, these determinations are based on measurements of the hyperfine interactions on the muon and nuclear spins which are sensitive to the unpaired electron spin density distribution and the neighboring nuclei. Two distinct neutral centers, one stable and the other metastable, are observed in Si, Ge, diamond, GaAs, and GaP (Patterson, 1988; Kiefl and Estle, 1991). The labels $Mu_T^0$ and $Mu_{BC}^0$ are now commonly

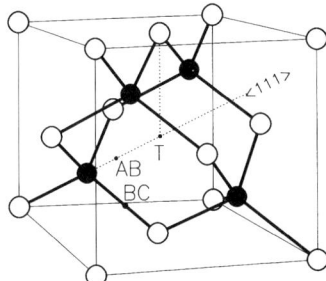

FIG. 1. Interstitial sites in the zinc-blende structure of a III–V semiconductor, such as GaAs or GaP. The tetrahedral (T) and bond-center (BC) sites are shown, as well as the antibonding (AB) site. The shaded circles represent one type of host nuclei, while the open circles the other type. This figure will also be valid for a group IV semiconductor such as Si or diamond, where both the shaded and open circles represent the same host atoms.

used where the superscript reflects the charge and subscript letter identifies the interstitial site (Fig. 1). $Mu_T^0$ is characterized by a large muon–electron hyperfine interaction that is isotropic, indicating that it is at or very near the tetrahedral (T) interstitial site. At least in Si, GaAs, and Ge, there is good evidence that $Mu_T^0$ is diffusing rapidly, even at low temperatures (see Patterson, 1988 and references therein; Kadono et al., 1990a, 1994a; Schneider et al., 1992b). The isotropic hyperfine parameter $A_\mu$ is roughly half that of muonium in vacuum (4463.302 MHz), indicating that the unpaired electron is centered on the muon in an orbital with a large 1s component. In the case of $Mu_{BC}^0$, the muon hyperfine interaction is axially symmetric about a $\langle 111 \rangle$ crystalline axis (i.e., bond direction) and is described by two parameters $A_\parallel$ and $A_\perp$, which are about an order of magnitude smaller than for $Mu_T^0$. Measurements of both the muon and surrounding nuclear hyperfine interactions have established that in the case of $Mu_{BC}^0$ the muon is located close to the bond-center (BC) position, with the two nearest neighbors atoms pushed substantially outward (Kiefl et al., 1987, 1988; Schneider et al., 1993a). $Mu_{BC}^0$ appears to be immobile over the relevant timescale in a $\mu$SR experiment (about 10 $\mu$s after implantation).

There have been two important developments in this field over the last few years. The first is the work on charged muonium (i.e., $Mu^+$ or $Mu^-$). Since in this case there is no unpaired electronic spin and associated hyperfine interactions, information on the structure and dynamics must be deduced from measurements of the muon-nuclear magnetic dipolar interaction. These couplings are much weaker and less distinctive than the large hyperfine interactions that occur in paramagnetic centers. Nevertheless, using the full complement of advanced $\mu$SR methods described in this chapter, one can obtain structural information on charged muonium that is equivalent to what is known about neutral paramagnetic centers. For example, it is now established that isolated Mu$^-$ is formed in heavily doped n-type GaAs and that it is located at the tetrahedral interstice surrounded by four nearest-neighbor Ga atoms, with only a small distortion in the positions of these nearest-neighbor atoms (Chow et al., 1995) (i.e., Mu$_T^-$).

A second important development is in the area of muonium dynamics. This includes studies of muonium diffusion, charge–spin exchange cycling with free carriers, and interconversion between the various muonium states. Until recently, little was known about these processes. A comprehensive picture of how muonium diffuses and interacts with charge carriers is beginning to emerge, at least in the case of Si and GaAs, and very similar behaviour is expected for hydrogen. One of the striking results is the dramatic difference in the diffusion rates of Mu$_T^-$ and Mu$_T^0$. For example, in GaAs, the diffusion rate of Mu$_T^-$ at room temperature appears to be about 10 orders of magnitude smaller than its neutral counterpart, Mu$_T^0$. Furthermore, it appears that Mu$_T^-$ in n-type GaAs is a deep electron-hole recombination center (Chow et al., 1996a). Another important result is that in silicon with low or intermediate doping levels, a rich and complicated set of dynamics involving transitions between all the muonium states appears to be ocurring on the timescale of the muon lifetime.

Before describing the experimental techniques and presenting relevant examples, it is important to review the extent to which $\mu$SR experiments on muonium in a semiconductor will model the behaviour of atomic hydrogen. The close relationship between muonium and hydrogen is due to the fact that the positive muon and the proton have the same spin and charge and also because it has a mass equal to about $\frac{1}{9}$ that of a proton or about 200 times that of an electron. Table I lists some of the physical properties of the muon and muonium and compares these with the proton and hydrogen atom. Note the reduced mass of the electron for a free muonium atom along with all other quantities related to electronic structure, such as binding energy, are almost identical to those of a hydrogen atom. There is good experimental evidence that this close correspondence between muonium and hydrogen, particularly with regard to electronic structure, carries over into the corresponding centers formed in semiconductors. For example, in the single case where isolated atomic hydrogen has been characterized by EPR (the AA9 center in Si), the observed hyperfine parameters agree within about 10% of those measured for the analogous muonium center (after correcting for a trivial factor due to fact that the magnetic moment of the muon is 3.18 times that of the proton). The nontrivial differences are likely due to the enhanced zero point motion of the much lighter muonium atom.

## TABLE I
### SOME PHYSICAL PROPERTIES OF THE MUON, PROTON, VACUUM MUONIUM AND VACUUM HYDROGEN

| Physical Properties | Muon | Proton |
|---|---|---|
| Charge[a] ($e$) | +1 | +1 |
| Spin[b] ($\hbar$) | $\frac{1}{2}$ | $\frac{1}{2}$ |
| Mass (MeV/c$^2$) | 105.66 | 938.27 |
| Gyromagnetic ratio $\bar{\gamma}$ (MHz/T) | 135.54 | 42.58 |
| Lifetime $\tau$ ($\mu$s) | 2.19703 | Stable |
| | Muonium | Hydrogen |
| Reduced electron mass ($m_e$)[c] | 0.995187 | 0.999456 |
| Binding energy (eV) | 13.54 | 13.60 |
| Hyperfine parameter (MHz) | 1402[d] | 1420.4 |

[a]The unit $e$ corresponds to the unit of electronic charge and has a value of $+1.602 \times 10^{-19}$ C.
[b]The unit $\hbar$ is $h/2\pi$, where $h$ is Planck's constant.
[c]The electron mass $m_e$ is 0.51100 MeV/c$^2$.
[d]The actual vacuum muonium hyperfine parameter is 4463.302 MHz. The value of 1402 MHz stated in the table is after correcting for the trivial factor due to the fact that the magnetic moment of the muon is 3.183 times that of the proton. The small difference with the stated hyperfine parameter of vacuum hydrogen is due to the non-equal spin densities at the nucleus because of the different nuclear masses.

Thus, we expect only modest differences in quantities related to the electronic structure, such as the local electronic density, location of neighboring atoms, and the energy level scheme within the semiconducting gap. On the other hand, more substantial differences are expected when comparing diffusion rates, since enhanced zero point vibrational energy reduces the effective energy barrier to the next site. Furthermore, even if the energy barriers were identical, one would expect more rapid quantum mechanical tunneling in the case of the lighter muonium atom. Theoretical work on Si indicates that the enhanced zero point motion of muonium can influence the relative stability of different sites under special circumstances (Ramirez and Herrero, 1994).

In addition to these fundamental differences arising from the mass difference, it is also important to keep in mind that the experiments on muonium and hydrogen are often done under different circumstances. For example, muons are studied in the extreme dilute limit, within a short timescale of about 10 $\mu$s after implantation. This short time window, which is dictated by the muon lifetime of 2.2 $\mu$s, along with the fact that there are at most only a few muons in the sample at a time, implies that muonium is almost always studied in an isolated form: that is, there is little chance for the muon to interact with another point defect and virtually no chance to interact with another muon. On the other hand, hydrogen is normally introduced at high concentrations and is studied over a much longer time window, ranging from a few minutes to several days after its introduction. As a result, H is most easily studied as part of a complex with other point defects, including itself (e.g., $H_2^*$; see Holbech et al., 1993).

Another important difference is that muons are implanted at high energies at any temperature, whereas most often hydrogen in diffused into the sample at high temperatures. The short time window of observation in the case of $\mu$SR means that muonium may be observed in non-equilibrium charge states or metastable sites. This is especially relevant at low temperatures, where the equilibration time is long, and in low-doped materials, where the lack of free charge carriers impedes the drive toward equilibrium. The most obvious example of this is the co-existence of two neutral muonium states (i.e., $Mu_T^0$ and $Mu_{BC}^0$), in GaAs (Section III.1.a; Kiefl et al., 1985). Clearly one of these must be metastable. This situation should be contrasted with studies on hydrogen, where it is very unlikely that the hydrogen would be observed in a metastable site or in a non-equilibrium charge state.

In summary, the main usefulness of muonium with regard to hydrogen in semiconductors is that it can serve as an experimental model for isolated interstitial hydrogen $H^0$, $H^+$, and $H^-$. Studies of muonium are the main source of information on these centers, which are difficult to study directly. In some instances, the information gained from muonium studies is expected to be transferable to hydrogen in a very direct way. For other kinds of information, the link between muonium and hydrogen is less direct. Nevertheless, even when significant differences in the behaviour of muon and hydrogen are anticipated, the muonium results should provide a stringent test for any theoretical model for hydrogen.

The remainder of this chapter will be divided as follows: Section II discusses the basic principles behind the production of muons and their inherent muon spin polarization. In any type of $\mu$SR experiment, the muon spin polarization is the observable from which one extracts physical information on the structure and dynamics of muonium. The basic principles behind the generation of spin-polarized muons are described in this section. Also, we present some methods for calculating the muon spin polarization and illustrate these techniques for free stationary muonium centers. The main body of this chapter is in Section III, which describes the various experimental techniques with relevant experimental examples in diamond, silicon, GaP, and GaAs. Muon spin rotation in a transverse magnetic field

(TF-$\mu$SR) and muon level-crossing resonance ($\mu$LCR) are treated first, since their application to semiconductors was first and they provide most of the information on electronic structure. Other closely related methods are used to measure the muon spin relaxation rate in zero applied field (ZF-$\mu$SR) or in a longitudinal applied field (LF-$\mu$SR). These along with radio-frequency muon spin resonance (RF-$\mu$SR) are natural complements to TF-$\mu$SR and $\mu$LCR since the former are primarily sensitive to dynamical processes. This section illustrates how one obtains information on the electronic structure, diffusion, spin-exchange scattering, cyclic charge-state changes, and the interconversion between the various muonium centers. The relevance to hydrogen is discussed throughout Section III and is summarized in Section IV.

## II. Fundamentals of μSR in Semiconductors

Studies of muonium in semiconductors may be classified under one of two categories: *electronic structure* or *dynamics*. Experiments on the former are concerned with the crystalline site of the muon, any possible rearrangement of the neighboring atoms, and the surrounding electronic density (or unpaired electronic spin density in the case of paramagnetic centers). As in related forms of magnetic spin resonance, information on structure is extracted from the relevant spin interactions (i.e., the muon and nuclear hyperfine interactions, muon-nuclear dipolar interactions, and nuclear-electric quadrupole interactions, etc.). *Muonium dynamics* refers to the diffusive or tunneling motion of muonium, spin exchange, or charge-changing reactions with free carriers and any site changes. Generally speaking, the spin interactions and resulting spin-Hamiltonian are a function of the electronic structure. We will show how time-independent spin-Hamiltonians invariably lead to a set of coherent muon spin precession frequencies that characterize the structure. Stochastic processes, which cause sudden changes in the spin interactions (e.g., such as diffusion or scattering with free carriers), result in muon spin relaxation. Both muon spin precession and muon spin relaxation are observable using the appropriate experimental technique.

In this section, we begin with a short review of how the spin-polarized muons are produced and then describe how the time evolution of the muon spin polarization $\vec{P}(t)$ is measured. We then show how $\vec{P}(t)$ is related to the quantities of interest, such as the hyperfine parameters.

## 1. Production and Decay of Spin-Polarized Muons

The most intense beams of spin-polarized muons are produced at meson factories such as TRIUMF (Canada) and PSI (Switzerland), and at spallation neutron sources such as ISIS (UK). These facilities provide intense beams (100 µA or more) of moderate energy protons ($\approx 500$ MeV), which are directed at suitable production targets (such as carbon or beryllium). The resulting energetic collisions produce a large flux of positive pions. Pions have zero spin and decay via the weak interaction, with a mean lifetime of 26.03 ns, through the dominant decay mode: $\pi^+ \to \mu^+ + \nu_\mu$. In the rest frame of the pion, both the muon and neutrino are created with negative helicity; that is, their spins are aligned perfectly antiparallel with their momentum, as illustrated in Fig 2. Pions that stop near the surface of the production target give rise to low-energy muons (4 MeV), which, when collected in a secondary beamline, are nearly 100% spin polarized in the lab frame. Most µSR experiments today use these so-called surface muons,

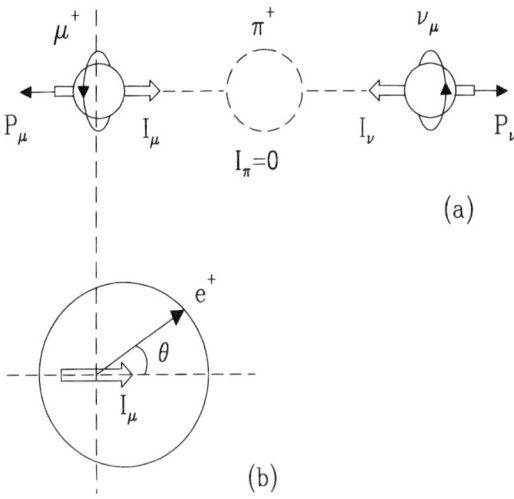

FIG. 2. (a) Pions decay into a muon and a neutrino in the center-of-mass frame. Conservation of angular momentum implies that the muon spin direction is *antiparallel* to the muon momentum. The symbols $P_\mu$ and $P_\nu$ denote the muon and neutrino momentum, respectively. The symbols $I_\mu$, $I_\pi$, and $I_\nu$ denote the muon, pion, and neutrino spins respectively. (b) The positron emission probability, averaged over all positron energies, as a function of angle in muon decay. The distance from the origin to a point on the curve is proportional to the decay rate at angle $\theta$ with respect to the muon spin direction. Note that the direction in which the decay positron is emitted is correlated with the muon spin polarization.

which have a mean range (calculated as the product of the density and the thickness of the material) of $\approx 120 \, \text{mg/cm}^2$ in carbon. This value is similar for the materials of interest in this chapter.

There are basically two types of muon facilities, depending on the temporal distribution of muons in the beam. TRIUMF and PSI provide continuous beams of muons: the muons arrive essentially at random. ISIS and KEK in Japan are pulsed facilities, such that the muons arrive at the sample in short pulses separated by a time that is long compared to the muon lifetime. The influence of this time structure on a $\mu$SR experiment is discussed in Section III. An ideal sample for a typical experiment would have a cross-sectional area of $\approx 10 \, \text{cm}^2$ and a thickness of several hundred $\mu$m. However, special methods have been developed (see Section III.1.b; Schneider et al., 1993b) that allow measurements on much smaller samples.

Positive muons have a mean lifetime of 2.197 $\mu$s (see Table I) and decay into a positron, an electron neutrino, and a muon anti-neutrino:

$$\mu^+ \to e^+ + \nu_e + \bar{\nu}_\mu \tag{1}$$

The positron detectors are made from a plastic scintillator coupled by plastic light guides to a photomultiplier (see Section III). The energy of the positron varies continuously from zero (if the two neutrinos are emitted in opposite directions and carry away all the kinetic energy) to a maximum of $\approx 53 \, \text{MeV}$ (if the two neutrinos travel together and antiparallel to the positron). Muon decay properties are calculated from the theory of weak interactions (see Schenck, 1985, and references therein). The most important feature of the decay is that the positron is emitted *preferentially in the direction of the muon spin* (see Fig. 2b). The large natural spin polarization in muon beams along with this asymmetry in muon decay are the basis of the $\mu$SR technique.

2. CALCULATION OF THE MUON SPIN POLARIZATION FUNCTION $\vec{P}(t)$

In this section, we discuss how one calculates the muon spin polarization (the observable quantity) given a spin-Hamiltonian. A few examples are then given that are relevant to the rest of this chapter.

Consider the spin-Hamiltonian (e.g., Kiefl and Estle, 1991) for a stationary paramagnetic muonium center surrounded by nuclear spins $\mathbf{J}^i$ in a magnetic field $\mathbf{B}$ (magnitude $B$). This may be written

$$\mathcal{H} = \mathcal{H}_{\text{Mu}} + \mathcal{H}_n \tag{2}$$

with

$$\mathcal{H}_{\text{Mu}} = h\tilde{\gamma}_e \mathbf{B}\cdot\mathbf{S} - h\tilde{\gamma}_\mu \mathbf{B}\cdot\mathbf{I} + \mathbf{S}\cdot\tilde{\mathbf{A}}^\mu\cdot\mathbf{I} \qquad (3)$$

and

$$\mathcal{H}_n = \sum_i [-h\tilde{\gamma}_n^i \mathbf{B}\cdot\mathbf{J}^i + \mathbf{S}\cdot\tilde{\mathbf{A}}^i\cdot\mathbf{J}^i + \mathbf{I}\cdot\tilde{\mathbf{D}}^i\cdot\mathbf{J}^i + \mathbf{J}^i\cdot\tilde{\mathbf{Q}}^i\cdot\mathbf{J}^i] \qquad (4)$$

$\mathcal{H}_{\text{Mu}}$ describes the Zeeman interactions of the electron spin **S** and the muon spin **I** in a magnetic field **B** and the hyperfine interaction between the muon and electron spin, and $\mathcal{H}_n$ contains additional terms involving the nuclear spins. Planck's constant is symbolized by $h$ and $\tilde{\gamma}_e$ (28024.95 MHz/T), $\tilde{\gamma}_\mu$ (see

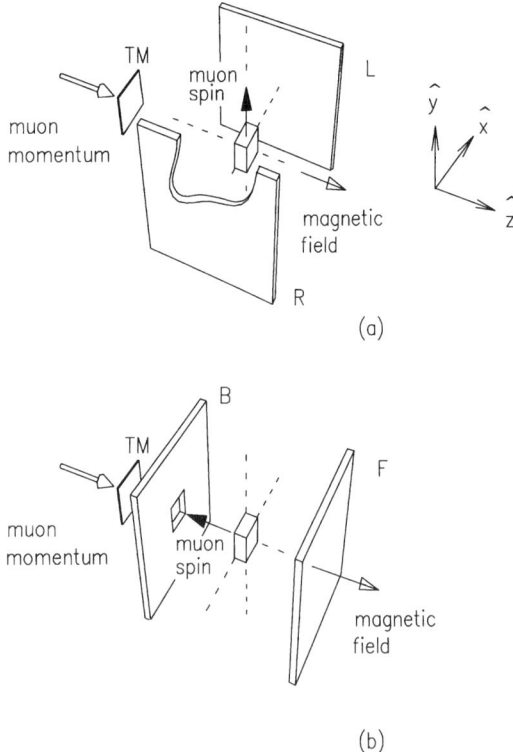

FIG. 3. The two standard configurations for a $\mu$SR experiment. In (a) the muon spin is initially perpendicular to the applied magnetic field, while in (b) the muon spin is initially parallel to the applied magnetic field. The designation TM refers to the "thin muon" counter. Positrons are detected by the thicker L and R (left and right) and F and B (forward and backward) scintillators.

Table I) and $\tilde{\gamma}_n^i$ are the electron, muon, and nuclear gyromagnetic (magnetogyric) ratios, respectively. The couplings between the muon, electron, and nuclei are defined as follows: $\tilde{\mathbf{A}}^\mu$ is the muon–electron hyperfine tensor and $\tilde{\mathbf{A}}^i$ is the tensor describing the hyperfine interaction between the electrons and the $i$th neighboring nucleus. $\tilde{\mathbf{D}}^i$ is the dipolar tensor between the muon and nucleus $i$ and $\tilde{\mathbf{Q}}^i$ is the nuclear quadrupole tensor for nucleus $i$.[1] The $g$ tensors are all assumed to be isotropic (Blazey et al., 1986), hence the gyromagnetic ratios are scalars ($\gamma \propto g$). The very weak dipolar interactions between different nuclear spins are neglected.

The two most common detector configurations in a $\mu$SR experiment are shown in Fig. 3. Note that they are distinguished by the orientation between the initial muon spin polarization vector and the applied magnetic field **B**. In a transverse-field (TF) experiment (see Fig. 3a), the initial muon spin is perpendicular to the applied magnetic field. The component(s) of the muon spin polarization vector perpendicular to the applied field is monitored with positron detectors in the plane of precession. In a longitudinal-field (LF) experiment the initial muon polarization direction is directed along the magnetic field direction ($\hat{z}$) and the axis of the counters (see Fig. 3b). Using these counter arrangements, one can measure the time dependence of the various components of the muon spin polarization, $\vec{P}(t)$, which are a function of the spin-Hamiltonian. Next, we describe the density matrix approach for calculating $\vec{P}(t)$ from $\mathcal{H}$, which is useful for Hamiltonians involving a small number of nuclear spins.

In a density matrix notation (Weissbluth, 1978; Roduner and Fischer, 1982),

$$\vec{P}(t) = \langle \vec{\sigma}^\mu(t) \rangle = Tr[\rho(t)\vec{\sigma}^\mu] \qquad (5)$$

Here, **B** is assumed, without loss of generality, to be applied along the $\hat{z}$ direction, and $\sigma_r^\mu$ are the Pauli matrices corresponding to the $r$th component of the muon spin ($r = x, y, z$):

$$\sigma_x^\mu = \begin{pmatrix} 0 & 1 \\ 1 & 0 \end{pmatrix}; \quad \sigma_y^\mu = \begin{pmatrix} 0 & -i \\ i & 0 \end{pmatrix}; \quad \sigma_z^\mu = \begin{pmatrix} 1 & 0 \\ 0 & -1 \end{pmatrix} \qquad (6)$$

The spin operators in Eq. (6) operate only on the muon spin.

In the Heisenberg representation, the density matrix $\rho(t) = e^{i\mathcal{H}t/\hbar}\rho(0)e^{-i\mathcal{H}t/\hbar}$. Straightforward quantum mechanical manipulation of Eq. (5) results in a form that is particularly useful for analytical or numerical calculation. If the initial muon spin is 100% polarized in the $\hat{r}$th direction

---

[1] Only nuclei with $J > \frac{1}{2}$ have a quadrupole moment (Weissbluth, 1978).

($r = x, y, z$) and the electron and nuclei are unpolarized, then the $s$th ($s = x, y$, or $z$) component of $\vec{P}(t)$ is given by,[2]

$$P_s(t) = 1 - \frac{2}{K} \sum_{i,j>i} |\langle \varepsilon_i | \sigma_r^\mu | \varepsilon_j \rangle|^2$$
$$+ \frac{1}{K} \sum_{i,j>i} [e^{-i\omega_{ij}t} \langle \varepsilon_i | \sigma_r^\mu | \varepsilon_j \rangle \langle \varepsilon_j | \sigma_s^\mu | \varepsilon_i \rangle + e^{i\omega_{ij}t} \langle \varepsilon_j | \sigma_r^\mu | \varepsilon_i \rangle \langle \varepsilon_i | \sigma_s^\mu | \varepsilon_j \rangle] \quad (7)$$

where $|\varepsilon_i\rangle$ denotes the energy eigenstates of $\mathscr{H}$ with energy eigenvalues $\varepsilon_i$, and $K$ is the dimension of the matrix representation of $\mathscr{H}$ (hence $K$ is also the number of energy levels). For example, in a system consisting of the $\mu^+$ and $N$ nuclei of spin $J$, $K = 2(2J + 1)^N$, while $K = 4(2J + 1)^N$ for a system consisting of neutral muonium $Mu^0$ ($\mu^+ + e^-$) and $N$ nuclei. Futhermore,

$$\omega_{ij} = \frac{\varepsilon_i - \varepsilon_j}{\hbar} \quad (8)$$

and $\hbar$ is $h/2\pi$.

Often, it is sufficient to consider the component of the polarization that is along the initial muon spin direction (i.e., $s = r$). This would be the case if the other components are either zero or if they only differ by the initial phase. For example, in the transverse-field arrangement shown in Fig. 3a one is normally interested in $P_x(t)$ (i.e., $s = r = x$) or $P_y(t)$ (i.e., $s = r = y$), while in the longitudinal-field arrangement shown in Fig. 3b, a calculation of $P_z(t)$ (i.e., $s = r = z$) is usually desired. Equation (7) can then be simplified to give:

$$P_r(t) = 1 - \frac{2}{K} \sum_{i,j>i} |\langle \varepsilon_i | \sigma_r^\mu | \varepsilon_j \rangle|^2 \times [1 - \cos(\omega_{ij}t)] \quad (9)$$

Specific examples illustrating the uses of the density matrix approach are discussed in Section II.2.a and Section II.2.b.

Another relevant observable is the time-integrated polarization, defined as follows:

$$\bar{P}_r = \frac{\int_{t_1}^{t_2} e^{-t/\tau_\mu} P_r(t) dt}{\int_{t_1}^{t_2} e^{-t/\tau_\mu} dt} \quad (10)$$

[2] $\rho(0) = (1/K)(1 + \sigma_r^\mu)$, where 1 is the unity operator.

As shall be discussed in Section III.2, $\bar{P}_z$ is the observable quantity in a $\mu$LCR experiment. In the special case where $t_1 = 0$ and $t_2 = \infty$, $\bar{P}_z$ is proportional to the total integrated positron asymmetry (the choice of $t_1 \neq 0$ and $t_2 \neq \infty$ enables one to enhance the signal (Leon, 1992):

$$\bar{P}_z = 1 - \frac{2}{K} \sum_{i,j>i} |\langle \varepsilon_i | \sigma_z^\mu | \varepsilon_j \rangle|^2 \times \frac{\omega_{ij}^2}{\omega_{ij}^2 + 1/\tau_\mu^2}, \qquad (11)$$

which is the sum of Lorentzians.

In summary, to calculate $\vec{P}(t)$ from $\mathcal{H}$, the following approach can be used: (1) Calculate the eigenvectors and eigenvalues of $\mathcal{H}$. (2) Apply Eq. (7) or Eq. (10) [or Eq. (11)] depending on whether one is interested in $\vec{P}(t)$ or the integrated polarization. Clearly, the complexity of the calculated $\vec{P}(t)$ depends on the number of terms that are considered in the Hamiltonian Eq. (2). It is instructive to first discuss muonium in the absence of nuclear spins (i.e., setting $\mathcal{H}_n = 0$ in Eq. (2)). In addition to its pedagogical value, such a simplification is relevant for interpretation of data in a large number of situations. For example, the approximation is an excellent one for muonium in Si where the majority of the host nuclei have zero spin ($^{28}$Si and $^{30}$Si occur with isotopic abundances of 92.21% and 3.09%, respectively) and only a small number have spin $\frac{1}{2}$ ($^{29}$Si occurs with 4.7% isotopic abundance).

*a. Diamagnetic Centers in the Absence of Nuclear Spins*

Suppose a diamagnetic muonium center is located in a system where the dipole and quadrupole interactions with the surrounding nuclei are negligible ($\mathcal{H}_n = 0$). Then the expression for a free muon in a magnetic field (which is assumed to be applied along the $\hat{z}$ axis) is recovered from Eq. (2):

$$\mathcal{H}/h = -\tilde{\gamma}_\mu B I_z \qquad (12)$$

This Hamiltonian is appropriate when there is no unpaired electron spin. Examples include Mu$^+$, Mu$^-$, or any diamagnetic complex containing muonium.

The muon spin polarization can be calculated following the just described prescription. First, the eigenstates and eigenvalues are determined. The basis set labeled by the muon magnetic quantum number $m_I$ are eigenstates of Eq. (12). These can be labeled as $|+\rangle$ (i.e., $m_I = +\frac{1}{2}$) and $|-\rangle$ (i.e., $m_I = -\frac{1}{2}$) and correspond to spin-up and spin-down, respectively. In the matrix

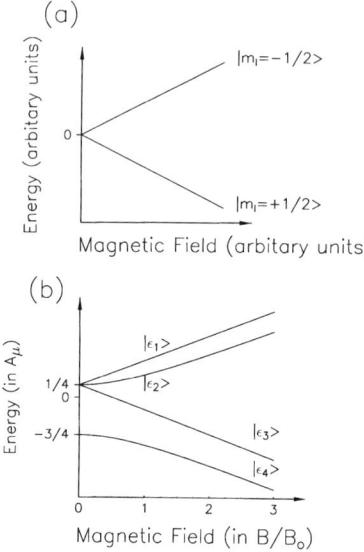

FIG. 4. The energy versus field for (a) free diamagnetic muonium and (b) muonium with an isotropic hyperfine parameter. At higher magnetic fields than shown in Fig. b, there is a crossing in the energy levels corresponding to $|\varepsilon_1\rangle$ and $|\varepsilon_2\rangle$. The eigenvectors marked in the figure are described in more detail in the text.

representation,

$$|+\rangle \equiv \begin{pmatrix} 1 \\ 0 \end{pmatrix}; \quad |-\rangle \equiv \begin{pmatrix} 0 \\ 1 \end{pmatrix} \quad (13)$$

The energy eigenvalues are $-\tilde{\gamma}_\mu B/2$ and $+\tilde{\gamma}_\mu B/2$ for the $|+\rangle$ and $|-\rangle$ states, respectively. The plot of energy versus magnetic field is shown in Fig. 4a.

The next step is to calculate $\vec{P}(t)$ for the transverse-field (see Fig. 3a) and longitudinal-field (see Fig. 3b) arrangements. In the TF arrangement shown in Fig. 3a, the initial muon spin is in the $\hat{y}$ direction, while the quantity of interest is $P_x(t)$. Eq. (9), with $r = y$, gives $P_y(t) = 1 - [1 - \cos(\omega_\mu t)] = \cos(\omega_\mu t)$ where $\omega_\mu = 2\pi \tilde{\gamma}_\mu B$. Similarly, $P_x(t)$ can be calculated directly from Eq. (7) to be $\sin(\omega_\mu t)$ while $P_z(t) = 0$. Note that $P_x(t)$ and $P_y(t)$ only differ by the value of the initial phase. This is the familiar result of the Larmor precession of a spin $\frac{1}{2}$ particle in an applied magnetic field (Cohen-Tannoudji et al., 1977), indicating that the muon precesses in the $\hat{x}\hat{y}$ plane. On the other hand, in a longitudinal-field (LF) experiment, Eq. (9) with $r = z$ gives a time-independent muon spin polarization along $\hat{z}$ of

$P_z(t) = 1$ ($P_x(t) = P_y(t) = 0$). This implies that the initial muon spin direction is preserved.

### b. Paramagnetic Centers in the Absence of Nuclear Spins

Recall that all the muonium centers observed to date in semiconductors have either an isotropic hyperfine interaction ($Mu_T^0$) or one with axial symmetry ($Mu_{BC}^0$). The spin-Hamiltonian describing the neutral $Mu_{BC}^0$ state in the absence of nuclear spins is as follows:

$$\mathcal{H}_{Mu_{BC}^0}/h = \tilde{\gamma}_e BS_z - \tilde{\gamma}_\mu BI_z + A_\parallel S_{z'} I_{z'} + A_\perp (S_{x'} I_{x'} + S_{y'} I_{y'}) \tag{14}$$

The magnetic field is assumed to be directed along $\hat{z}$. The last three terms represent an anisotropic, axially symmetric hyperfine interaction with symmetry axis $\hat{z}'$ (tilted at an angle $\theta$ from $\hat{z}$). (The hyperfine interaction of $Mu_{BC}^0$ is axially symmetric about the $\langle 111 \rangle$ crystalline axis.) The parallel and perpendicular hyperfine constants are labeled $A_\parallel$ and $A_\perp$ respectively.[3] The eigenvectors and energy eigenvalues are orientation dependent and their determination generally requires the solution of a fourth-order polynomial. Analytical expressions for the eigenvectors, energy eigenvalues and time-dependence of the muon polarization can be found for special values of $\theta$ (0° and 90°), (see Patterson, 1988). In addition, useful approximate solutions can be obtained using the "effective-field" treatment described in the next section.

---

[3] The $s$ and $p$ electron-spin densities can be calculated by (Morton and Preston, 1978; Kiefl and Estle, 1991; Schneider et al., 1993a)

$$\eta_s^2 = \frac{A_\parallel + 2A_\perp}{3A_s^f}, \quad \eta_p^2 = \frac{A_\parallel - A_\perp}{3A_p^f}$$

where the free atom values $A_s^f$ and $A_p^f$ are given by

$$A_s^f = \frac{8\pi}{3} \frac{\mu_o h^2}{4\pi} \tilde{\gamma}_e \tilde{\gamma}_\mu \rho_{s,f}(0)$$

$$A_p^f = \frac{\mu_o h^2}{4\pi} \tilde{\gamma}_e \tilde{\gamma}_\mu \int d^3r \rho_{p,f}(\vec{r}) \frac{3\cos^2\alpha - 1}{2r^3}$$

where $\mu_o = 4\pi \times 10^{-7}$ Vs A$^{-1}$ m$^{-1}$ is the permeability of vacuum, $\rho_{s,f}$ is the $s$ spin density on the muon, $\rho_{p,f}(\vec{r})$ is the $2p$ spin density at position $\vec{r}$ with respect to the muon, and $\alpha$ is the angle between $\hat{z}$ and $\vec{r}$. Note that the $1s$ ($2p$) spin density is equal to the square of the normalized valence $1s$ ($2p$) wave function of the muon.

TABLE II

THE EIGENVECTORS AND EIGENVALUES OF MUONIUM WITH AN ISOTROPIC HYPERFINE PARAMETER

| Level | Eigenvectors | Eigenvalues |
|---|---|---|
| 1 | $\|\varepsilon_1\rangle = \|++\rangle$ | $\nu_1 = \varepsilon_1/h = \frac{1}{4}A_\mu + \frac{1}{2}(\tilde{\gamma}_e - \tilde{\gamma}_\mu)B$ |
| 2 | $\|\varepsilon_2\rangle = \cos\alpha\|-+\rangle + \sin\alpha\|+-\rangle$ | $\nu_2 = \varepsilon_2/h = -\frac{1}{4}A_\mu + \frac{1}{2}\sqrt{A_\mu^2 + (\tilde{\gamma}_e + \tilde{\gamma}_\mu)^2 B^2}$ |
| 3 | $\|\varepsilon_3\rangle = \|--\rangle$ | $\nu_3 = \varepsilon_3/h = \frac{1}{4}A_\mu - \frac{1}{2}(\tilde{\gamma}_e - \tilde{\gamma}_\mu)B$ |
| 4 | $\|\varepsilon_4\rangle = -\sin\alpha\|-+\rangle + \cos\alpha\|+-\rangle$ | $\nu_4 = \varepsilon_4/h = -\frac{1}{4}A_\mu - \frac{1}{2}\sqrt{A_\mu^2 + (\tilde{\gamma}_e + \tilde{\gamma}_\mu)^2 B^2}$ |

We use the basis set $|m_I m_S\rangle$, consisting of the muon ($m_I = \pm\frac{1}{2}$) and the electron ($m_S = \pm\frac{1}{2}$) magnetic quantum numbers in the $\hat{z}$ direction. The symbol + indicates $m_I m_S = +\frac{1}{2}$ and − indicates $m_I, m_S = -\frac{1}{2}$ and $\cos 2\alpha = x/\sqrt{1 + x^2}$ where $x = B/B_o$ is a dimensionless quantity with $B_o = A_\mu/(\tilde{\gamma}_e + \tilde{\gamma}_\mu)$.

If $A_\perp = A_\parallel = A_\mu$, Eq. (14) reduces to the spin-Hamiltonian for Mu$_T^0$, which is characterized by an isotropic muon-electron hyperfine interaction. The eigenvectors and energy eigenvalues for the Mu$_T^0$ center are listed in Table II. The magnetic field parameter $x$ in Table II is $x = B/B_o$ where the dimensionless quantity $B_o = A_\mu/(\tilde{\gamma}_e + \tilde{\gamma}_\mu)$. The field dependence of the energy levels constitutes a so-called Breit-Rabi diagram, an example of which is shown in Fig. 4b.

In an LF-$\mu$SR experiment, Eq. (9), with the eigenvectors and eigenvalues given in Table II, yields:

$$P_z(t) = \frac{1 + 2x^2}{2(1 + x^2)} + \frac{1}{2(1 + x^2)} \cos(\omega_{24} t) \quad (15)$$

where $\omega_{24} = 2\pi A_\mu \sqrt{1 + x^2}$. [Note that $P_x(t) = P_y(t) = 0$.] Similarly, a straightforward calculation shows that the time dependence of the muon spin polarization in a TF-$\mu$SR experiment is

$$P_x(t) = \tfrac{1}{2}(\cos^2\alpha \cos\omega_{12} t + \sin^2\alpha \cos\omega_{23} t + \cos^2\alpha \cos\omega_{34} t + \sin^2\alpha \cos\omega_{14} t)$$

(16)

At high fields, $\cos^2\alpha \to 1$, while $\sin^2\alpha \to 0$, and only the $\cos\omega_{12} t$ and $\cos\omega_{34} t$ terms have significant amplitude (see Table II for the definition of $\alpha$).

3. Effective-Field Approximation

In a high magnetic field such that the electron Zeeman interaction greatly exceeds the hyperfine interactions (i.e., $\tilde{\gamma}_e B \gg |A_{\parallel,\perp}^\mu|$), an approximate spin-

Hamiltonian describing the muon subsystem can be obtained (Slichter, 1989; Kiefl and Estle, 1991) from Eq. (2) and Eq. (14). For each value of the electron magnetic quantum number $m_s$, this approximate Hamiltonian gives rise to an effective magnetic field acting on the muon spin, which in general has components both parallel and perpendicular to the applied magnetic field **B**. The two effective fields are:

$$\begin{aligned} \mathbf{B}^+ &= (B - B_\parallel, -B_\perp) \quad \text{if } m_s = +1/2 \\ \mathbf{B}^- &= (B + B_\parallel, +B_\perp) \quad \text{if } m_s = -1/2 \end{aligned} \quad (17)$$

where the first term in the parentheses is the parallel component of the effective field, while the second represents the perpendicular component. If $\theta$ is the angle between the applied magnetic field and the hyperfine symmetry axis of $\text{Mu}_{BC}^0$, then in first order,

$$\begin{aligned} B_\parallel &= \frac{(A_\perp \sin^2\theta + A_\parallel \cos^2\theta)}{2\tilde{\gamma}_\mu} \\ B_\perp &= \frac{(A_\perp - A_\parallel)\sin 2\theta}{4\tilde{\gamma}_\mu} \end{aligned} \quad (18)$$

Note that, in general, the effective fields are *not* parallel to the applied magnetic field. Thus in high magnetic fields, the total muon polarization is the sum of precession signals about $\mathbf{B}^+$ and $\mathbf{B}^-$. For example, $v^+ = \tilde{\gamma}_\mu B^+$ and $v^- = \tilde{\gamma}_\mu B^-$ approximate the high-field precession frequencies $v_{12}$ and $v_{34}$ for $\text{Mu}_T^0$ ($A_\parallel = A_\perp = A_\mu$) and $\text{Mu}_{BC}^0$.

### 4. INFLUENCE OF NUCLEAR SPINS

As mentioned, the results that are obtained by neglecting the influence of nuclear spins provide an important framework for understanding and interpreting muonium experiments in semiconductors. It should be emphasized, however, that it is the interactions with the neighboring nuclear spins that enable one to obtain detailed information on the electronic structure and to monitor the diffusion of muonium. Next, we briefly outline the influence of nuclear spins on the observables in $\mu$SR experiments. These effects will be expanded upon in Section III, where we describe the $\mu$SR techniques and illustrate their applications with relevant experimental examples.

In the case of the diamagnetic centers, where the unpaired electron spin and the corresponding hyperfine interactions are absent, the primary coupling between the muon and the surrounding nuclear spins is through the muon-nuclear dipole interaction. As a result of this coupling, the muon spin observables are, in general, also sensitive to the nuclear Zeeman interaction and the muon-induced quadrupole interaction for nuclei with spin 1 or greater. Typically, the nuclear dipolar coupling leads to a spread in internal magnetic fields at the muon site. In a TF-$\mu$SR experiment, where the muon undergoes Larmor precession about the total magnetic field (applied plus internal), there will be a loss in phase coherence, and the precession signal will be damped (Section III.1). Furthermore, if the nearest neighbor nuclei have an electric quadrupole interaction, the observed line broadening will, in general, be a function of the magnitude of the applied magnetic field and its orientation with respect to the crystalline axes. In zero field (ZF), the nuclear dipolar fields will also lead to a decay of the muon spin polarization (see Section III.3). On the other hand, in an LF-$\mu$SR experiment, where the applied magnetic field is significantly larger than the nuclear dipole fields, the muon spin polarization is time-independent, except near energy-level crossings. As discussed in more detail in Section III.2, such crossings can give rise to the muon-nuclear cross relaxation that can be observed as resonances in the time-integrated longitudinal polarization as a function of magnetic field. The study of muon level-crossing resonances, or $\mu$LCR, provides information regarding the electronic structure of muonium. If the muon is hopping, it experiences fluctuating dipolar fields. This affects the linewidth of the precession signal in TF-$\mu$SR (see Section III.1) and the relaxation of the muon polarization in ZF-$\mu$SR (see Section III.3) and in weak longitudinal fields (Hayano et al., 1979).

An analogous situation exists for the paramagnetic centers. The aforementioned dipole interactions still exist, but the hyperfine interactions between the muon, the unpaired electron spin, and the host nuclei now dominate. Nuclear hyperfine interactions (also called superhyperfine interactions) cause additional level splittings in the Breit-Rabi diagram. In a TF-$\mu$SR experiment, these nuclear hyperfine interactions can lead to splittings of the muonium precession frequencies which are field-dependent. In the limit of high transverse magnetic fields (Section II.3 and Section III.1), the line splittings due to the nuclear hyperfine interactions are totally suppressed. Under these conditions, only two frequencies are observed that depend solely on the muon hyperfine interaction and the applied magnetic field. In lower applied magnetic fields, where the muon and nuclear spins are coupled, the *nuclear* hyperfine interactions lead to line splittings and a much more complicated frequency distribution. As with the diamagnetic centers, the combination of the nuclear hyperfine and quadrupolar interactions can also produce level-crossing resonances in an LF-$\mu$SR experiment (Section III.2). The $\mu$LCR technique is a well-established method for investigating such interactions and hence determining the electronic structure of static neutral muonium centers. As in the case of diamagnetic centers, diffusion and the resulting spin fluctuations lead to $1/T_1$ spin relaxation. The investigation of such dynamics is described in Section III.4.

Spin dynamics for processes such as cyclic charge-state changes (Section III.4), spin exchange scattering (Section III.4), and interconversion between the various muonium states (Section III.5) are often modeled *neglecting* the influence of nuclear spins. This assumption makes the theory for the spin dynamics manageable and is a very good approximation in cases such as Si. In other materials, such as GaAs, the conditions for which such an approximation is valid must be examined more carefully. However, in studying such dynamics, effects due to nuclear interactions are often of "higher order" and can be ignored.

## III. Experimental Techniques and Examples

In the previous section, it was discussed in general terms how the muon spin polarization is measured and how one calculates the (time evolution of the) polarization given a spin-Hamiltonian. In this section, the main body of the chapter, we discuss the experimental aspects of $\mu$SR in more detail. The applications of $\mu$SR to the study of muonium in semiconductors is illustrated with some experiments in crystalline group IV and group III–V materials. We describe how the choice of technique depends on the physical phenomena being investigated and also why, in some cases, one muon facility is better suited than another facility for a certain experiment. Studies of muonium in semiconductors have benefited a great deal from this diversity and in many situations the information obtained from the various techniques (and facilities) are used together to form a coherent picture of the underlying physics.

Figure 3 shows schematics of the two basic arrangements for $\mu$SR experiments. Spin-polarized muons are detected by a thin muon (TM) counter before being implanted into the sample.[4] The decay positrons are detected by additional thicker scintillation counters in the vicinity of the sample. These detectors are made of plastic scintillators that emit flashes of light whenever they are traversed by an ionizing particle. The light is guided to a photomultiplier, which generates a voltage pulse. These pulses are then carried by coaxial cables to the electronics in the "counting room" for processing.

[4]The reason for the thickness (on the order of $\approx 250\ \mu$m) is to minimize multiple scattering, which will cause the muons to miss the sample and add to the background signal.

The number of positron counters and their location depend on the type of experiment. These are labeled in Fig. 3 as left (L), right (R), forward (F), and backward (B). [Occasionally, up (U) and down (D) counters are also used.] Note that the labeling scheme used assumes a viewpoint in the sample looking in the beam direction. The B counter shown in Fig. 3b has a hole in order to avoid the path of the muon beam, and the F counters are often "split" to accommodate axial cryostats or ovens.

The various $\mu$SR techniques are often classified by (1) whether time information is stored or not and (2) the orientation between the magnetic field and the muon spin. The terminology *time-differential* or *time-integral* is commonly used regarding the storage of time information. The former refers to data in which such information is recorded; that is, the time-dependence of the muon spin polarization is monitored. There are expriments in which determination of the time-dependence of the muon polarization function is unnecessary. Instead, one measures its integrated value over some length of time that is long relative to the muon lifetime [see Eq. (10), for example]: this is the basis for time-integral $\mu$SR experiments. In time-differential experiments at continuous beam facilities, muons are detected by the TM counter, shown in Fig. 3, which provides the signal to start the clock. The stop signal comes from one of the positron counters placed around the sample. Because of the requirement that each decay positron should be correlated with its parent muon, one is restricted to having only one muon in the sample at a time. In the integral mode, this restriction is removed and there is no theoretical upper limit on the incoming muon rate. Subsequently, the count rates, and hence the sensitivity, are greatly enhanced. At pulsed muon sources, "time zero" is determined by the arrival of a bunch of muons, called a "muon pulse," rather than by the arrival of individual muons. Hence, a TM counter is not needed. At these facilities, all the data collection is done in the time-differential mode without a penalty in the count rate. This allows added flexibility during data analysis; for example, the integration limits $t_1$ and $t_2$ of the integrated polarization [see Eq. (10)] can be selected to enhance the signal (Leon, 1992).

As already mentioned in Section II.2, Fig. 3a describes a *transverse-field* (TF) $\mu$SR experiment; that is, the initial muon spin and the applied magnetic field **B** are perpendicular to each other. On the other hand, Fig. 3b is appropriate for a *longitudinal-field* (LF) experiment, in which the magnetic field is applied along the initial muon spin polarization direction. A *zero-field* (ZF) experiment, as the name implies, is one in which no magnetic field is applied. Finally, a *radio-frequency* (RF) $\mu$SR experiment has the same arrangement of **B** and initial muon spin as an LF-$\mu$SR experiment, but a small oscillating magnetic field is also applied perpendicular to the large static field.

4 μSR ON MUONIUM IN SEMICONDUCTORS IN RELATION TO HYDROGEN

The remainder of this section will be organized as follows: in Section III.1, the TF-μSR technique (time-differential) is described. This is useful in structural determinations of muonium and provides qualitative information on certain dynamical processes. Next, in Section III.2, we discuss the muon level-crossing resonance technique (μLCR), which is a natural complement to TF-μSR for structural determinations. For example, in the case of paramagnetic centers, TF-μSR is used primarily to determine the muon hyperfine parameters, whereas μLCR is much more powerful in determining hyperfine interactions with surrounding nuclear spins. ZF-μSR (time-differential) is discussed in Section III.3. This technique has been used to investigate muonium dynamics in semiconductors. Its usefulness stems from its sensitivity to the diffusion of diamagnetic centers in semiconductors. Then, in Section III.4, we discuss the LF-μSR technique, which is well suited for investigations of dynamics—diffusion and spin and charge exchange. We also briefly describe the "repolarization" technique, which is used to characterize muonium centers in polycrystalline or amorphous semiconductors. RF-μSR, discussed in Section III.5, is a complementary technique to both TF-μSR and LF-μSR. For example, it has been used to provide a detailed model of the interconversion between the various muonium states in silicon. During the discussion of each of these five related techniques, we will also introduce experimental examples illustrating how various aspects of the physics of muonium in semiconductors are studied. In Section III.6, we compare pulsed and continuous-source muon facilities.

1. MUON SPIN ROTATION IN A TRANSVERSE MAGNETIC FIELD (TF-μSR)

Transverse-field (TF) μSR is commonly used to study muonium in semiconductors. Although more detailed information can sometimes be obtained by other techniques, TF-μSR is often used to first identify and characterize muonium in a semiconductor. In a TF-μSR experiment, one observes precession of the muon spin polarization, which occurs at a number of discrete frequencies. The pattern of precession frequencies provides a signature of the muonium center since they are a function of the muon hyperfine interaction, although under certain circumstances (low magnetic fields), these frequencies also depend on the nuclear hyperfine interactions. The linewidth also provides information on a variety of physical phenomena. For example, the strength and symmetry of the muon-nuclear dipolar and quadrupolar interactions of diamagnetic muonium centers can be determined from the dependence of the linewidth on the applied field. In addition to providing information on electronic structure, diffusion and charge-changing reactions of muonium can also be studied. We shall discuss examples of such uses in Section III.1.a to Section III.1.c for GaAs and diamond.

The experimental arrangement for an ideal time-differential transverse-field experiment is shown schematically in Fig. 3a. The magnetic field is usually applied parallel to the muon momentum, while the muon spin is rotated perpendicular to **B**.[5] The raw spectra, (i.e., the number of positron events recorded for the left (L) and right (R) counters), is given by

$$N_L(t) = N_L^0 \exp(-t/\tau_\mu)[1 + A_{x,L}P_x(t)] + b_L$$
$$N_R(t) = N_R^0 \exp(-t/\tau_\mu)[1 - A_{x,R}P_x(t)] + b_R \quad (19)$$

where $A_{x,i}$ is a constant whose value depends on factors such as the probability distribution of the emitted positron and the solid angle of the $i$th counter (and typically has values $\approx 0.25$), while $b_i$ is due to the uncorrelated background events counted by the $i$th counter and is typically assumed to be time-independent. An example of such a TF histogram is shown in Fig. 5a for a heavily doped n-type GaAs:Si sample (net donor concentration $4.5 \times 10^{18}$ cm$^{-3}$). Superimposed on the exponential muon decay is the signal of interest, in this case a spin precession close to the Larmor frequency of the muon. The data are often displayed and/or analysed as "asymmetries" $A(t)$ of two opposing matched counters (such as L and R). In the experimental arrangement shown in Fig. 3a,

$$A(t) = \frac{[N_L(t) - b_L] - [N_R(t) - b_R]}{[N_L(t) - b_L] + [N_R(t) - b_R]} \quad (20)$$

One common approach is to adopt the notation $\alpha = N_R^0/N_L^0$ and assume $A_{x,R} = A_{x,L} = A_x$, leading to

$$A(t) = \frac{(1 - \alpha) + A_x P_x(t)(1 + \alpha)}{(1 + \alpha) + A_x P_x(t)(1 - \alpha)} \quad (21)$$

Note that the preceding definition of $A(t)$, also referred to as the "raw asymmetry," eliminates the random background and exponential decay of the muon from the spectrum, thereby showing the polarization function $P_x(t)$ more clearly. For example, in an ideal situation, where $\alpha = 1$, $A(t) = A_x P_x(t)$. Usually, one fits $\alpha$ to obtain the "corrected asymmetry"

---

[5]Alternatively, the muon spin is not rotated, and a weak field can be applied perpendicular to the muon spin. This allows for transverse-field studies on beamlines where no spin rotator is available.

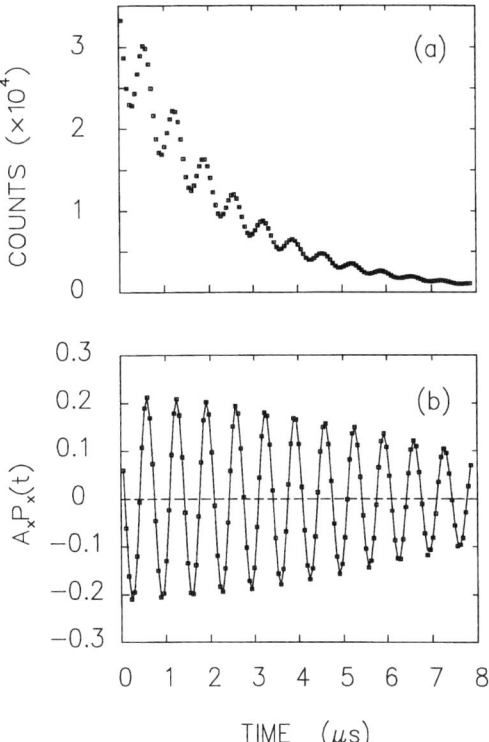

FIG. 5. (a) Raw spectrum in one of the counters from a TF-$\mu$SR experiment showing the exponential decay of muon modulated by the muon spin precession. The sample is heavily doped n-type GaAs:Si with net donor concentration in the range 2.5–5.0 × $10^{18}$ cm$^{-3}$. The temperature is ≈295 K, and a magnetic field of ≈11 mT is applied parallel to a $\langle 110 \rangle$ crystalline axis. (b) The corresponding asymmetry plot. The solid line is the best fit to the data, assuming a Gaussian damping of the precession signal.

$A_x P_x(t)$. Figure 5b shows the corrected asymmetry plot for the TF example in Fig. 5a. The signal is relatively simple, consisting of an oscillation at the Larmor frequency, implying that a large fraction of the implanted muons form diamagnetic centers.[6]

---

[6] It is interesting to make an analogy with nuclear magnetic resonance (NMR) at this point. The $\mu$SR precession signal and experimental arrangement is similar to that required to observe "free induction decay." Note, however, that the muon spin is produced 100% polarized, whereas the excess of moments initially pointing along the applied field in NMR is a thermal equilibrium distribution. As in NMR, the $\mu$SR precession decays due to interaction of the probe with the surroundings.

a. *Detection of $Mu^0_{BC}$, $Mu^0_T$, and the Diamagnetic Center in High-Resistivity GaAs*

The most stable equilibrium charge state of muonium (or hydrogen) in a semiconductor depends on the location of the Fermi level $E_F$ relative to the muonium level (which is assumed to be in the gap). In particular, we expect $Mu^+$ ($H^+$) to be the stable state in heavily doped p-type semiconductors where $E_F$ is close to the valence band, whereas $Mu^-$ ($H^-$) is predicted to be the most stable state in n-type materials (see Van de Walle, 1991, for example). This hypothesis is supported by the fact that the fraction of diamagnetic to paramagnetic centers is much larger in heavily doped semiconductors as compared to the lightly doped materials. Furthermore, in some systems such as the III–V semiconductors, the diamagnetic state in p-type materials (i.e., $Mu^+$) behaves differently than the diamagnetic center in n-type samples (i.e., $Mu^-$). However, when studying high-resistivity materials, non-equilibrium conditions can occur because of the low carrier concentrations and the short lifetime of the muon (see Section I). In this case, it is possible for $Mu^0_T$, $Mu^0_{BC}$, and a diamagnetic center to co-exist.

This is illustrated in Fig. 6, which shows the Fourier Transform of the TF-$\mu$SR spectrum at low temperatures in high-resistivity GaAs (Kiefl et al., 1985). In addition to the diamagnetic center labeled as $v_{\mu^+}$, signatures for $Mu^0_T$ and $Mu^0_{BC}$ are also evident. The frequency labeled as $v_{\mu^+}$ occurs at the Larmor frequency of the muon and corresponds to a diamagnetic center.

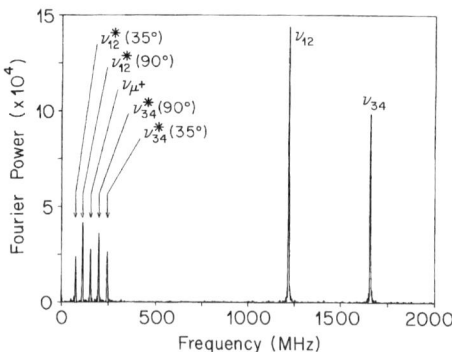

FIG. 6. The $\mu$SR frequency spectrum in GaAs at 10 K in an external field of 1.15 T applied along a $\langle 110 \rangle$ crystalline direction. The upper two frequencies are due to $Mu^0_T$, which has an isotropic hyperfine interaction. The starred frequencies are due to $Mu^0_{BC}$, which has an axially symmetric hyperfine interaction. The angles in parentheses refer to the direction of the external field with respect to the $Mu^0_{BC}$ symmetry axis. The frequency labeled $v_{\mu^+}$ is due to a diamagnetic center. (From Kiefl et al., 1985.)

These centers are characterized by the absence of an unpaired electron spin and the accompanying hyperfine interactions; hence, either $Mu^+$ or $Mu^-$, or any diamagnetic complex containing muonium, contributes to this signal. The remaining frequencies are due to $Mu_{BC}^0$ and $Mu_T^0$. In these high-transverse magnetic fields, the polarization of these centers is given by $P_x(t) = \frac{1}{2}(\cos\omega_{12}t + \cos\omega_{34}t)$, as discussed in Section II.2.b and Section II.3. The two frequencies labeled $v_{12}$ and $v_{34}$ in Fig. 6 are due to $Mu_T^0$. The

TABLE III

Hyperfine Parameters of the Muon and Nearest Neighbor Nuclei on the Symmetry Axis for $Mu_{BC}^0$ in Group IV and III–V Semiconductors

| Center | Nucleus | $A_\parallel^1$ (MHz) | $A_\perp^1$ (MHz) | Q (MHz) | $\eta_s^2$ | $\eta_p^2$ |
|---|---|---|---|---|---|---|
| $Mu_{BC}^0$ Si | Muon | −16.82(1) | −92.59(5)[a] | — | −0.0151 | — |
|  | $^{29}$Si | −137.5(1) | −73.96(5)[b] | — | +0.0207 | +0.185 |
| AA9 Si | "Proton" | 19.7(3.0) | 99.9(3.0)[c] | — |  |  |
|  | $^{29}$Si | 140.1(1.0) | 72.9(1.0)[c] | — |  |  |
| $Mu_{BC}^0$ Ge | Muon | −27.27(1) | −131.04(3)[a] | — | −0.0216 | — |
| $Mu_{BC}^0$ C | Muon | +167.98(6) | −392.59(6)[a] | — | −0.0461 | — |
|  | $^{13}$C | +218(10) | +80(5)[d] | — | +0.0334 | +0.428 |
| $Mu_{BC}^0$ GaAs | Muon | +218.54(3) | +87.87(5)[e] | — | +0.0294 | — |
|  | $^{75}$As | +563.1(4) | +128.4(2)[e] | +18.8(2) | +0.0186 | +0.433 |
|  | $^{69}$Ga | +1052(2) | +867.9(3)[e] | +1.08(33) | +0.0761 | +0.301 |
| $Mu_{BC}^0$ GaP | Muon | +219.0(2) | +79.48(7)[f] | — | +0.0282 | — |
|  | $^{31}$P | +620.2(4) | +249.7(1)[g] | — | +0.0280 | +0.337 |
|  | $^{69}$Ga | +1017.8(1) | +787.4(1)[h] | +3.94(3) | +0.0708 | +0.377 |

The results on the hydrogen AA9 center in Si are included for comparison (Gorelkinskii and Nevinnyi, 1987). The "proton" hyperfine parameters are multiplied by the ratio of muon and proton magnetic moments (3.183) to aid in the comparison. The s and p densities are calculated from $\eta_s^2 = \frac{1}{3}(A_\parallel + 2A_\perp)/A_s^{free}$ and $\eta_p^2 = \frac{1}{3}(A_\parallel - A_\perp)/A_p^{free}$, where the free atom values are from Morton and Preston (1978).

[1]The sign of $A_\parallel$ relative to $A_\perp$ for the muon (or nucleus) is a measured quantity. Also, for any given center, the signs of the muon hyperfine parameters relative to those of a nucleus are measured quantities. Finally, the signs of the muon hyperfine parameters in diamond have been measured relative to those of the normal muonium center (Odermatt, 1988). Thus, the absolute signs, where given, follow from the assumption that the large spin densities for normal muonium in diamond and the nearest-neighbor nuclei in Si, GaAs, and GaP must be positive.

[a]From Blazey et al. (1983).
[b]From Kiefl et al. (1988).
[c]From Gorelkinskii and Nevinnyi (1987).
[d]From Schneider et al. (1993b).
[e]From Kiefl et al. (1987).
[f]From Kiefl et al. (1985).
[g]From Kiefl and Estle (1991).
[h]From Schneider et al. (1993a).

positions of the frequencies do not depend on the orientation between **B** and the crystalline directions, confirming the isotropic nature of the hyperfine interaction. This should be contrasted with the $\text{Mu}_{BC}^0$ frequencies (indicated by $v_{ij}^*$ in the figure). Since **B** is applied parallel to the $\langle 110 \rangle$ direction, two of the $\text{Mu}_{BC}^0$ centers are at an angle of $\theta = 90°$ and two centers are at $\theta = 35.3°$. As noted in Section II.3, $\theta$ is the angle between **B** and the $\text{Mu}_{BC}^0$ hyperfine symmetry axis (also the bond axis). A knowledge of $v_{12}$ and $v_{34}$ enables the muon hyperfine parameter $A_\mu$ for $\text{Mu}_T^0$ to be estimated ($v_{12} + v_{34} = A_\mu$). Similarly, knowing $v_{12}^*$ and $v_{34}^*$ allows the anisotropic hyperfine parameters $A_\parallel$ and $A_\perp$ for $\text{Mu}_{BC}^0$ to be determined, provided $\theta$ is known, although in practice, the field dependence of the frequencies is also used. The hyperfine parameters of the muonium centers in a number of semiconductors are listed in Table III and Table IV—most of these were determined by similar methods.

Note that the $\mu$SR spectrum shown in Fig. 6 is relatively simple. In particular, the lines are sharp, indicating that in the high magnetic fields used in the experiment, the nuclear hyperfine interactions are totally suppressed (Patterson, 1988; Kiefl et al., 1984). The muon spin is said to be decoupled from the nuclear spins: the muonium precession frequencies are independent of the *nuclear* terms in the spin-Hamiltonian and depend solely on the *muon* hyperfine interaction and the applied magnetic field. As the magnetic field decreases, the effect of nuclear spins is to broaden and split the lines. An investigation of this additional structure is discussed in the next subsection.

TABLE IV

The Isotropic Hyperfine Parameter for $\text{Mu}_T^0$ in Covalent Group IV and III–V Semiconductors

| Center | $A_\mu$ (MHz) | $\eta_s^2$ | Temperature (K) |
|---|---|---|---|
| $\text{Mu}_T^0$ Si | 2006.3(2.0)[a] | 0.450 | T → 0 |
| $\text{Mu}_T^0$ Ge | 2359.5(2)[a] | 0.529 | T → 0 |
| $\text{Mu}_T^0$ C | 3711(21)[a] | 0.831 | T → 0[b] |
| $\text{Mu}_T^0$ GaAs | 2883.6(3)[c] | 0.646 | 10 |
| $\text{Mu}_T^0$ GaP | 2914(5)[c] | 0.653 | 10 |

A more complete list, which includes group II–VI and I–VII materials, is given by Kiefl and Estle (1991). The s-density ($\eta_s^2$) is equal to the reduced hyperfine parameter $A_\mu/A_{\text{free}}$, where $A_{\text{free}} = 4463.302$ MHz.

[a]From Holzschuh (1983).
[b]Extrapolated to T = 0.
[c]From Kiefl et al. (1985).

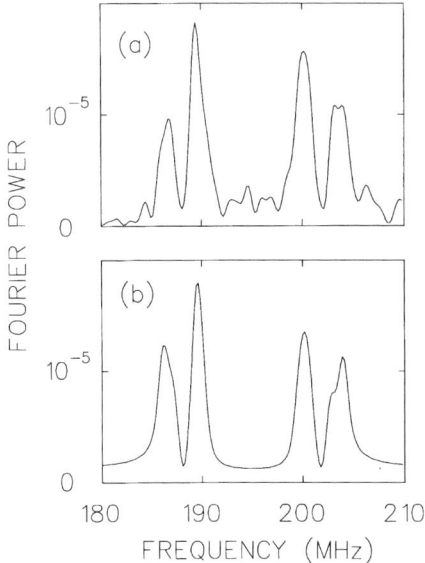

FIG. 7. TF-$\mu$SR spectrum for Mu$_{BC}^0$ center in $^{13}$C-enriched diamond at room temperature, where the field of 45.2 mT is applied parallel to a $\langle 110 \rangle$ axis. This is one of the six measured spectra used by Schnedier et al. (1993b) to extract the nuclear hyperfine parameters for the two nearest-neighbor $^{13}$C nuclei. Only the $\theta \approx 90°$ lines are shown. (b) The calculated spectrum using the parameters given in Table III.

b. *Resolution of the Nuclear Hyperfine Interaction in $^{13}$C Enriched Diamond*

If the host nuclei have nonzero spins, and are in low applied magnetic fields, the TF-$\mu$SR spectra consists of a multitude of small amplitude lines that cannot be resolved. However, at intermediate fields, one can reach conditions where the structure from the largest nuclear hyperfine parameters is resolvable (Kiefl et al., 1987; Kiefl and Estle, 1991). An example of such an experiment is the study of the Mu$_{BC}^0$ center in 99% $^{13}$C-enriched single-crystal diamond (Schneider et al., 1993b) at room temperature. Figure 7 shows the resulting TF-$\mu$SR spectrum when a magnetic field of 45.2 mT was applied parallel to a $\langle 110 \rangle$ direction. Only the $\theta \approx 90°$ lines are shown. At this field, and in the absence of nuclear hyperfine interactions with the $^{13}$C neighbors, two narrow lines [corresponding to $v_{12}$ and $v_{34}$ (Section II.2 and Section II.3)] at 190 MHz and 202 MHz are expected. The additional interactions result in the observed splitting and broadening of the lines and enable the nuclear hyperfine parameters to be estimated (see Table III).[7] Another way of measuring hyperfine and nuclear hyperfine parameters, often with greater accuracy, is by using the $\mu$LCR technique described in Section III.2.

*c. Structure and Dynamics of the $Mu^-$ Center in Heavily-Doped n-Type GaAs*

Figure 5 shows the TF-$\mu$SR spectrum for heavily doped n-type GaAs:Si (with Si concentrations in the range of 2.5–5.0 × $10^{18}$ cm$^{-3}$) at room temperature (Chow et al., 1996a). The large amplitude of the signal close to the Larmor frequency of the free muon indicates that a diamagnetic state is formed. From the preceding discussion of equilibrium considerations, it is most likely $Mu^-$. The neutral $Mu_T^0$ is quickly converted to $Mu^-$ by electron capture. The decay of the precession signal is attributed to the presence of additional interactions with the surrounding nuclear spins, in this case, the magnetic dipolar and muon-induced quadrupolar interactions with the nearest-neighbor Ga and As, all of which have a spin of $\frac{3}{2}$. The large abundance of nuclear moments in III–V semiconductors makes them ideal for investigations of the structure and diffusion of charged muonium. By contrast, analogous investigations of the diamagnetic centers in Si would be much more difficult, because only 4.7% of the host nuclei have a nonzero spin ($^{29}$Si, spin $\frac{1}{2}$).

If the $Mu^-$ center is not moving, the surrounding nuclear spins, which are unpolarized, give rise to a spread in the internal fields experienced by the muon. This leads to a damping of the precession signal, which is well described by a Gaussian envelope function $\exp(-\sigma^2 t^2)$. The parameter $\sigma$ is a measure of the strength of nuclear magnetic dipolar fields at the muon (Schenck, 1985). A Gaussian damping function is a good model for the data in Fig. 5b (Chow et al., 1996a). The line broadening is a strong function of the magnitude of **B** and its direction with respect to the crystallographic directions, as demonstrated in Fig. 8 (Chow et al., 1995). The dependence of $\sigma(B)$ on the nuclear quadrupole interaction was first discussed by Hartmann (1977).[8] There are two extreme field regions. At low magnetic fields, the nuclear Zeeman interaction is smaller than the nuclear electric quadrupole interaction, and the stationary spin states of the nucleus are determined by

---

[7]The dimensions of the sample were 3.5 × 3.5 × 3 mm$^3$. The authors were able to measure a sample of such small cross-sectional area because of the construction of a new $\mu$SR apparatus, which reduced the background events dramatically.

[8]The polarization can also be calculated directly from Eq. (9). However, this approach can be quite cumbersome for a system containing many spins.

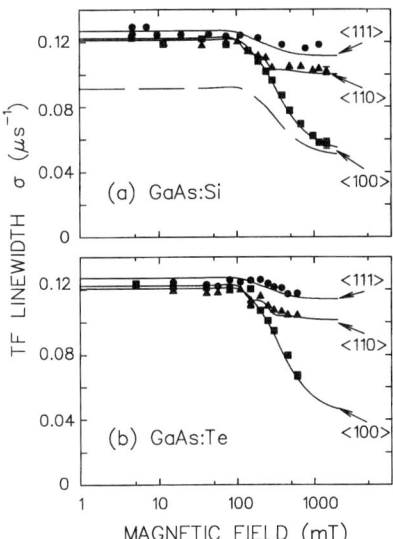

FIG. 8. The muon TF-μSR depolarization rate σ as a function of the magnetic field in heavily doped n-type (a) GaAs:Si and (b) GaAs:Te. The magnetic field is applied parallel to the crystallographic directions shown. The solid curves are for a fit assuming a $T_{Ga}$ muon site and allowing symmetric lattice distortions. The long-dashed curve in (a) is for $\mathbf{B} \parallel \langle 100 \rangle$ in an undistorted lattice.

the quadrupole interaction associated with that nucleus. In the high-field limit, the appropriate quantization axis defining the stationary states of the nuclear spin is along **B**, and σ reduces to the more familiar Van-Vleck expression for the dipolar linewidth (Slichter, 1989). The Van-Vleck width is field-independent and vanishes for $3\cos^2\theta - 1 = 0$, where $\theta$ is the angle between **B** and the vector distance from the muon to the nucleus $\mathbf{r}_{\mu n}$. Hence, the dramatic decrease when $\mathbf{B} \parallel \langle 100 \rangle$ indicates that $\mathbf{r}_{\mu n}$ and the $\langle 100 \rangle$ direction are at an angle $\theta = 54.7°$ (i.e., the nuclei dominating σ lie on the $\langle 111 \rangle$ axis). The magnetic field at which σ decreases indicates a crossover from one regime to another, and hence provides a rough measure of the strength of the quadrupole interaction of the nucleus. Note that the data for GaAs:Si is virtually identical to GaAs:Te ($4.5 \times 10^{18}$ cm$^{-3}$), implying that the muon is *not* closely associated with a dopant atom. The Mu-Si and Mu-Te complexes would have different signatures, since the Si donor substitutes at a Ga site, whereas Te substitutes at an As site. The discussion of the local electronic structure of the Mu$^-$ center will be continued in Section III.2, where we discuss the μLCR technique. In that section, we will

TABLE V

Fitted Quadrupole Parameters $Q^i$ and the Muon-Nuclear Distances for Mu$^-$ at the $T_{Ga}$ Site in GaAs:Si and GaAs:Te

|         | $|Q(^{71}Ga)|/h$ (MHz)$^a$ | $|Q(^{75}As)|/h$ (MHz)$^a$ | $r_{Ga}$ (Å) | $r_{As}$ (Å) |
|---------|---------------------------|----------------------------|--------------|--------------|
| GaAs:Si | 1.472(7)                  | 0.636(9)                   | 2.199(7)     | 2.72(5)      |
| GaAs:Te | 1.532(7)                  | 0.70(1)                    | 2.188(12)    | 2.82(14)     |

The $r_{Ga}$ and $r_{As}$ distances are 2.45 Å and 2.83 Å, respectively, for an undistorted T site.

$^a q(^{69}Ga)/q(^{71}Ga) = 0.178/0.112$. Errors are statistical.

describe why the $T_{Ga}$ site is the one most consistent with existing experimental and theoretical data. By assuming that the muon is located at this site (Chow et al., 1995), the strength of the dipolar interactions, and hence the muon-nucleus distances, could be estimated by global fits of the TF-$\mu$SR data, taking into account the nearest neighbor Ga and next-nearest neighbor As nuclei (solid lines in Fig. 8). The results are given in Table V. Note that the effective Ga-muon distance is $\approx 10\%$ shorter than expected for an undistorted GaAs lattice. It is not clear whether this is due to the enhanced zero point motion of the muon or a slight contraction of the tetrahedral cage about the muon.

The temperature-dependence of the linewidth parameter $\sigma$ is shown in Fig. 9 and gives information on the dynamics of Mu$^-$. The large temperature-independent linewidth below 500 K (region I) indicates that Mu$^-$ remains static. At higher temperatures (region II), the linewidth decreases, indicating that Mu$^-$ starts to diffuse on the timescale of $1/\sigma$. Once the muon is hopping at a rate $v$, which is fast compared to $\sigma$, the muon-nuclear dipolar interaction is motionally averaged. In this region, the damping function should cross over to an exponential, but this is difficult to verify because the relaxation rate is small. The increase in $\sigma$ beginning at $\approx 700$ K (region III) is attributed to interactions with free charge carriers, which are present at high concentrations at these temperatures. In particular, the cyclic charge-changing reaction Mu$^- \leftrightarrow$ Mu$^0$ is postulated.

As mentioned previously, TF-$\mu$SR is often used to complement other related techniques. For example, $\mu$LCR provides more detailed information about the local electronic structure (region I). Similarly, ZF-$\mu$SR is a more sensitive measure of the diffusion in region II, whereas LF-$\mu$SR is the technique of choice for studying charge-state cycling in region III. These will all be discussed in Section III.2, Section III.3, and Section III.4, respectively. Thus, while more precise information can often be obtained by other

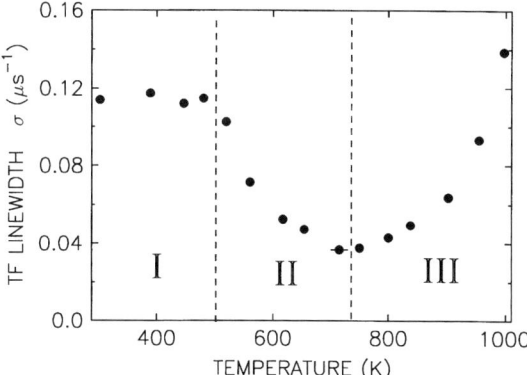

FIG. 9. Temperature-dependent of the TF-$\mu$SR Gaussian linewidth $\sigma$ in heavily doped n-type GaAs:Si. The three regions are discussed in more detail in the test.

techniques, TF-$\mu$SR is the method first used to define the various regions of interest and identify which type of muonium is present.

## 2. Muon Level-Crossing Resonance ($\mu$LCR)

In this section, we discuss a relatively new method used to characterize muonium in semiconductors. This method has been labeled as muon level-crossing resonance ($\mu$LCR), avoided level-crossing (ALC), and also cross-relaxation (CR). The reasons for all of these labels will become clear when we discuss the principles of the method in more detail; we (arbitrarily) choose the designation $\mu$LCR. The appropriate experimental arrangements for a $\mu$LCR experiment is shown in Fig. 3b. The magnetic field is applied along the initial muon spin direction, and the relevant polarization information can be derived from the count rates in the F and B counters. The total number of counts in some time interval from $t_1$ and $t_2$ is recorded—$\eta_i = \int_{t_1}^{t_2} v_\mu N_i(t)dt$, where $v_\mu = 1/\tau_\mu$ is the inverse muon lifetime and $N_i(t)$ is the number of positron events recorded in counter $i$. It is common to have $t_1 = 0$ and $t_2 = \infty$, since then one does not have to keep track of any time information; that is, one simply records the total positron rate $\eta_i$ in each counter. If both F and B counters are used, a count rate asymmetry can also be defined, such as $(\eta_B - \eta_F)/(\eta_B + \eta_F)$. Regardless of whether one counter or a pair of counters are used, one is interested in isolating a quantity that is proportional to the time-integrated muon spin polarization $\bar{P}_z$ (see Section II.2).

μLCR is a powerful probe of the electronic structure of muonium in semiconductors and has been used to investigate both the diamagnetic and paramagnetic centers, with studies of the latter being considerably more abundant. Specifically, the Mu$_{BC}^0$ center in Si (Kiefl et al., 1988), GaAs (Kiefl et al., 1987), GaP (Schneider et al., 1993a), and the two normal muonium centers (Mu$^I$ and Mu$^{II}$) in CuCl (Schneider et al., 1990a, 1990b) have all been investigated with μLCR. This technique has also been applied to the Mu$^-$ center in heavily doped n-type GaAs (Chow et al., 1995). In general, μLCR allows one to determine much more accurate information on the parameters of the spin-Hamiltonian involving the nuclear spins (e.g., nuclear electric quadrupole and nuclear hyperfine interactions) than any other technique for studying the muon. This in turn provides key information on the muon site and the location of neighboring atoms. For example, these studies established that Mu$_{BC}^0$ in Si, GaAs, and GaP is located near the center of a covalent bond with unpaired electron spin density residing primarily on the two nearest neighbors that have relaxed appreciably away from the muon. The similarity between the hyperfine parameters of the muon and the nearest-neighbor $^{29}$Si nuclei and the corresponding parameters for the AA9 hydrogen center in Si (Gorelkinskii and Nevinnyi, 1987; 1991) is notable (see Table III). This proves that the AA9 center is the hydrogenic analog of Mu$_{BC}^0$ and confirms that studies of muonium yield direct microscopic information on isolated hydrogen in semiconductors. Furthermore, it suggests that many of the other muonium centers observed in other semiconductors should have hydrogenic counterparts.

In the next subsections, we describe the principle of the μLCR technique and illustrate its application to two recent investigations: (1) the Mu$^-$ center in n-type GaAs:Si and (2) the Mu$_{BC}^0$ center in semi-insulating GaP.

*a. The Principle of μLCR*

The principles of level-crossing resonance (LCR) applied to muons and nuclei have been reviewed elsewhere [e.g., Kiefl and Estle (1991) and Kiefl and Kreitzman (1992)]. Here, we discuss μLCR for a diamagnetic center, such as Mu$^-$ in n-type GaAs:Si and GaAs:Te. The approach follows the theoretical discussion in Section II.2: the eigenvalues and eigenvectors are numerically calculated from the appropriate spin-Hamiltonian and then Eq. (10) is used to calculate $\bar{P}_z$. A diamagnetic center such as Mu$^-$ or Mu$^+$ has no unpaired electron spin density at the muon. Setting $S = 0$ in Eq. (2), the spin-Hamiltonian has terms describing the Zeeman interactions of the muon and neighboring nuclear spins with **B** ($\mathcal{H}^Z$), the magnetic dipolar coupling ($\mathcal{H}^D$) between the muon and nuclear spins, and the nuclear electric quadrupole interaction ($\mathcal{H}^Q$). If the magnetic field **B** is assumed to be applied along the $\hat{z}$ direction, the spin-Hamiltonian can be expressed as follows:

$$\mathcal{H}/h = \mathcal{H}^Z/h + \mathcal{H}^D/h + \mathcal{H}^Q/h \qquad (22)$$

where

$$\mathcal{H}^Z/h = -\tilde{\gamma}_\mu BI_z + \sum_i -\tilde{\gamma}_n^i BJ_z^i$$

$$\mathcal{H}^D/h = \sum_i D^i(-2I_{z'}J_{z'}^i + I_{x'}J_{x'}^i + I_{y'}J_{y'}^i) \qquad (23)$$

$$\mathcal{H}^Q/h = \sum_i Q^i\left(J_{z'}^{i2} - \frac{J^i(J^i+1)}{3}\right).$$

Since an adequate description of the data does not require more generality, the nuclear electric quadrupole tensors are asumed to be axially symmetric about the $\hat{z}'$ axis, which points along the muon-nucleus direction. This axis is tilted by an angle $\theta$ from **B**. The summation extends over all the nearest-neighbor nuclei for which $\mathcal{H}^D$ is appreciable. The strengths of the dipole (i.e., $D^i$) and quadrupole (i.e., $Q^i$) interactions are related to other microscopic quantities via the following equations:

$$D^i = \frac{\mu_o h \tilde{\gamma}_\mu \tilde{\gamma}_n^i}{4\pi r_i^3}; \qquad Q^i = \frac{3V_{z'z'}^i eq^i}{4J^i(2J^i-1)} \qquad (24)$$

where $Q^i$ scales with the electric field gradient $V_{z'z'}^i$ which, in the case of Mu$^-$ in GaAs, is assumed to be induced mainly by the muon. The nuclear electric quadrupole moment $eq^i$ is known for any particular nucleus. Note that the magnetic dipole parameters $D^i$ vary inversely with the cube of the distance from the muon to the $i$th nucleus ($r_i$), so only the nearest-neighbor nuclei need to be considered.

In a $\mu$LCR experiment, the field dependence of $\bar{P}_z$ is recorded. One is searching for specific magnetic field values for which the muon polarization along **B** is transferred to the neighboring nuclei. This causes a resonant-like dip in $\bar{P}_z$ as a function of field. Recall from Section II.2a that in the absence of muon-nuclear interactions, the initial muon spin state is a good eigenstate of $\mathcal{H}$, and hence $P_z(t) = 1$ and $\bar{P}_z = 1$ [Eq. (10)]. When nuclear spin terms are included in $\mathcal{H}$, this is still correct at most magnetic fields. However, at specific magnetic fields, a resonant transfer of polarization occurs when the muon Zeeman energy splitting is matched to the appropriate energy splitting of a neighboring nucleus. In our example of the Mu$^-$ center, the nuclear spin splittings are determined by the quadrupole and nuclear

Zeeman interactions. In a plot of the magnetic field dependence of the energy levels, this would correspond to a crossing of two energy levels corresponding to two spin states of the combined muon-nucleus system if there were no muon-nuclear dipole interaction [$\mathscr{H}^D = 0$ in Eq. (22)]. However, an avoidance of the levels occurs if one includes the muon-nuclear dipole interaction. Although this interaction is usually small compared to the muon Zeeman interaction and the nuclear Zeeman and quadrupole interactions, it completely "mixes" the two spin states involved, leading to a resonant transfer of muon polarization to the nuclei (and vice versa). A similar technique was developed for $\beta$-NMR, where it is referred to as "cross-relaxation" (Fujura et al., 1980; Jäger et al., 1990).

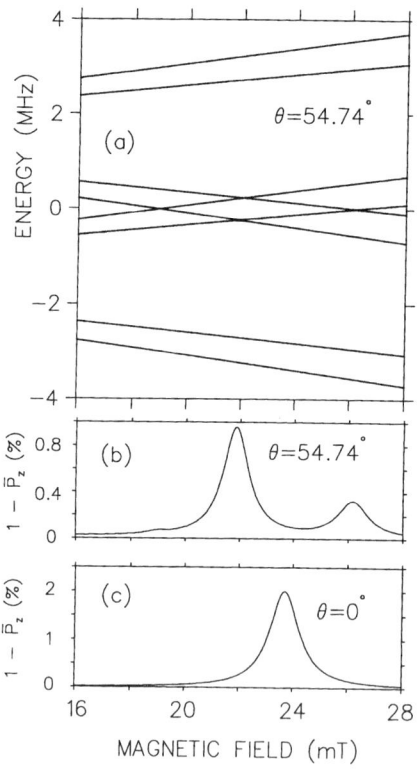

FIG. 10. (a) The field-dependence of the energy levels of a system consisting of a muon and one spin $\frac{3}{2}$ nucleus ($^{71}$Ga) with $\theta = 54.74°$, $Q = -1.472$ MHz, and $D = 2.1 \times 10^{-2}$ MHz. All four of the "crossings" shown are, in reality, AICs. The corresponding $\mu$LCR spectra for (b) $\theta = 54.74°$ and (c) $\theta = 0°$.

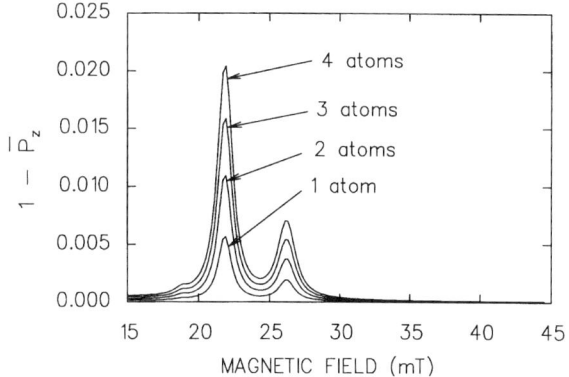

FIG. 11. The simulated μLCR spectra of up to four spin $\frac{3}{2}$ $^{71}$Ga nuclei at $\theta = 54.74°$. The parameters $Q = -1.472$ MHz and $D = 1.0 \times 10^{-2}$ MHz were used.

A system containing the muon and one spin $\frac{3}{2}$ neighbor, in this example taken specifically to be $^{71}$Ga, gives rise to eight energy levels (see Section II.2). The behaviour of these levels in the vicinity of a resonance is illustrated in Fig. 10a, where it is assumed that the applied magnetic field **B** and the muon-nucleus direction $\hat{z}'$ is at an angle of $\theta = 54.74°$. There are four μLCRs. The size of the energy gap $E_{\text{gap}}$ due to the avoidance of two levels is on the order of the dipole interaction. As can be seen from Eq. (9), a fraction of the muon polarization will oscillate at $E_{\text{gap}}/\hbar$. This (slow) oscillation(s) in $P_z(t)$ leads to $\bar{P}_z < 1$, and hence an apparent resonance.[9] The magnetic fields at which the resonances appear are determined primarily by $Q^i$ and $\theta$, while their absolute widths and intensities are governed by $D^i$ and the muon lifetime. Figure 10b shows the calculated field-dependence of $\bar{P}_z$ for $\theta = 54.74°$ and should be contrasted with the situation in which $\theta = 0°$. In Figs 10a, 10b, and 10c, the same dipole and quadrupole parameters are used. Note that the resonances are plotted as $1 - \bar{P}_z$; hence, they appear as peaks rather than dips.

Several important features of μLCR spectroscopy should be pointed out:

1. Groups ("shells") of equivalent nuclei contribute to the same resonance. More specifically, for diamagnetic centers, if $D^i \ll 1/\tau_\mu$, additional equivalent nuclei produce more intense and only slightly broader resonances, as shown in Fig. 11. Moreover, the intensity is roughly proportional to the number of equivalent nuclei $N_{\text{eq}}$.

---

[9]This discussion also illustrates that the resonant-like behaviour occurs when there is an ALC; hence, the common reference of this technique as "level-crossing resonance" is somewhat of a misnomer.

2. The resonances attributed to one shell of nuclei are largely unaffected by resonances from other shells. This makes it much easier to identify neighboring nuclei and the electronic structure around them. This is one of the most powerful features of $\mu$LCR.

*b. The Electronic Structure of $Mu^-$ in n-Type GaAs*

The $\mu$LCR spectrum for $Mu^-$ in the heavily doped n-type GaAs:Si sample discussed in Section III.1 is shown in Fig. 12 (Chow et al., 1995). The magnetic field was applied parallel to a $\langle 100 \rangle$ crystalline direction. These data were obtained at room temperature, at which the muon is static on a timescale much longer than the muon lifetime (region I in Fig. 9). A rough estimate of the strength of the quadrupole interaction, and hence the positions of the resonances, was obtained from the TF-$\mu$SR data (see Fig. 8), as discussed in Section III.1.c. A total of five resonances are present in the field range shown, and no additional ones were observed up to 120 mT. The four upper resonances may be assigned with certainty to the two isotopes of gallium ($^{69}$Ga and $^{71}$Ga) and show unambiguously that the muon and nearest-neighbor(s) Ga lie on the same $\langle 111 \rangle$ axis. This data supports the conclusion of the TF-$\mu$SR measurements described in Section III.1.c and also confirms that the Ga nuclei are responsible for the decrease in $\sigma$. Each of the gallium spin $\frac{3}{2}$ isotopes contributes a pair of lines.[10] The structure of each doublet (c.f. Fig. 10) is consistent with **B** and $r_i$ forming an angle of 54.7°. Furthermore, the relative positions and intensities of the two doublets define a unique value of the field gradient $V_{z'z'}$. The relative positions scale with the quadrupole moments $eq_i$, while their relative intensities scale with the gyromagnetic ratios and natural abundance of the two isotopes. The broader fifth resonance at lower field is attributed to $^{75}$As—the position of this resonance indicates a smaller field gradient at the $^{75}$As neighbors for which the structure of the resonance is not resolved. Furthermore, very similar results were obtained for n-type GaAs:Te, again confirming that $Mu^-$ is isolated in this material. The relevant parameters are listed in Table V.

Despite extensive knowledge of the interactions with the nearest neighbors, the site of the muon is still ambiguous. However, with a few reasonable assumptions based on theory and common sense, one can rule out all but one site. There are three candidate sites compatible with the experimental observation of a Ga nucleus along $\langle 111 \rangle$. These are the tetrahedral interstice surrounded by four nearest-neighbor Ga ($T_{Ga}$), the bond-center

---

[10]Natural Ga is an isotopic mixture, consisting of 60.2% $^{69}$Ga and 39.8% $^{71}$Ga, while natural As is 100% pure $^{75}$As.

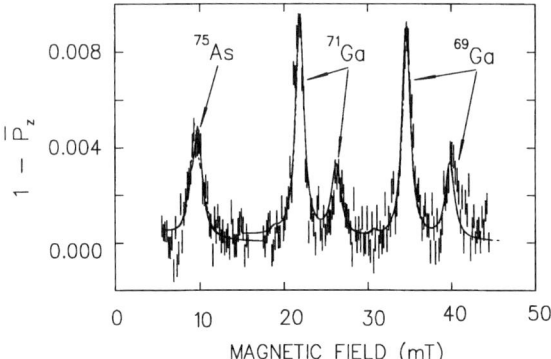

FIG. 12. Room-temperature μLCR spectra for heavily doped n-type GaAs:Si with a net donor concentration in the range 2.5–5.0 × $10^{18}$ cm$^{-3}$. The magnetic field is applied parallel to a $\langle 100 \rangle$ axis.

(BC), and an antibonding site next to a Ga (AB$_{Ga}$). These sites are illustrated in Fig. 1. Since it is not possible to determine $N_{eq}$ and $D^i$ separately, because the strength of a given resonance is given by the product $N_{eq}D^i$ (see Section III.2.b), the spectrum does not in itself distinguish between the BC (one Ga neighbor) and T$_{Ga}$ (four equivalent Ga neighbors). However, the electron-rich BC site is highly unlikely since this site is theoretically the highest energy location for Mu$^-$ (Pavesi and Giannozzi, 1992), whereas Mu$^-$ is lowest in energy when occupying regions of low valence charge, such as the interstitial T site. Furthermore, calculations of the adiabatic potential energy surface (Pavesi and Giannozi, 1992; Adams et al., 1995) place the overall minimum off-center from T$_{Ga}$ but at only slightly lower energy with small barriers. The large muon zero-point energy then yields a state spread out over the entire tetrahedral cage, centered at T$_{Ga}$, and including all four AB$_{Ga}$ sites. Hence, the most reasonable explanation for these data is that the muon is at a tetrahedral interstice surrounded by four nearest-neighbor Ga. This is the basis of the assumption for the muon site made in Section III.1.

Thus far, there is no direct experimental information on the structure of isolated H$^-$ (or H$^+$) in any semiconductor. The presence of the isolated H$^-$ center in n-type GaAs has been inferred from reverse-bias annealing experiments (Yuan et al., 1991; Cho et al., 1991; Roos et al., 1991; Leitch et al., 1991). However, some researchers still debate the existence of H$^-$, and alternative explanations for some of these data have been proposed (Morrow, 1993). Except for zero-point energy differences, the results for Mu$^-$ should model negatively charged isolated hydrogen in GaAs.

### c. The $Mu_{BC}^0$ Center in High-Resistivity GaP

Another use of the μLCR technique was to resolve the nuclear hyperfine structure of $Mu_{BC}^0$ in GaP (Schneider et al., 1993a). The spin-Hamiltonian Eq. (22) now becomes more complicated because of the inclusion of the muon and nuclear hyperfine interactions, and more resonances will occur. Nevertheless, the same principles apply. A resonance is possible when the muon energy-splitting matches those of a neighboring nucleus and if some additional interaction is present that "mixes" the muon and nuclear spin levels. If one considers only $Mu_{BC}^0$ and one neighboring nucleus, $n$, then the spin-Hamiltonian can be written as

$$\mathcal{H} = \mathcal{H}_\mu + \mathcal{H}_n \quad (25)$$

where $\mathcal{H}_\mu$ is given by Eq. (14) and

$$\mathcal{H}_n = -\tilde{\gamma}_n B J_z + A_\parallel^n S_{z'} J_{z'} + A_\perp^n (S_{x'} J_{x'} + S_{y'} J_{y'}) + Q \left[ J_{z'}^2 - \frac{J(J+1)}{3} \right] \quad (26)$$

The same notation as in Section II.2 is used. In addition, $A_\parallel^n$ and $A_\perp^n$ are the parallel and perpendicular hyperfine parameters, respectively, for the nearest-neighbor nuclei on the $\langle 111 \rangle$ axes, and $Q$ is the strength of the muon-induced quadrupole interaction (see Section III.2.a). As noted, because the μLCRs from one nucleus are effectively independent of other non-equivalent nuclei, it is sufficient to consider a spin-Hamiltonian involving a single nucleus.

The measured spectrum in an undoped GaP sample is shown in Fig. 13. The magnetic field is applied parallel to a $\langle 110 \rangle$ axis, and the temperature is 9 K. The spectrum is plotted as an "asymmetry difference." This derivative-like quantity is defined as

$$\Delta A = A^+ - A^- \quad (27)$$

where $A^\pm = A(B \pm B_f)$ and

$$A(B) = \frac{\int_0^\infty [N_B(t, B) - N_F(t, B)] \, dt}{\int_0^\infty [N_B(t, B) + N_F(t, B)] \, dt} \quad (28)$$

The small flip field $B_f$ of about 0.5 mT is applied alternatively along and opposite to **B** to smooth out small systematic effects due to the fluctuations of the beam intensity. All the resonances that appear in Fig. 13 are from

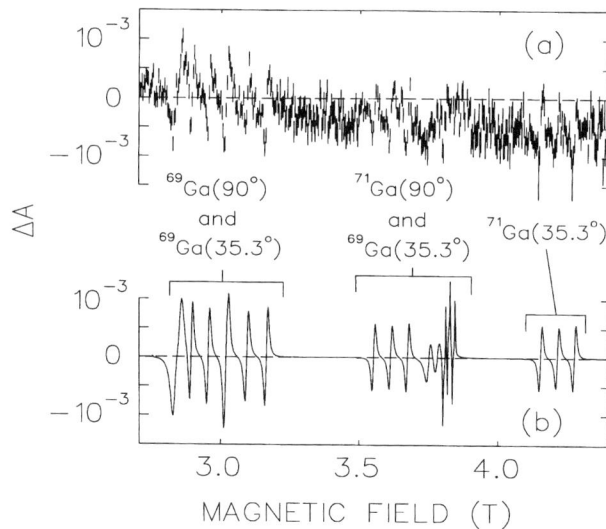

FIG. 13. (a) Measured μLCR spectrum for $Mu^0_{BC}$ in high-resistivity GaP, with a crystalline $\langle 110 \rangle$ axis aligned with the magnetic field and taken at a temperature of 9 K. All resonances are from $^{69}$Ga or $^{71}$Ga nuclei. (b) Computer simulation using the Ga hyperfine parameters given in Table III.

$^{69}$Ga or $^{71}$Ga nuclei. The positions of these resonances and a study of their orientation dependence (with $\mathbf{B} \parallel \langle 100 \rangle$) enables the hyperfine parameters for the two nearest neighbors (one Ga and the other P) to be determined (see Table III). Figure 13b is an exact numerical calculation of the μLCR spectrum using these parameters, following the prescription outlined in Section II.2, and demonstrates the excellent agreement between theory and experiment. The precise nuclear hyperfine parameters for the nearest-neighbor Ga and P atoms allow estimates of the $s$ and $p$ character of the unpaired spin density on Ga and P (see Section II.2.b, for example) and, consequently, the displacement of each of these atoms due to the presence of $Mu^0_{BC}$. Schneider et al. (1993a) estimated that the Ga and P atoms have moved $\approx 0.24$ Å and $\approx 0.54$ Å, respectively, away from their unrelaxed positions. The high precision with which the muon and the nearest-neighbor hyperfine parameters were determined in GaP, and also in GaAs (Kiefl et al., 1987), provides a data set that can be used as a stringent test of *ab initio* calculations. For example, the parameters in GaAs are in good agreement with state-of-the-art first-principles spin-density-functional calculations for neutral hydrogen or muonium (Van de Walle and Pavesi, 1993). A similar calculation for GaP is not yet available. The complete set of measured muon and nearest-neighbor nuclear hyperfine parameters for $Mu_{BC}^0$ in GaP and GaAs also enabled a detailed comparison of the spin density among the group III and group V nearest neighbors. It was found that for GaP there is about 43% of the spin density on the group III element and 35% on the group V element, while for GaAs the corresponding numbers are 36% and 44%. This result seems incompatible with the trends predicted by a qualitative theoretical model for $Mu_{BC}^0$ in compound semiconductors (Estricher et al., 1989; Maric et al., 1989).

These examples demonstrate that $\mu$LCR is indeed a powerful probe of the electronic structure of static neutral and charged muonium. The size and symmetry of the couplings with the neighboring spins can be accurately determined. For example, in the case of the paramagnetic centers, the electronic spin density on the nearest-neighbor atoms, and sometimes the next-nearest neighbors, can be measured. This information can be used as a stringent test of state-of-the-art theoretical calculations. In addition, key experimental information on the locations of the muon and neighboring nuclear spins can be obtained.

### 3. Zero-Field Muon Spin Relaxation/Rotation (ZF-$\mu$SR)

The counter arrangement shown in Fig. 3b is also appropriate for experiments in ZF-$\mu$SR. Counters along the initial muon spin direction are used to obtain the relevant polarization information [i.e., $P_z(t)$]. The count rates in each histogram and the asymmetries for the F and B counters are given by expressions analogous to Eq. (19) to Eq. (21). Helmholtz coils are usually used to cancel out stray magnetic fields at the sample. Typically, it is sufficient to reduce any residual stray field to be $\approx 0.01$ mT. This is smaller than any relevant internal magnetic field, typically the muon-nuclear dipolar field ($\approx 0.1$ mT).

The application of the ZF-$\mu$SR technique to semiconductors is still relatively new, and its main use so far has been for investigating the diffusion of the diamagnetic centers in group III–V semiconductors. ZF-$\mu$SR is more sensitive to the slow hopping motion of muonium than TF-$\mu$SR (Hayano et al., 1979). In Section III.3.a, we discuss the principles of ZF-$\mu$SR as it pertains to the diffusion of diamagnetic centers. In Section III.3.b, we discuss the example of Mu$^-$ diffusion in heavily doped n-type GaAs:Si.

*a. Studying Diffusion—the Kubo-Toyabe Functions*

Consider the situation in which the muon is stationary in a crystal and surrounded by nuclear spins that are unpolarized. To a first approximation,

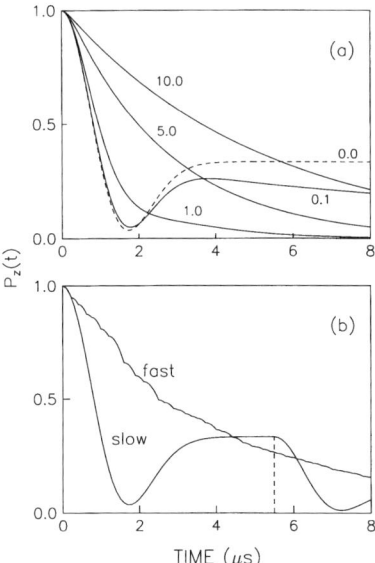

FIG. 14. (a) Static and dynamic Kubo-Toyabe functions. The dashed line is the static Kubo-Toyabe, while the others are the dynamic Kubo-Toyabe function, assuming the muon is hopping at a rate $v = 0.1$, $1.0$, $5.0$, and $10\,\mu s^{-1}$. The value of $\Delta = 1.0\,\mu s^{-1}$ was assumed for this simulation. (b) A schematic representation of the dynamization of the static Kubo-Toyabe function. The dashed line indicates the time that the muon hops for the slow-hopping regime. The labels "slow" and "fast" indicate that the muon hop rate $v$ is much slower and much faster than $\Delta$, respectively.

the nuclei exert a Gaussian field distribution at the muon, which is the same for all three components of **B**:

$$F(B_i) = \frac{\gamma_\mu}{\sqrt{2\pi}\Delta} \exp\left[\frac{-\gamma_\mu^2 B_i^2}{2\Delta^2}\right] \qquad (29)$$

where the dipolar width $\Delta$ characterizes the distribution of local fields and is given by $\Delta^2 = \gamma_\mu^2 \langle B_x^2 \rangle = \gamma_\mu^2 \langle B_y^2 \rangle = \gamma_\mu^2 \langle B_z^2 \rangle$ and $\gamma_\mu = 2\pi\tilde{\gamma}_\mu$. In zero applied magnetic field, $P_z(t)$ is then described by the static Kubo-Toyabe function (Hayano et al., 1979). This function, labeled as $p_z(t)$, has the following form:

$$p_z(t) = \frac{1}{3} + \frac{2}{3}(1 - \Delta^2 t^2)\exp\left[-\frac{\Delta^2 t^2}{2}\right]. \qquad (30)$$

The notable features of $p_z(t)$, shown in Fig. 14a, is the minimum that occurs at $t = \sqrt{3}/\Delta$ and the asymptotic value of $\frac{1}{3}$ at long times. The effect of muon diffusion on $P_z(t)$ can be approximated by a stochastic model for the time evolution of a local field by the migration of a muon under a Markovian process (strong collision model). This approach was first adopted by Hayano et al. (1979), and the resulting dynamic Kubo-Toyabe function is given by:

$$P_z(t) = p_z(t)e^{-vt} + v \int_0^t p_z(\tau)e^{-v\tau} P_z(t-\tau)d\tau. \tag{31}$$

The muon is assumed to be hopping randomly between equivalent sites at an average frequency $v$. The time spent during a jump is taken to be much shorter than the mean residence time at a site. The observed muon polarization $P_z(t)$ is made up of the sums of contributions from muons that have not hopped at all, those that have hopped once, twice, and so on. Figure 14b shows two extreme examples, illustrating the situation in which the muon has jumped 38 times and once within 8 $\mu$s, respectively. The former is in the regime of fast hopping ($v \gg \Delta$) and the latter is an example of slow hopping ($v \ll \Delta$). Between jumps, the evolution of the muon polarization is given by the static Kubo-Toyabe function, which after each hop starts with an initial amplitude that is consequently smaller. Equation (31) is the mathematical description of such a random process, taking into account weighting factors due to the average lifetime $1/v$ between hops. This equation can be solved numerically for $P_z(t)$, and examples of the results are shown in Fig. 14a.

b. *Diffusion of $Mu^-$ in n-Type GaAs*

One application of ZF-$\mu$SR is to determine precisely the muon hop rate in region II, shown in Fig. 9. Typical zero-field data are shown in Fig. 15a. First, the static Kubo-Toyabe function is used to fit the data below $\approx 400$ K, where the muon is static (giving $\Delta = 0.165(5)\mu s^{-1}$). This value is then fixed, and the data from 475 to 625 K are fitted by the dynamic Kubo-Toyabe function to obtain $v$. The temperature dependence of the hop rates, shown in Fig. 15b, has an activated form over nearly two decades in $v$, yielding an activation energy of 0.73(1) eV and a prefactor of $5.6(5) \times 10^{12} s^{-1}$. The hop rate can also be related to the diffusion constant; for example, in a simple model, in which the muon makes random jumps in three dimensions between the nearest *equivalent* tetrahedral interstitial sites in the zinc-blende lattice, the muon diffusion constant is $D_\mu = a^2 v/12$, where $a$ is the lattice constant ($= 5.64$ Å in GaAs). Hence, the value of $D_\mu = 1.48 \times 10^{-3} \exp(-0.73 \text{eV}/k_B T) \text{cm}^2/\text{s}$.

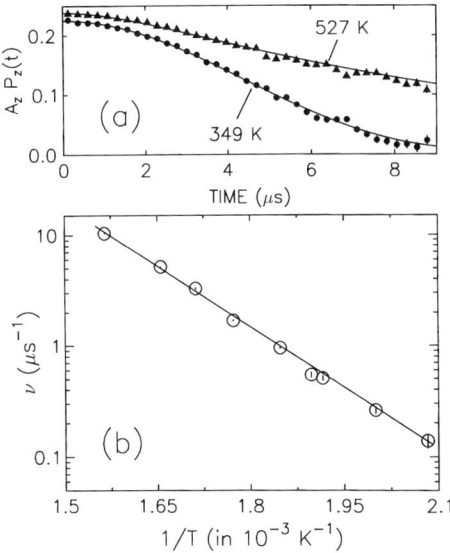

FIG. 15. (a) Examples of the ZF-$\mu$SR spectra for n-type GaAs:Si. The solid lines are the best fits to the data with the static and dynamic Kubo-Toyabe functions. (b) The temperature-dependence of the hop rate $v$. The solid line is the best fit, assuming an Arrhenius functional form.

As a comparison with results on hydrogen, Chevallier et al. (1992) reported a diffusion constant of deuterium in heavily doped n-type GaAs of $4 \times 10^{-12}$ cm$^2$/s at 513 K, while $D_\mu(513K) = 1.1 \times 10^{-10}$ cm$^2$/s. The ratio of the diffusion constants is considerably larger than predicted by classical mass scaling arguments, assuming the same processes. This could suggest that quantum or nonadiabatic effects are important. However, one should keep in mind that the bulk diffusion constant of deuterium may be limited by the trapping and detrapping at dopant atoms, whereas measurements of $Mu_T^-$ diffusion reported here are on a microsecond timescale and thus insensitive to trapping effects. One of the most remarkable results to come out of these measurements is the enormous difference in the diffusion rates for $Mu_T^0$ and $Mu_T^-$, as we shall see in Section III.4.a. In particular, the diffusion constant for $Mu_T^0$ is about $10^{10}$ times larger than $Mu_T^-$ at room temperature.

At this point, it is worthwhile pointing out that the diffusion depth profiles of hydrogen are usually complicated. In Si and GaAs, the time dependences depart significantly from that expected for the diffusion of a single species. This difficulty may stem from the fact that hydrogen can exist in multiple charge states, and because it appears to be present in a number of configurations, including atomic, diatomic, and as a complex (see Herring and Johnson, 1991 for a discussion). For example, Chevallier et al. (1992) speculate that both $H^0$ and $H^-$ contribute to the diffusion profile of hydrogen in n-type GaAs. This view is certainly reasonable if an analogy is made with measurements of muonium, such as those discussed in Section III.1.c. We shall return to these points in later sections, where we describe the studies of spin and charge exchange dynamics and the interconversion between the various muonium states.

4. Muon Spin Relaxation in a Longitudinal Magnetic Field (LF-$\mu$SR)

In a time-differential longitudinal-field experiment, the magnetic field is applied along the initial muon spin direction. As with $\mu$LCR and ZF-$\mu$SR, the counter arrangement for LF-$\mu$SR is shown in Fig. 3b. LF-$\mu$SR is often used to measure the $1/T_1$ spin relaxation rate of muonium, which is quite sensitive to muonium dynamics—diffusion, spin-exchange scattering, and cyclic charge-state changes of muonium. By spin-exchange scattering, we mean the processes whereby a neutral muonium center and a free conduction electron scatter, leading to a spin-flip of the bound muonium electron. Changes in the charge state of muonium include the familiar example in which a neutral center is ionized and hence becomes positively charged. Under certain conditions, such as when there are a significant number of free charged carriers, the muonium charge state can cycle many times during the lifetime of the muon. Experimental examples of spin and charge dynamics are abundant in semiconductors, and the most detailed studies so far have been in Si and GaAs.

The following experimental examples are used to illustrate the uses of the LF-$\mu$SR technique: (1) the diffusion of $Mu_T^0$ in high-resistivity GaAs (Section II.4.a), (2) charge-state cycling of muonium at high temperatures in nearly intrinsic Si (Section III.4.c) and in heavily-doped n-type GaAs (Section III.4.d), and (3) the spin-exchange scattering of muonium in n-type Si (Section III.4.e). In addition, Section III.4.b contains a discussion on how spin and charge exchange influence the $1/T_1$ rates in a LF-$\mu$SR experiment. Finally, in Section III.4.f, we briefly describe the "repolarization" technique, although in these experiments one is not actually interested in the $1/T_1$ relaxation. This technique is sometimes used to characterize muonium centers in semiconductors when it is difficult to obtain the same information by other methods.

Recall from Section I that the lighter mass of the muon compared to the proton implies that significant differences are expected when comparing diffusion rates of muonium with hydrogen. On the other hand, the mass difference should only have a minor influence on processes such as spin-exchange scattering or charge-state cycling. Studies of muonium dynamics are thus particularly important since there is only very limited experimental information on such dynamics for hydrogen in semiconductors. For example, the existence of interconversion between the various charge states of hydrogen has been suggested to play an important role for hydrogen diffusion in semiconductors (see Section III.3). In another example, Johnson et al. (1994) have studied the rates of charge-state changes of atomic hydrogen ($H^+ + 2e^- \leftrightarrow H^-$) in n-type Si:P by capacitance techniques. The results lead the authors to suggest that the hydrogen acceptor level is lower in energy than the donor level, indicating that hydrogen is a negative $U$ impurity in Si,[11] as indicated by theory.

*a. Quantum Diffusion of $Mu_T^0$ in High-Resistivity GaAs*

As summarized in Section III.1.a, there are two neutral muonium centers, $Mu_{BC}^0$ and $Mu_T^0$, that co-exist at low temperatures in high-resistivity GaAs. Recall that $Mu_{BC}^0$ is stationary on the timescale of the muon lifetime, whereas $Mu_T^0$ is highly mobile. This is believed to be true in Si, Ge, and GaAs. The most detailed measurements of the diffusion of $Mu_T^0$ are in the case of GaAs.

The Hamiltonian for the stationary isotropic muonium $Mu_T^0$ or hydrogen center surrounded by nuclei with spins $J^i$ can be written as:

$$\mathcal{H} = \tilde{\gamma}_e B S_z - \tilde{\gamma}_\mu B I_z + A_\mu \mathbf{I} \cdot \mathbf{S}$$
$$+ \sum_i [-\tilde{\gamma}_n^i B J_z + A_\parallel^{n,i} J_{z'}^i S_{z'} + A_\perp^{n,i} (J_{x'}^i S_{x'} + J_{y'}^i S_{y'})] \quad (32)$$

The same notation is used as in Section III.2.c, where we discussed the $\mu$LCR studies of the $Mu_{BC}^0$ center in high-resistivity GaP. The first three terms (see Section II.2.b) describe the free $Mu_T^0$ center, while the remaining terms describe the interactions involving the nuclei ($\mathcal{H}_n$). The summation is over all the neighboring nuclear spins, and the small nuclear electric quadrupole interactions are neglected. Now, suppose that $Mu_T^0$ is hopping between equivalent sites. Its unpaired electron experiences a fluctuating nuclear hyperfine interaction originating from the randomly oriented nuclear spins. The electron spin relaxes and consequently results in muon spin relaxation due to the muon hyperfine interaction. Celio (1987) and Yen

---

[11] In a negative $U$ system, two $H^0$ centers can lower their energy by changing to $H^+$ and $H^-$.

(1988) calculated $P_z(t)$ by assuming that the nuclear terms are replaced by

$$\mathscr{H}_n = \delta_{ex} \mathbf{S} \cdot \mathbf{T}(t) \tag{33}$$

where $\delta_{ex}$ is the effective nuclear hyperfine interaction strength and $\mathbf{T}(t)$ is the unit vector randomly fluctuating with correlation time $\tau_c$. Hence, the fluctuation rate $1/\tau_c$ can be roughly interpreted as the "hop rate." They found that the theoretical $P_z(t)$ is a sum of four expontentials, and very often, the spectrum can be well fitted by an average single exponential with decay rate, $1/T_1$, which is a function of $\mathbf{B}$, $A_\mu$, $\delta_{ex}$, and $\tau_c$. A useful approximate equation for $1/T_1$ (Kadono, 1990b) is

$$1/T_1 \approx \left(1 - \frac{x}{\sqrt{(1+x^2)}}\right) \frac{\delta_{ex}^2 \tau_c}{1 + \omega_{12}^2 \tau_c^2} \tag{34}$$

where $\omega_{12} = 2\pi\nu_{12}$ and $x = B/B_0$ are defined in Section II.2.b.

Figure 16a shows the temperature dependence of the average relaxation rate at three fields applied parallel to a $\langle 100 \rangle$ crystalline direction (Kadono et al., 1990a, 1994a; Schneider et al., 1992b) of a high-resistivity GaAs sample. Notice that the relaxation rates are still quite fast in high magnetic fields. This should be contrasted with studies of the diamagnetic centers, where the relevant interaction (muon-nuclear dipolar interaction) is much weaker (see Section III.3). The inverse correlation time $1/\tau_c$, shown in Fig. 16, is obtained from simultaneous fits of the Celio and Yen model to the three-spectra-per-temperature point. Note that the hop rate is a minimum at a crossover temperature $T_X \approx 90$ K. Above this temperature, one is in the regime where thermally activated motion, primarily phonon-assisted tunneling, is dominant. Below $T_X$, one observes that the $1/\tau_c$ *increases* (with an approximately $T^{-3}$ law) with decreasing temperature as a result of incoherent quantum tunneling. Immediately below $T_X$, the motion is still hopping-like. The tunneling rate is controlled by phonon scattering, which decreases with temperature. Below about 10 K, there is a crossover to coherent bandlike motion where the mean free path is longer than a lattice constant. In this region, $k_B T$ is still much greater than the bandwidth for muonium, so the average velocity of muonium, and hence $1/\tau_c$, is temperature-independent (Kagan and Prokof'ev, 1990; MacFarlane et al., 1994).

A comparison with the diffusion rate of $Mu_T^-$ in heavily doped n-type GaAs (see Section III.3.b) shows the dramatic influence of the charge state. $Mu_T^-$ is essentially static in the low temperature range, whereas $Mu_T^0$ exhibits aspects of quantum diffusion. At room temperature, $Mu^-$ is moving more slowly by a factor of $\approx 10^{10}$. The implication for hydrogen is that the $H^-$ ion is also far less mobile than a neutral (T-site) hydrogen atom.

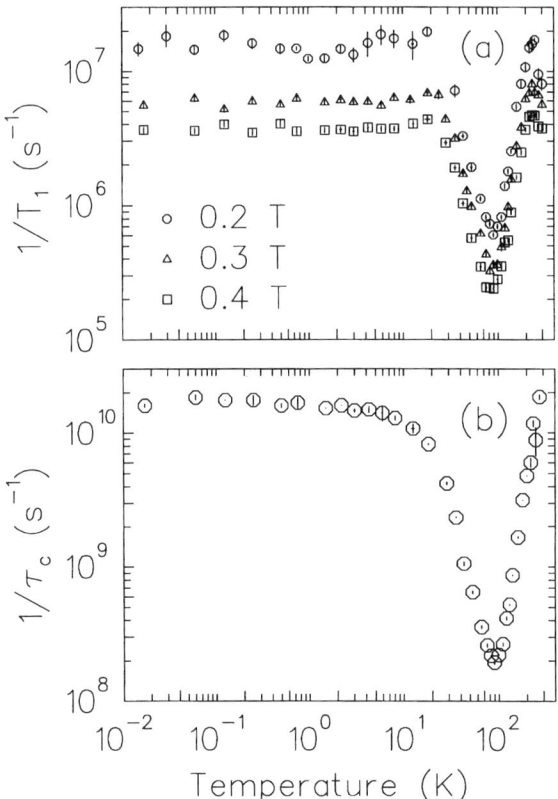

FIG. 16. Temperature dependence of (a) the average $1/T_1$ rate at three different applied fields and (b) the corresponding inverse correlation time $1/\tau_c$ for $Mu_T^0$ to hop in high-resistivity GaAs.

### b. *Influence of Spin and Charge Exchange on $1/T_1$*

Thus far, we have been concerned primarily with the electronic structure and diffusion of muonium. We now consider the role of free carriers, which depend sensitively on external factors such as temperatue and the concentration of dopants present. In general, the interaction with free carriers is strong and produces significant $1/T_1$ spin relaxation. Recent experiments emphasize that two situations are common; which one dominates depends on the experimental conditions. These processes are (1) cyclic charge-state changes of muonium and (2) spin-exchange scattering of muonium with free electrons. The former has also been called "charge exchange" (CE) and describes the situation in which the charge state of muonium fluctuates between two (or more) charge states. Typical examples are $Mu^0 \leftrightarrow Mu^+ + e^-$ and $Mu^0 + e^- \leftrightarrow Mu^-$. Several conditions must be met for CE to dominate. The first is that the temperature must be sufficiently high so that any relevant ionization processes can take place. The second is that a significant number of free carriers should be present in the system in order for capture processes to take place at an adequate rate. Note that unidirectional conversion such as $Mu^0 \rightarrow Mu^-$ or $Mu^0 \rightarrow Mu^+$ can be considered special cases of CE. Thus far, spin-exchange scattering has been observed only in n-type materials and involves neutral muonium and electrons. This process is important at low temperatures where muonium ionization does not occur but where there are many free electrons in the system. This condition will be experimentally realized in doped n-type semiconductors at temperatures at which there is ionization of the donor levels. When conduction electrons scatter off muonium, the unpaired electron on the muon can flip its spin accompanied by a flipping of the conduction electron spin. Hence, the muonium electron cycles at a rate $v_{SE}$ between its $m_s = +\frac{1}{2}$ and $m_s = -\frac{1}{2}$ states. The electron dynamics are transmitted to the muon via the muon-electron hyperfine interaction. Although CE and spin exchange are fundamentally very different processes, they lead to similar field dependences of the $1/T_1$ spin relaxation rate, and hence can be interchanged for purposes of qualitative discussions (Senba, 1990b).

We now give a qualitative presentation of how CE leads to $1/T_1$ relaxation in order to provide an intuitive understanding of the parameters that are important in such dynamics. This will be especially helpful when the behavior of $Mu^0_{BC}$ undergoing such dynamics is considered. Consider muonium with an *isotropic* hyperfine parameter, such as $Mu^0_T$ in LF. In the absence of CE, the muon polarization for $Mu^0_T$ can be written as [see Eq. (15)]

$$P_z(t) = \frac{1 + 2x^2}{2(1 + x^2)} + \frac{1}{2(1 + x^2)} \cos(\omega_{24} t) \tag{35}$$

and consists of the sum of a constant and a term oscillating at the field-dependent $\omega_{24}$ frequency. In contrast, the spin of a diamagnetic center $Mu^+$ or $Mu^-$ is "locked" along the direction of the applied field (see Section II.2.a):

$$P_z(t) = 1 \tag{36}$$

Now consider the situation in which there is rapid cycling between $Mu^0$ and $Mu^+$. The time-dependence of the muon polarization is shown schematically in Fig. 17. At the end of each cycle, part of the initial muon polarization is lost. The observed polarization is an average of many such cycles and decreases as a function of time (i.e., spin relaxation occurs). As in the case of NMR, this is referred to as $1/T_1$ relaxation. Spin-exchange scattering can be understood in a similar manner. Roughly speaking, in the case of spin-exchange scattering, cyclic transitions are taking place between the following two states: (1) muon and electron spin parallel and (2) muon and electron spin antiparallel. The former is an eigenstate of Eq. (14), with $A_\parallel = A_\perp = A_\mu$, and hence the muon polarization behaves similarly as a diamagnetic center (i.e., independent of time). The muon polarization of the latter (state 2) is qualitatively similar to $Mu_T^0$, consisting of oscillating and non-oscillating components. Therefore, the qualitative conclusions drawn from the discussion on CE also apply to spin-exchange scattering.

Theoretical calculations show that CE and spin-exchange involving either $Mu_{BC}^0$ or $Mu_T^0$ lead to a decay of the muon polarization, which is well described by a single exponential function $e^{-t/T_1}$. Several of these calculations are described by Nosov and Yakovleva (1963), Ivanter and Smilga

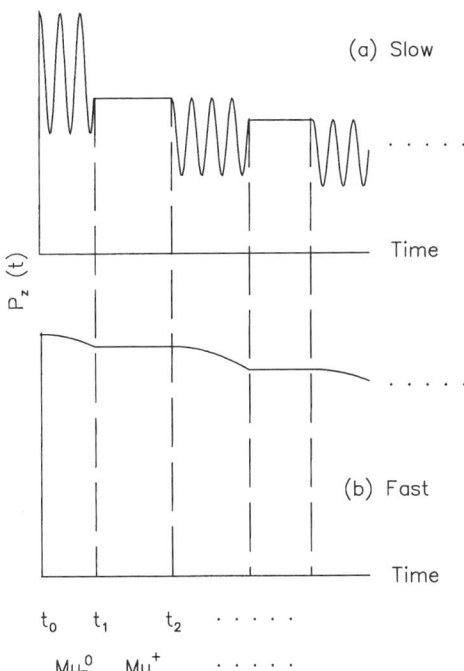

FIG. 17. Isotropic muonium undergong CE in the limits where the conversion rate from $Mu_T^0$ to the diamagnetic center is (a) slow and (b) fast compared to the $\omega_{24}$ frequency.

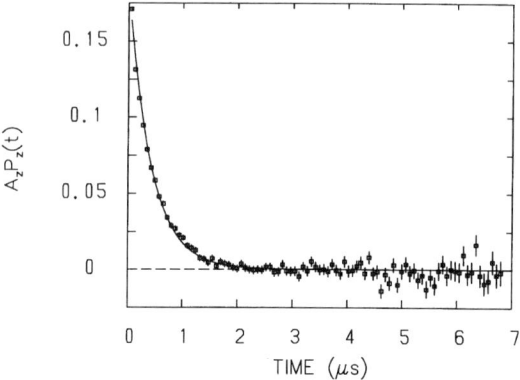

FIG. 18. A typical asymmetry plot, $A_zP_z(t)$, for nearly intrinsic Si at 805 K. The LF of $\approx 3.0$ Tesla is applied parallel to a $\langle 110 \rangle$ crystallographic direction.

(1971), Odermatt (1988), Senba (1990a, 1990b, 1991), Chow et al. (1993), and Kreitzman et al. (1995). This is illustrated in Fig. 18, which shows an example of the LF asymmetry $A_zP_z(t)$ for muonium in nearly intrinsic Si at $\approx 800$ K, where cyclic charge-state changes of muonium are taking place. This well-compensated sample, also called P11 in Section III.5, is slightly p-type with a nominal room-temperature resistivity of $33,000 \,\Omega \cdot$ cm (net acceptor concentration $\approx 5 \times 10^{11}$ cm$^{-3}$).

The field-dependence of the $1/T_1$ rates behave very differently, depending on the nature of the hyperfine interaction. If muonium has an *isotropic* hyperfine parameter, the decay rate is well approximated by (Chow et al., 1993)

$$1/T_1 \approx \frac{\lambda_{0D}\lambda_{D0}}{\lambda_{0D} + \lambda_{D0}} \left( \frac{\omega_o^2}{\lambda_{0D}^2 + \omega_{24}^2} \right) \frac{1}{2}. \tag{37}$$

The preceding equation is valid for CE involving Mu$^0$ and a diamagnetic center Mu$^D$. The symbols $\lambda_{0D}$ and $\lambda_{D0}$ represent the conversion rates for Mu$^0 \to$ Mu$^D$ and Mu$^D \to$ Mu$^0$, respectively. $1/T_1$ is flat at "low" magnetic fields (where $\omega_{24} \ll \lambda_{0D}$) and eventually shows a $B^{-2}$ behavior at "high" fields (where $\omega_{24} \gg \lambda_{0D}$). Such a field-dependence is a characteristic signature for charge and spin dynamics involving isotropic muonium.

On the other hand, the highly anisotropic hyperfine interaction of Mu$_{BC}^0$ produces a dramatically different field-dependence of the $1/T_1$ rates, especially in the case of slow spin or CE. The most prominent feature is a *peak* in

$1/T_1$, which occurs close to $B_p = (A_\perp \sin^2\theta + A_\parallel \cos^2\theta)/2\tilde{\gamma}_\mu$, where $\theta$ symbolizes the angle between **B** and the hyperfine symmetry axis (parallel to the $\langle 111 \rangle$ direction). The value of the $1/T_1$ rate at the peak is approximately the electron spin-flip rate $v_{SE}$ (Chow et al., 1994a). Observation of this peak is evidence for the existence of $Mu_{BC}^0$ under conditions in which it is not observable via other methods. The origins of this pronounced feature in $1/T_1$ can be understood qualitatively within the effective field description presented in Section II.3. Recall that the total muon polarization is the average over the precession of muon spins about two effective fields, $\mathbf{B}^+$ and $\mathbf{B}^-$ [Eq. (17)]. The muon precesses in a cone of fixed angle about each effective field and the projection of its polarization onto the axis parallel to the applied field **B** consists of a constant and an oscillating component. As shown in Eq. (17), when $B = B_\parallel$, the muon polarization for states with $m_s = +\frac{1}{2}$ precesses about an effective field that is perpendicular to the applied field, since the parallel component of the effective field is zero. At this field, the amplitude of the oscillating component is largest. As $B$ moves away from $B_\parallel$, this amplitude rapidly goes to zero. In analogy with the discussion on $Mu_T^0$, a useful approximate picture of LF depolarization is that for small $v_{SE}$, the amount of muon polarization lost per spin-exchange cycle is proportional to the amplitude of these oscillations. Consequently, when $B = B_\parallel$, one expects the maximum $1/T_1$ rate.

### c. Charge-State Dynamics in Si at High Temperatures

Figure 19 shows the high-temperature behavior of the $1/T_1$ rates in the well-compensated Si sample discussed previously (see Section III.4.c and Fig. 18). At the elevated temperatures for which the data (Chow et al., 1993) were obtained, (1) ionization of the neutral state is possible and (2) there are a large number of thermally generated free carriers. In high field, $1/T_1$ falls as $B^{-2}$, demonstrating that an *isotropic* neutral center is involved in cyclic charge-state changes. In fact, the data are well described by a two-state model involving thermal ionization of $Mu^0$ followed by electron capture: $Mu^0 \leftrightarrow Mu^+ + e^-$. Within this model, the average muon-electronic hyperfine parameter of the neutral state is consistent with that of $Mu_T^0$, indicating that when muonium is in the neutral charge state, it spends a significant fraction of its time near T sites. This result is surprising since most theoretical calculations (Van de Walle, 1991; Myers et al., 1992; Chu and Estreicher, 1990) predict that the BC site is the global minimum in the total energy for the neutral charge state. Furthermore, there are predictions that the BC state is at least $\approx 0.3$ eV below that of the T site, implying that the BC site should be occupied much more frequently.

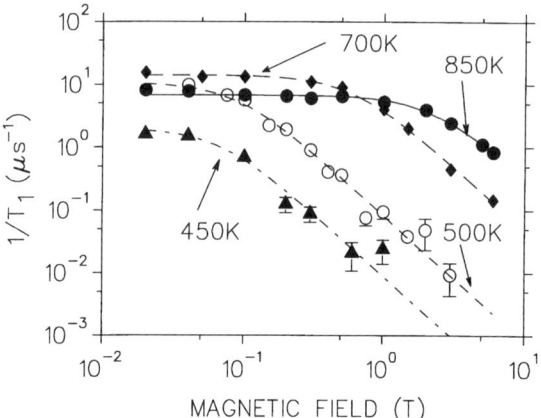

FIG. 19. The field-dependence of the $1/T_1$ rates in nearly intrinsic Si. The curves are fits to the data as described in Chow et al. (1993).

One possible explanation for this result is that quantum effects can significantly modify the conclusions of previous calculations. Ramirez and Herrero (Ramirez and Herrero, 1994; Herrero and Ramirez, 1995) claim that although hydrogen at the BC site remains the stable state, the difference in mass between the muon and the proton can be enough to make muonium located at the same site metastable due to the enhanced zero-point energy ($\approx$ three times larger for muonium compared to hydrogen). Their simulations also show that the lifetime of $\text{Mu}_{BC}^0$ decreases with increasing temperature, and it eventually converts to $\text{Mu}_T^0$.

Nevertheless, most theoretical calculations predict that $\text{Mu}_{BC}^0$ and $\text{H}_{BC}^0$ are both the *stable* states, provided that the neighboring atoms undergo significant relaxation. These calculations apply equally to both hydrogen and muonium since both are treated as classical point particles. If this result is still correct after taking into account zero-point motion, then the foregoing experimental result could be a consequence of the nature of the lattice response to the motion of muonium. At the high temperatures, where these experiments were performed, it is likely that theoretical calculations based on the assumption that the lattice can relax fully in the presence of the muon are no longer valid. This is supported by molecular dynamics calculations of the interstitial proton in Si (T $\gg$ 1200 K), which show that the surrounding host atoms, which are much more massive, do not have time to respond to the fast-moving proton (Buda et al., 1989). Hence, sites such as the BC, which were originally low in energy, are no longer so because the surrounding lattice atoms do not relax fully. Within this model, the muon spends a significant fraction of its time near the T site since a much smaller lattice relaxation is required there than for muonium occupying the BC cite. Note that since the proton is much more massive and less mobile than the muon, the inability of the lattice to respond to the muon should occur at lower temperatures. The suggestion that the host nuclei cannot respond quickly enough to a rapidly moving muonium center has been invoked previously to explain the existence of $Mu_T^0$ at low temperatures in silicon (Van de Walle, 1991), the disagreement between the theoretical and experimentally determined isotropic hyperfine parameter for normal muonium in GaAs (Van de Walle and Pavesi, 1993), and the local tunneling motion of the $Mu^I$ center in CuCl (Schneider *et al.*, 1992a).

### d. Evidence for Muonium as a Deep Recombination Center in GaAs

Recall from Section III.1 that at high temperatures in heavily doped n-type GaAs:Si, there is a dramatic increase in the TF linewidth of the $Mu^-$ precession signal (region III in Fig. 9). This is attributed to cyclic charge-state changes involving $Mu_T^0$ and $Mu^-$. A study of the LF relaxation in this temperature region has yielded the most detailed information to date on the nature of one of the muonium charge cycles that have been observed in many semiconductors (Chow *et al.*, 1996a). The field-dependence of the $1/T_1$ rates at 900 K and 1000 K (not shown) is characteristic of a neutral center with a large isotropic hyperfine interaction. Hence, the authors assumed that only $Mu_T^-$ and $Mu_T^0$ are involved in the charge dynamics (i.e., $Mu_T^0 \leftrightarrow Mu_T^-$). This also represents the simplest and most plausible cycle, since both states should be located near a $T_{Ga}$ site. The experimental $1/T_1$ rates at a low magnetic field of 48 mT and a high magnetic field of 2.9 T are displayed in Fig. 20a. By assuming that $A_\mu = 2884$ MHz, Eq. (37) can be used at each temperature to calculate the transition rate $\lambda_{0-}$ for $Mu_T^0 \to Mu_T^-$ and the rate $\lambda_{-0}$ for the reverse process. These are displayed in Fig. 20b and Fig. 20c.

The rate $\lambda_{0-}$ is nearly temperature-independent. Since the conduction electron concentration $n_e$ is essentially constant below 1000 K in this heavily doped n-type crystal, these results imply that $Mu_T^0$ captures an electron to form $Mu_T^-$. (The observed weak temperature-dependence of $\lambda_{0-}$ is inconsistent with an activated process, as would be expected for the alternate reaction $Mu_T^0 \to Mu_T^- + h^+$). Fitting $\lambda_{0-}$, assuming the process $Mu_T^0 + e^- \to Mu_T^-$, and assuming that the transition rate has the form $\lambda_{0-} = \sigma_e n_e v_e$, where $v_e$ is the electron velocity, yields an average capture cross-section $\sigma_e$ of 16.0(6) Å$^2$. The reverse rate $\lambda_{-0}$, as shown in Fig. 20b, fits well to an activated process with an activation energy ($\approx 1.67$ eV) comparable to the bandgap $E_{gap}$.[12] The most plausible explanation is that $Mu_T^-$ converts to $Mu_T^0$ via hole capture (i.e., $Mu_T^- + h^+ \to Mu_T^0$). Since the

---

[12]One explanation is that the $0/-$ level of $Mu_T$ is near the valence band edge and the process is thermal ionization of $Mu_T^-$ ($Mu_T^- \to Mu_T^0 + e^-$). However, the prefactor ($\approx 10^{17}$ s$^{-1}$) seems too large for this explanation.

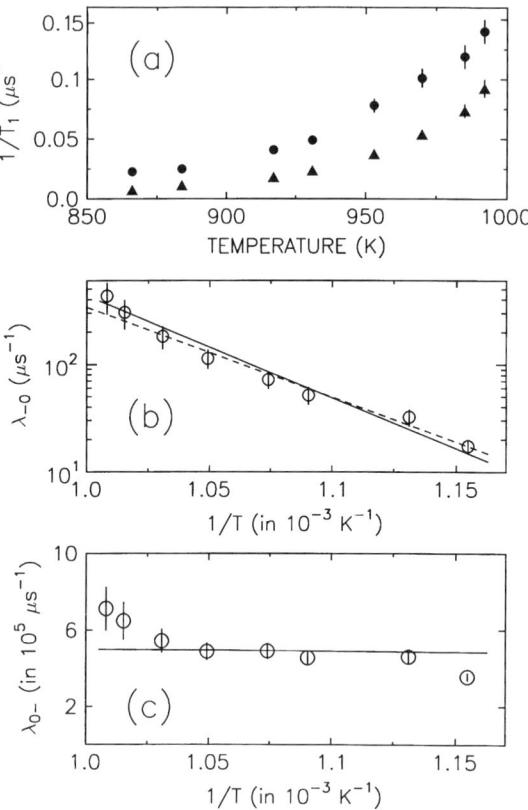

FIG. 20. (a) The temperature-dependence of the $1/T_1$ rates at $B = 2.9$ T (triangles) and $B = 48$ mT (circles). (b) The temperature-dependence of the rate for $Mu_T^- \to Mu_T^0$. The dashed line is the best fit to the data, assuming an Arrhenius function, while the solid line is the best fit to the data, assuming that $Mu_T^- + h^+ \to Mu_T^0$. (c) The rate for $Mu_T^0 + e^- \to Mu_T^-$ and the corresponding fit.

electron concentration is nearly constant, the hole concentration $n_p$ is proportional to $\exp[-E_{gap}/k_B T]$. Fitting $\lambda_{-0}$, assuming the functional form $\sigma_p n_p v_p$, yields a cross-section $\sigma_p$ of $2.6(2) \times 10^3$ Å$^2$. This value is much larger than the capture cross-section of $Mu_T^0$ because of the Coulomb attraction between the hole and $Mu_T^-$. Consequently, $Mu_T^-$ acts as a recombination center since the charge cycle involves alternate trapping of electrons and holes: $Mu_T^- + h^+ \to Mu_T^0$ and $Mu_T^0 + e^- \to Mu_T^-$. (This is in contradistinction to the behaviour of muonium in Si at high temperatures; i.e., Section III.4.c). Such electron-hole recombination is commonly observed for deep-level defects in semiconductors and would imply that the $Mu_T(0/-)$ level is not shallow. Since the mass difference between the muon and a proton should not alter cross-sections by much, we expect that H in GaAs will also function as a recombination center, and the corresponding level is deep, in disagreement with some interpretations of the hydrogen data (Chevallier *et al.*, 1992).

Finally, we note that $Mu_T^-$ diffuses relatively slowly at these temperatures, as discussed in Section III.3.b (see also Fig. 15). Although muonium spends much more time as the charged species at high temperatures (i.e., $1/\lambda_{0-} \ll 1/\lambda_{-0}$), extrapolation of the diffusion data on $Mu_T^0$ below room temperature (see Section III.4.a and Fig. 16) implies that the diffusion is likely to be dominated by the neutral center. By analogy, similar dynamics should also occur for H. Although a quantitative statement about H diffusion rates based on muonium is not possible, one is led to the possibility that the diffusion of hydrogen in n-type GaAs at high temperatures is controlled by the transient $H^0$ species present during charge cycling.

*e. Spin-Exchange Scattering in n-Type Silicon*

Contrast Fig. 19 with Fig. 21, which shows the field-dependence of the $1/T_1$ rates in n-type $10^{14}$ cm$^{-3}$ Si:P with the magnetic field applied parallel to a $\langle 111 \rangle$ crystal direction (Chow *et al.*, 1994a). The pronounced peak is evidence that the $Mu_{BC}^0$ center is responsible for the $1/T_1$ relaxation and is undergoing either spin-exchange scattering or cyclic charge-state changes (see Section III.4.b). The location of the peak at 0.32 T is in excellent agreement with the theoretically expected value [using $A_\parallel = -16.82$ MHz, $A_\perp = -92.59$ MHz and $\theta = 70.53°$ in the expression for $B_\parallel$ as given by Eq. (18)]. At the temperatures at which these data were taken, ionization of $Mu_{BC}^0$ is not significant, but the electron concentration is high. Hence, spin-exchange scattering dominates. Fits to the data shown in Fig. 21 are obtained by using a phenomenological model developed by Nosov and Yakovleva (1963) to describe the spin dynamics. From the fitted spin-exchange rates, an average cross-section of 44(5) Å$^2$ is obtained. These results demonstrate that the $Mu_{BC}^0$ state is metastable in n-type silicon below 100 K. The resistance to forming a negative ion implies that there is a substantial energy barrier for the electron capture/site change reaction $Mu_{BC}^0 + e^- \rightarrow Mu^-$. Similar peaks have been observed in more heavily doped n-type Si and also in n-type GaAs at low temperatures. In contrast, in p-type Si, hole capture occurs rapidly and $Mu_{BC}^0$ is converted to $Mu^+$; see Chow *et al.* (1994b) and Estle and Lichti (1996) for a more complete discussion.

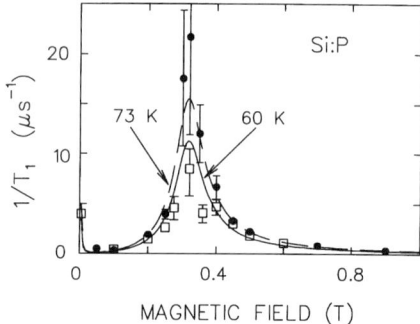

FIG. 21. Field-dependence of the $1/T_1$ rates at 60 K (open squares) and 73 K (closed circles) in $\approx 10^{14}$ cm$^{-3}$ Si:P. The solid [$v_{SE} = 11.2(5)$ $\mu$s$^{-1}$] and dashed [$v_{SE} = 15.4(5)$ $\mu$s$^{-1}$] lines are the best fits to the data using the Nosov-Yakovleva theory for 60 K and 73 K, respectively. The relaxation is assumed to be due to Mu$^0_{BC}$ in Si with $\theta = 70.53°$.

The detection of Mu$^0_{BC}$ in n-type materials suggests that H$^0_{BC}$ may also exist as a metastable (in the sense that Mu$^-$ or H$^-$ is expected to be the equilibrium state in n-type materials) species in these samples. Recall from Section I that there have been only two spectroscopic observations of isolated hydrogen in a semiconductor, both via EPR (Gorelkinskii and Nevinnyi, 1987, 1991; Bech Nielsen et al., 1994). These experiments, which show that the hydrogen center detected is H$^0_{BC}$, were both performed in high-resistivity silicon. If H$^0_{BC}$ exists in n-type Si, it would undoubtedly be undergoing rapid spin-exchange scattering with conduction electrons and therefore would be difficult to identify with EPR.

### f. Repolarization of Muonium

Finally, we mention briefly the "repolarization" technique. The experimental arrangement is that of LF-$\mu$SR, and the quantity of interest is the apparent initial amplitude $\langle P_z(0) \rangle$ of the LF-$\mu$SR signal as a function of magnetic field. Such experiments are most useful when dynamics are *not* important, and yet the conventional methods for identifying muonium do not work well.

In general, the function $P_z(t)$, describing the various muonium centers, consists of a constant plus oscillating terms (see Section II). The apparent initial amplitude $\langle P_z(0) \rangle$ is different from $P_z(0)$ if there are frequencies (or relaxations) that are too fast to observe experimentally and hence appear as a "missing fraction." Note that the name *repolarization* comes from the fact that for any muonium center, $P_z(t) = 1$ (and hence $\langle P_z(0) \rangle = 1$) at high magnetic fields. A simple example illustrates the technique. Consider free muonium with an isotropic hyperfine parameter. $P_z(t)$ is given by Eq. (15). In most experiments, the timing resolution is usually not good enough to enable observation of the very rapidly oscillating component proportional to $\cos(\omega_{24}t)$. Hence, this component is averaged to zero, and the field dependence of $\langle P_z(0) \rangle$ is given by the first term in Eq. (15), which increases with field monotonically. This should be contrasted with $\langle P_z(0) \rangle$ for the $Mu_{BC}^0$ center, which shows more complicated behavior because of the highly anisotropic hyperfine interaction (Patterson, 1988; Cooke et al., 1994a; Section II.3). Even in the case of polycrystalline samples, distinct features can still be present in the field dependence of $\langle P_z(0) \rangle$ for $Mu_{BC}^0$. The application of the repolarization technique to polycrystalline Si is discussed by Meier (1994) and Cooke et al. (1994a).

The repolarization technique has several potential applications. First, it can be used at pulsed muon facilities to identify muonium states and estimate hyperfine parameters in situations in which the TF-$\mu$SR frequencies usually associated with paramagnetic centers are too fast to be observable (see Section III.6). Second, in principle, the field-dependence of $\langle P_z(0) \rangle$ can still be used to estimate the average muon and nuclear hyperfine interactions in situations in which there is a distribution of hyperfine parameters that broadens conventional TF-$\mu$SR signals beyond detection. For example, one would naturally expect such broad lines in disordered materials such as amorphous Si.

5. Muon Spin Resonance in an RF Magnetic Field (RF-$\mu$SR)

The arrangement of the counters for RF-$\mu$SR experiments is again that of the standard LF-$\mu$SR geometry (i.e., Fig. 3b, in which the main magnetic field direction, counter axis, and initial spin polarization are collinear along $\hat{z}$). In addition, a small RF field oscillating at angular frequency, $\omega_{osc}$, is applied perpendicular to the main field **B**. The resonance condition (i.e., when the RF $\omega_{osc}$ equals (one of) the muonium precession frequencies in the applied magnetic field) can be achieved experimentally by varying either the main field at a fixed RF or the RF at a fixed **B**.

Consider the diamagnetic state. Since the muon spin is initially parallel to the applied magnetic field, the muon spin polarization is retained (i.e., $P_z(t) = 1$), except near a resonance. The resonance condition is simply $\omega_{osc} = 2\pi\bar{\gamma}_\mu B$. On resonance, the muon spin polarization precesses at a frequency determined by the magnitude of the RF field. Hence, as in $\mu$LCR, the time-integrated polarization will be reduced from unity. Similarly, for the neutral states in longitudinal applied field, the amplitude of the nonoscillating component of the polarization ($a_\parallel$) is preserved (i.e., $P_z(t) = a_\parallel$), unless $\omega_{\text{osc}}$ matches one of the hyperfine transitions of muonium $\varepsilon_i - \varepsilon_j$ (see Section II). A mathematical treatment of the principles of RF-$\mu$SR is given in Section III.5.a.

The resonance condition implies that one can selectively study one muonium center at a time. This is unlike the time-differential TF-$\mu$SR or LF-$\mu$SR techniques, which detect signals from all muonium states simultaneously. Although the technique yields similar information to time-differential $\mu$SR with regard to electronic structure, it has an important advantage if there is a transition occurring between two muonium states. In particular, in time-differential $\mu$SR, the final state can be observed only if the transition rate is much faster than the frequency difference between the two states. In RF-$\mu$SR, which is not phase-sensitive, the final state can be observed if the transition rate is on the order of $v_\mu = 1/\tau_\mu$.

In principle, there are many possible transitions between the four muonium states, $\text{Mu}_T^0$, $\text{Mu}_{BC}^0$, $\text{Mu}^+$, and $\text{Mu}^-$, in semiconductors. Information about the interconversions between the various centers can be obtained by studying the temperature and doping dependences of the RF-$\mu$SR amplitudes of the muonium states. This is illustrated in Section III.5.b for the diamagnetic muonium center in silicon.

### a. Principles of RF-$\mu$SR: A More Mathematical Approach

Since we will be concerned with the study of the singly charged muonium state in Section III.5.b, we present a mathematical treatment of the principles of RF-$\mu$SR that emphasizes the diamagnetic centers. The extension of this treatment to paramagnetic centers is discussed in the paper by Kreitzman et al. (1995), which also describes the experimental technique in more detail.

As in conventional magnetic resonance (Slichter, 1989), it is convenient to treat the effect of the oscillating RF magnetic field on a spin system by breaking it into two counter-rotating components, each of magnitude $B_1$. Only the component rotating in the same sense as the muon is retained. The time-dependence of $B_1$ can then be eliminated by transformation into a frame rotating at the oscillator frequency $\omega_{\text{osc}}$ around the $\hat{z}$-axis, the direction of the main magnetic field. Denoting $\hat{x}_R$, $\hat{y}_R$, and $\hat{z}$ as the unit vectors in the rotating reference frame (RRF), the effective magnetic field in the RRF is then given by $\mathbf{B}_{\text{eff}} = [(\omega_0 - \omega_{\text{osc}})/2\pi\tilde{\gamma}_\mu]\hat{z} + [\omega_1/2\pi\tilde{\gamma}_\mu]\hat{\rho}$. The frequency $\omega_1$ is $2\pi\tilde{\gamma}_\mu B_1$, and $\omega_o$ is the precession frequency of the muon in the applied magnetic field **B** (i.e., $2\pi\tilde{\gamma}_\mu B$). Note that $\mathbf{B}_{\text{eff}}$ is composed of a modified $\hat{z}$ component and another component along $\hat{\rho}$ ($\hat{\rho} = \hat{x}_R \cos\phi + \hat{y}_R \sin\phi$) in the ($\hat{x}_R, \hat{y}_R$)-plane of the RRF.

Assuming unit initial polarization along $\hat{z}$, the time evolution of the $\hat{z}$ component of the polarization with the RF field on (+) is given by

$$P_z^+(t) = \cos^2\theta + \sin^2\theta \cos\omega_{\text{eff}} t \tag{38}$$

where $\omega_{\text{eff}} = \sqrt{(\omega_0 - \omega_{\text{osc}})^2 + \omega_1^2}$ is the precession frequency in the effective field, and $\cot\theta = (\omega_0 - \omega_{\text{osc}})/\omega_1$.

The quantity of interest in an RF-$\mu$SR experiment is the time-integrated muon decay asymmetry $A_{\text{rf}}$. This quantity can be defined as follows:

$$A_{\text{rf}} = \frac{N_F^+ - N_F^-}{N_F^+ + N_F^-} - \frac{N_B^+ - N_B^-}{N_B^+ + N_B^-} \tag{39}$$

where $N_F$ and $N_B$ are the integral number of counts in the forward and backward counters, and the superscripts designate whether the RF is on (+) or off (−) during data collection. The RF field is cycled between RF on to off (and vice versa) every few seconds. Combining the data in this manner serves to isolate the integral change in the LF asymmetry due to RF irradiation and also reduces unwanted systematic variations in the signal. The integral counts in Eq. (39) are given by

$$N_{F/B}^\pm = N_{F/B}^0 \int_0^\infty v_\mu e^{-v_\mu t}[1 \mp A_{F/B} P_z^\pm(t)] dt \tag{40}$$

where the subscript F(B) refers to the forward(backward) counter. $N_{F/B}^0$ is the initial positron flux and reflects the beam intensity and geometry of the counters, $v_\mu$ is the inverse muon life time, and $A_{F/B}$ is the asymmetry of the relevant counter. The background term is not included since it will be subtracted out in Eq. (39).

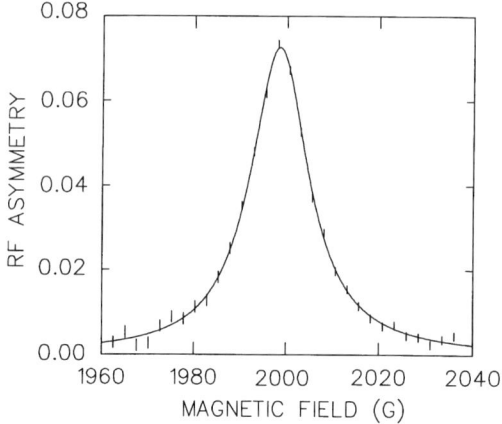

FIG. 22. Example of the diamagnetic RF-$\mu$SR resonance observed in Si (P11) at 310 K. The signal was measured at a fixed oscillator frequency of 27.1 MHz by varying the external magnetic field.

The total number of counts in the forward and backward counters in Eq. (40) are calculated with the RF on ($P_z(t) = P_z^+(t)$), and when the RF is off ($P_z(t) = P_z^-(t) = 1$) and then substituted in Eq. (39), which yields a Lorentzian lineshape

$$A_{rf} = A_{rf}(0) \frac{\delta\omega^2}{(\omega_o - \omega_{osc})^2 + \delta\omega^2} \qquad (41)$$

Here the on-resonance ($\omega_0 = \omega_{osc}$) amplitude is given by $A_{rf}(0) = \omega_1^2/(v_\mu^2 + \omega_1^2)$ and the half-width at half maximum is equal to $\delta\omega = \sqrt{v_\mu^2 + \omega_1^2}$.

Typical data are shown in Fig. 22 for a nearly intrinsic Si sample (P11) at 310 K. This resonance is due to the diamagnetic signal. The lineshape is clearly a Lorentzian and is fit to Eq. (41) to yield the on-resonance amplitude, width, and resonance frequency. Additional analysis is then required to extract the normalized RF amplitude. For diamagnetic centers, one way of normalizing the signal is to compare it to the signal in a reference material such as $CaCO_3$ measured under the same conditions.

### b. *Interconversion Between the Various Muonium States in Silicon*

RF-$\mu$SR has been used primarily to study interconversion of muonium centers in Si. The influence of dopants and crystal orientations have been

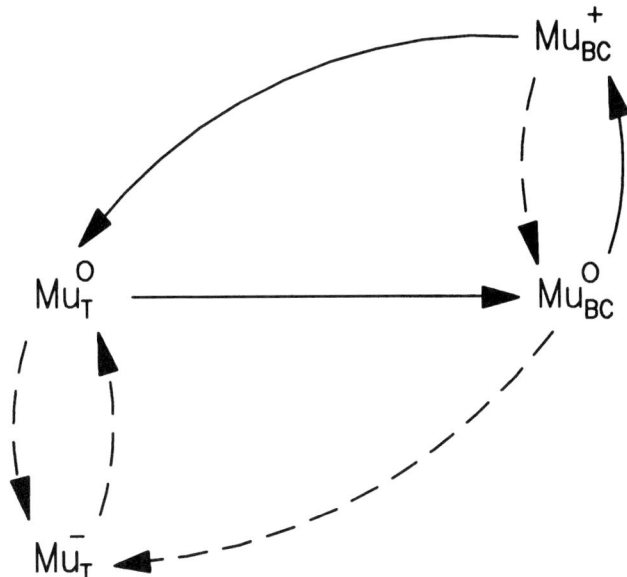

FIG. 23. The model of muonium dynamics used by Kreitzman et al. (1995) to explain the RF-$\mu$SR data. The meanings of the solid and dashed lines are discussed in more detail in the text.

investigated (Kreitzman et al., 1995; Scheuermann et al., 1995; Hitti et al., 1996). Here, we discuss primarily the results of Kreitzman et al. (1995). Based on the RF results, earlier TF data, and theoretical calculations, the authors developed a model that describes muonium dynamics in Si. From fitting the RF data to a strong collision model (Kreitzman et al., 1995), energy levels within the bandgap for some of the centers are extracted, and rate parameters (prefactors, energy barriers, and cross-sections), which govern interconversion between the centers, were estimated.

The transitions that are important within the model are summarized in Fig. 23. The solid lines represent transitions active in p-type samples, while the broken lines are for reactions that are *also* important in n-type samples. Some of these transitions involve charge changes, some involve site changes, and others involve both charge and site changes. The relative rates for competing transitions, which depend on the type and concentration of dopants, determine which of the transitions are important. We illustrate how such information can be obtained by using examples from slightly p-type and n-type Si samples. Very similar data for Si have been obtained by others (Scheuermann et al., 1995).

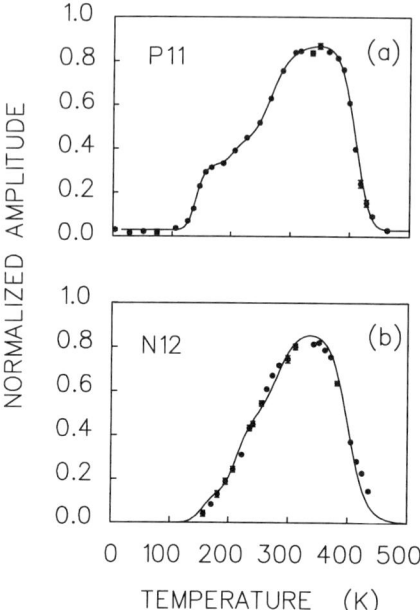

FIG. 24. The temperature-dependence of the normalized RF amplitude for diamagnetic muonium. (a) P11 is a well-compensated, slightly p-type Si sample, and (b) N12 is a low-doped n-type Si sample. The solid lines are the result of fitting to the model described by Kreitzman et al. (1995) and Hitti et al. (1996).

P11 is a well-compensated slightly p-type Si sample first discussed in Section III.4.c. At low temperature, the diamagnetic fraction is essentially zero, and approximately equal fractions of $Mu_{BC}^0$ and $Mu_T^0$ are present. As shown in Fig. 24, as the temperature is increased, a sharp rise in the diamagnetic amplitude occurs at about 140 K. This corresponds to the thermal ionization of $Mu_{BC}^0$ ($Mu_{BC}^0 \rightarrow Mu_{BC}^+ + e^-$: $E_{BC}^{0/+} = 0.21(1)$ eV). Since $Mu_{BC}^0$ has a highly anisotropic hyperfine interaction, the polarization transferred to the ionized product state has a complicated temperature-dependence. This anisotropy is also responsible for the small bump observed just above 200 K. The gradual increase in the amplitude of the diamagnetic signal centered around 270 K corresponds to the ionization of $Mu_T^0$. Since the positive charge state is expected to be located at the BC site, this process is thought to involve both site and charge changes. This process was modeled by assuming the process $Mu_T^0 \rightarrow Mu_{BC}^0 \rightarrow Mu_{BC}^+ + e^-$, involving first a site change of the neutral state followed by rapid ionization ($E_{T/BC}^0 =$

0.39(4) eV). At room temperature, most of the muons end up in the diamagnetic state $Mu_{BC}^+$. Above room temperature, the alternate ionization and capture of conduction electrons (i.e., an example of a charge exchange process described already in Section III.4.c) results in the loss of muon polarization. Parameters characterizing the previously described relevant processes can be obtained by fitting the steps at 270 K and 140 K. The ionization rates for $Mu_T^0$ and $Mu_{BC}^0$ can be extrapolated to higher temperatures, where the diamagnetic amplitude undergoes a sharp drop. It is found that the corresponding extrapolated rate for ionization of $Mu_{BC}^0$ is much faster than its precession frequencies; hence, little polarization should be lost (see Fig. 17 for an analogy). On the other hand, the extrapolated rates for $Mu_T^0$ are such that it is comparable to its hyperfine frequency, and hence significant loss of polarization will occur. Consequently, the sharp drop observed in the data above room temperature can be reproduced by the model only if the neutral center $Mu_T^0$ is involved in the charge cycle ($Mu_{BC}^0 + e^- \to Mu_T^0$: followed by $Mu_T^0$ ionization). This is also seen from the magnetic-field dependence of the $1/T_1$ rates at high temperatures, as discussed in Section III.4.c.

While the ionization of $Mu_{BC}^0$ near 140 K in the p-type samples is irreversible, the reverse reaction is possible in n-type samples because of the increased concentration of conduction electrons. This is illustrated in the sample called N12, a $\langle 111 \rangle$ Si:P sample with net donor concentration $N_D$ of $2.4 \times 10^{12}$ cm$^{-3}$. As shown in Fig. 24b, the rise in the diamagnetic amplitude is shifted to higher temperature compared to the P11 sample. The solid line in Fig. 24 was calculated by adding an extra reaction: electron capture at the BC site ($Mu_{BC}^0 \leftrightarrow Mu_{BC}^+ + e^-$: capture cross-section $\sigma_{BC}^{+/0} = 3300$ Å$^2$). Specifically, after $Mu_{BC}^0$ ionizes, it can capture an electron with either the same spin or the opposite spin as its pre-ionized electron. The effective magnetic field in which the muon precesses is different if the spin of the bound electron is parallel (spin-up) or antiparallel (spin-down) to the external field, as discussed in Section II.3. Therefore, the cyclic reaction at the BC site results in the loss of the diamagnetic amplitude until the ionization rate of $Mu_{BC}^0$ becomes rapid compared to its precession frequencies. In more heavily doped n-type samples, experimental evidence exists for the formation of $Mu_T^-$ and spin-exchange scattering (see Section III.4). Because of the increased capture rates compared to the ionization rates in the moderate to heavily doped ($N_D \geqslant 10^{15}$ cm$^{-3}$) n-type samples, it is reasonable to expect that the diamagnetic state is predominantly $Mu_T^-$ ($Mu_{BC}^0$ or $Mu_T^0 + e^- \to Mu_T^-$: $\sigma_T^{0/-} = 10$ Å$^2$). On the other hand, in samples with intermediate doping levels ($10^{15}$ cm$^{-3} \geqslant N_D \geqslant 10^{13}$ cm$^{-3}$), the low-temperature diamagnetic state is predominantly $Mu_T^-$, while at high temperatures, it is mostly $Mu_{BC}^+$. Spin-exchange scattering has not yet been included in the dynamical model described by Kreitzman et al. (1995). However, preliminary calculations indicate that it is responsible for the reduced amplitude observed in the intermediate doped n-type samples. Scheuermann et al. (1995) have treated spin-exchange processes as an integral part of their analysis and claim that such dynamics were essential for providing a quantitative description of the data in n-type Si.

The preceding dynamical model is expected to be at least qualitatively correct for hydrogen in Si. Comparison of the parameters for muonium transitions with the few analogous parameters for hydrogen show good agreement when expected differences related to zero-point energies are taken into account (see Table VI).

RF-$\mu$SR is an important complementary technique that is especially useful when there are interconversions between the various muonium centers. These RF-$\mu$SR results in silicon together with LF-$\mu$SR (Section III.4) have clearly demonstrated that muonium cycles rapidly (on the timescale of few microseconds, which is short compared to the measurement time of most experiments on hydrogen) between its various charge states and sites at elevated temperatures. These results suggest that atomic hydrogen will behave in a similar manner.

TABLE VI

COMPARISON OF THE DYNAMICAL PARAMETERS OBTAINED FROM RF-$\mu$SR WITH THE FEW EXISTING DATA FOR ISOLATED HYDROGEN IN Si

| Parameter | $\mu$SR | Hydrogen |
|---|---|---|
| $E_{BC}^{0/+}$ | $0.21 \pm 0.01$ eV | $0.164 \pm 0.011$ eV DLTS |
| $E_{T/BC}^{0}$ | $0.39 \pm 0.04$ eV | |
| $E_{BC/T}^{+/0}$ | $0.31 \pm 0.06$ eV | $0.44 \pm 0.01$ eV DLTS |
| | | $0.48 \pm 0.04$ eV EPR |
| $E_{BC/T}^{0/-}$ | $0.34 \pm 0.01$ eV | $0.293 \pm 0.003$ eV DLTS |
| $E_T^{-/0}$ | $\approx 0.56$ eV | |
| Prefactors | $\approx 10^{13}$ s$^{-1}$ | |
| $\sigma^{0/\pm}$ | $\approx 10$ Å$^2$ | |
| $\sigma^{+/0}$ | $\approx 3300$ Å$^2$ | |

In addition to the energy levels and activation energies, typical values of the prefactors and capture cross-sections ($\sigma$) are listed. In the notation used, the subscripts carry site information and the superscripts specify the charge state. For each transition, the initial state is listed first, followed by the final state in the appropriate subscript or superscript. All the hydrogen data involve $H_{BC}^0$, which is the only isolated H impurity directly observed and unambiguously identified. Most of the hydrogen results come from deep-level transient spectroscopy (DLTS) (Holm et al., 1991). The exception is an EPR study (Gorelkinskii and Nevinnyi, 1987, 1991).

## 6. COMPARISON OF TECHNIQUES AND FACILITIES

As was mentioned in Section II.1, $\mu$SR experiments can be carried out at several facilities worldwide. Two of these, TRIUMF and PSI, provide continuous muon beams, while another two, ISIS and KEK, are pulsed muon sources. Recall that at continuous facilities, the muons arrive essentially at random, while at pulsed facilities, muons arrive in short pulses. At continuous facilities, there are large differences in the data collection rate between a time-differential and an integral experiment. In the integral mode, the one-muon one-positron restriction is removed, and as high an incoming rate as possible is used. On the other hand, at pulsed facilities, no such restriction exists, and all the data collection is inherently time differential without any rate limitation. The differences in the previously discussed time structure of the muon beams imply that certain types of experiments are better suited for one type of facility or the other.

Experiments that involve studying fast oscillations or fast relaxation rates can be carried out only at continuous muon facilities. At pulsed facilities, the finite time width of each pulse of muons, on the order of several tens of nanoseconds, means that the timing resolution is severely curtailed. Presently, signals with precessions or relaxation rates greater than $\approx 5$ MHz or $\approx 5\,\mu s^{-1}$, respectively, are not resolvable at pulsed facilities.[13] Furthermore, at pulsed muon sources, it is possible to implant "too many" muons into the sample. If two or more decay positrons arrive at a counter too close together in time, they will not be recognizable as separate entities. This will produce distortions in the data at early times. In comparison, the limitations at TRIUMF or PSI are due primarily to the time resolution of the detectors, typically $\approx 1-2$ ns. Currently, frequencies up to $\approx 250$ MHz and relaxation rates up to $\approx 30\,\mu s^{-1}$ are easily resolvable. By using detectors that are specially designed for timing resolution, frequencies as high as 4.5 GHz can be resolved (Holzschuh, 1983; Kiefl et al., 1984). In addition, the ability to correlate one decay positron with its parent muon implies that certain technical advances, such as a "low-background" apparatus, are possible. With such an apparatus, one can discriminate against all muons that do not stop in the sample of interest, making it possible to measure very small samples that cannot otherwise be studied with $\mu$SR. The low-background apparatus was first used for studying the small $^{13}$C-enriched diamond sample discussed in Section III.1.b (see Schneider et al., 1993b). In principle, $\mu$LCR experiments can be performed at both types of facilities. However, for small resonances, it is important to have stable beam conditions as well as

---

[13]One way to get around this problem is being explored at ISIS. A short-lived transverse magnetic field is applied *after* all the muons are implanted into the sample (which rotates the muon spins by 90°). This bypasses the problem of timing resolution due to the finite width of the incoming beam and allows one to observe much higher TF frequencies.

high muon rates. Consequently, the majority of $\mu$LCR experiments have been performed at continuous beam facilities, where the counting rate is considerably higher.

On the other hand, pulsed sources are better for measuring very small relaxation rates because data can be measured to long times as a result of the very low number of random background events. This is a consequence of the "duty cycle" of the muon beam at pulsed facilities. For example, at ISIS, data collection only takes place for up to $\approx 30\,\mu$s every 20 ms. Reliable data can be taken to $\approx 16\,\mu$s as compared to $\approx 8-10\,\mu$s at continuous sources. Also, for RF-$\mu$SR experiments, working with pulsed beams has some advantages. At continuous muon facilities, the incoming count rates are maximized using an integral mode. Hence, relevant time information is normally lost. Furthermore, investigating a diamagnetic signal requires the generation of strong RF fields and power levels of up to several hundred watts (see Kreitzman *et al.*, 1995). At TRIUMF, the RF duty cycle can be as high as 50%, which can result in considerable RF heating of the sample. At pulsed sources, by timing the RF field to be coincident with the incoming muon pulse, a much lower RF duty cycle can be achieved.

A series of illumination experiments (Kadono *et al.*, 1994b) were carried out at KEK to investigate transitions in Si and Ge and involve carrier capture via athermal generation of charge carriers. The ability to shift the time of the light pulse with respect to the incoming muon pulse also enabled the carrier lifetimes to be monitored. Since the light source was synchronized with the incoming pulsed muon beam, the average heat influx on the samples was minimal. One of the conclusions of this study is that in Si at low temperatures, the interactions of $Mu_T^0$ with photoinduced carriers result in a rapid conversion to $Mu_{BC}^0$, confirming that the latter is the stable state. This result is in line with studies of the AA9 hydrogen atom in Si, where it was found that atomic hydrogen at the BC was observed only after illumination at lower temperatures ($<200$ K) (Gorelkinskii and Nevinnyi, 1987, 1991). The neutral centers are also found to undergo spin- and/or charge-exchange interactions with the photoexcited carriers.

## IV. Summary

In this chapter, we have discussed a group of complementary $\mu$SR techniques that can be applied to the study of muonium in semiconductors. TF-$\mu$SR is often the first method employed to identify muonium in a semiconductor and to characterize the electronic structue. The spectrum of muon spin precession frequencies in an external magnetic field is the most direct method to measure the muon hyperfine interaction in the case of paramagnetic centers and to confirm the existence of charged diamagnetic muonium centers. TF-$\mu$SR also provides key information on the interactions with neighboring nuclear spins: nuclear hyperfine interactions in the case of paramagnetic centers or nuclear dipolar interactions in the case of diamagnetic centers. A very powerful complementary method, called $\mu$LCR, can be used to obtain more detailed structural information than is possible with TF-$\mu$SR alone. $\mu$LCR is a form of double resonance that yields information similar to that obtained from electron-nuclear double-resonance (ENDOR) on conventional paramagnetic centers. The most important recent developments in the field of muonium in semiconductors have been on the diamagnetic charge states of muonium and on muonium dynamics. The latter category includes diffusion and interaction of muonium with free carriers (i.e., charge exchange and spin exchange). Muonium dynamics are most easily studied using $\mu$SR in ZF-$\mu$SR or in LF-$\mu$SR or RF-$\mu$SR.

Throughout this chapter we have emphasized that there are many similarities between muonium and hydrogen, but also some important differences. In particular, the lighter mass of the muon will enhance the zero-point energy of muonium relative to hydrogen. Provided these zero-point energy differences make only minor contributions to the structure, as all experimental evidence indicates, then hydrogenic analogs of many of the experimentally observed muonium centers are expected to exist and to have closely corresponding electronic structures. This is true for the one case in which comparison between isolated atomic hydrogen (AA9 center in Si) and muonium ($\text{Mu}_{BC}^0$) are possible. We anticipate that experimental studies of the electronic structure of muonium in its three charge states will continue to provide the best experimental information of the structure of isolated hydrogen.

Recent work on muonium has also revealed a rich and complicated dynamics, particularly at elevated temperatures, at which hydrogen is typically introduced into semiconductors. In a few cases (Si and GaAs), there is now a good understanding of the basic processes, such as diffusion, spin-exchange scattering, and charge- and site-changing reactions. With the exception of diffusion, we expect a very close correspondence between these dynamical processes for muonium and hydrogen. While these processes may not be directly observable in the case of hydrogen, due to the short timescale involved, they will strongly influence any time-averaged quantities of hydrogen and must be taken into account. These experiments show that any theory for hydrogen diffusion at high temperatures based on a single diffusing species is clearly oversimplified.

One of the most important aspects of hydrogen in semiconductors is the formation of hydrogen complexes and passivation of electrical activity. So far, attempts to detect the muonium analog of the passivation complexes have been unsuccessful (e.g., Maric et al., 1991; Cooke et al., 1994b), although recently, promising possibilities have been reported in some heavily doped III–V semiconductors (Chow, 1996b; Chow et al., 1996c; Lichti et al., 1996). The basic problem is the timescale for muonium experiments (only a few muon lifetimes), which is generally short compared to the time thought to be required for muonium to form a complex. $\mu$SR is ideally suited to studying muonium in an isolated form, whereas conventional methods are better at studying hydrogen complexes. Consequently, the most important impact of $\mu$SR experiments will likely continue to be on muonium in its isolated form.

In conclusion, the collection of $\mu$SR techniques has provided important information regarding the rich physics of muonium in semiconductors. $\mu$SR studies of these important materials are the primary source of information on the electronic structure and dynamics of isolated hydrogen in semiconductors, much of which cannot be obtained by other methods.

### Acknowledgments

We would like to thank the large team of scientist who have made research into this field possible, and especially to T. L. Estle, S. R. Kreitzman, and R. L. Lichti for their valuable contributions over the years. This work was partially supported by the National Sciences and Engineering Research Council of Canada.

### References

Adams, T. R., Roberson, M. A., and Lichti, R. L. (1995). *Phil. Mag. B* **72**, 183.
Bech Nielsen, B., Bonde Nielsen, K., and Byberg, J. R. (1994). *Mat. Sci. Forum* **143–147**, 909.
Blazey, K. W., Estle, T. L., Holzschuh, E., Odermatt, W., and Patterson, B. D. (1983). *Phys. Rev. B* **27**, 15.
Blazey, K. W., Estle, T. L., Holzschuh, E., Meier, P. F., Patterson, B. D., and Richner, M. (1986). *Phys. Rev. B* **33**, 1546.
Buda, F., Chiarotti, G. L., Car, R., and Parrinello, M. (1989). *Phys. Rev. Lett.* **63**, 294.
Celio, M. (1987). *Helv. Phys. Acta* **60**, 600.
Chevallier, J., Dautremont-Smith, W. C., Tu, C. W., and Pearton, S. J. (1985). *Appl. Phys. Lett.* **47**, 108.
Chevallier, J., Machayekhi, B., Grattepain, C. M., Rahbi, R., and Theys, B. (1992). *Phys. Rev. B* **45**, 8803.
Cho, H. Y., Min, S.-K., Chang, K. J., and Lee, C. (1991). *Phys. Rev. B* **44**, 13779.

Chow, K. H., Kiefl, R. K., Schneider, J. W., Hitti, B., Estle, T. L., Lichti, R. L., Schwab, C., DuVarney, R. C., Kreitzman, S. R., MacFarlane, W. A., and Senba, M. (1993). *Phys. Rev. B* **47**, 16004.

Chow, K. H., Lichti, R. L., Kiefl, R. F., Dunsiger, S., Estle, T. L., Hitti, B., Kadono, R., MacFarlane, W. A., Schneider, J. W., Schumann, D., and Shelley, M. (1994a). *Phys. Rev. B* **50**, 8918.

Chow, K. H., Kiefl, R. F., Schneider, J. W., Estle, T. L., Hitti, B., Lichti, R. L., Schwab, C., Kreitzman, S. R., DuVarney, R. C., Senba, M., Sonier, J., Johnston, T. M. S., and MacFarlane, W. A. (1994b). *Hyperfine Int.* **86**, 693.

Chow, K. H., Kiefl, R. F., MacFarlane, W. A., Schneider, J. W., Cooke, D. W., Leon, M., Paciotti, M., Estle, T. L., Hitti, B., Lichti, R. L., Cox, S. F. J., Schwab, C., Davis, E. A., Morrobel-Sosa, A., and Zavieh, L. (1995). *Phys. Rev. B* **51**, 14762.

Chow, K. H., Hitti, B., Kiefl, R. F., Dunsiger, S. R., Lichti, R. L., and Estle, T. L. (1996a). *Phys. Rev. Lett.* **76**, 3790.

Chow, K. H. (1996b). *Hyperfine Int.* **105**, 285.

Chow, K. H., Cox, S. F. J., Davis, E. A., Dunsiger, S. R., Estle, T. L., Hitti, B., Kiefl, R. F., and Lichti, R. L. (1996c). *Hyperfine Int.* **105**, 309.

Chu, C. H. and Estreicher, S. K. (1990). *Phys. Rev. B* **42**, 9486.

Cohen-Tannoudji, C., Diu, B., and Laloe F. (1977). *Quantum Mechanics*. Herman and John Wiley and Sons, New York.

Cooke, D. W., Leon, M., Paciotti, M. A., Meier, P. F., Cox, S. F. J., Davis, E. A., Estle, T. L., Hitti, B., Lichti, R. L., Boekema, C., Lam, J., Morrobel-Sosa, A., and Oostens, J. (1994a). *Phys. Rev. B.* **50**, 4391.

Cooke, D. W., Leon, M., Paciotti, M. A., Bennett, B. L., Rivera, O. M., Cox, S. F. J., Boekema, C., Lam, J., Morrobel-Sosa, A., Meier, P. F., Estle, T. L., Hitti, B., Lichti, R. L., Davis, E. A., Oostens, J., and Haller, E. E. (1994b). *Hyperfine Int.* **86**, 639.

Estle, T. L. and Lichti, R. L. (1996). *Hyperfine Int.* **97–98**, 171.

Estreicher, S. K., Chu, C. H., and Marynick, D. S. (1989). *Phys. Rev. B* **40**, 5739.

Fujara, F., Stöckman, H.-J., Ackermann, H., Buttler, W., Dörr, K., Grupp, H., Heitjans, P., Kiese, G., and Körblein, A. (1980) . *Z. Physik B* **37**, 151.

Gorelkinskii, Y. V. and Nevinnyi, N. N. (1987). *Sov. Tech. Phys. Lett.* **13**, 45.

Gorelkinskii, Y. V. and Nevinnyi, N. N. (1991). *Physica B* **170**, 155.

Hartmann, O. (1977). *Phys. Rev. Lett.* **39**, 832.

Hayano, R. S., Uemura, Y. J., Imazato, J., Nishida, N., Yamazaki, T., and Kubo, R. (1979). *Phys. Rev. B* **20**, 850.

Herrero, C. P. and Ramírez, R. (1995). *Phys. Rev. B* **51**, 16761.

Herring, C. and Johnson, N. M. (1991). *Hydrogen in Semiconductors*, ed. J. I. Pankove and N. M. Johnson. Academic Press, New York, Vol. 34, pp. 225–350.

Hitti, B., Kreitzman, S. R., Estle, T. L., Lichti, R. L., and Lightowlers, E. C. (1966). *Hyperfine Int.* **105**, 321.

Holbech, J. D., Bech Nielsen, B., Jones, R., Sitch, P., and Öberg, S. (1993). *Phys. Rev. Lett.* **71**, 875.

Holm, B., Bonde-Nielsen, K., and Bech-Nielsen, B. (1991). *Phys. Rev. Lett.* **66**, 2360.

Holzschuh, E. (1983). *Phys. Rev. B* **27**, 102.

Ivanter, I. G. and Smilga, V. P. (1971). *Sov. Phys. JETP* **33**, 1070.

Jäger, E., Ittermann, B., Sulzer, G., Bürkmann, K., Fischer, B., Frank, H.-P., Stöckmann, H.-J., and Ackermann, H. (1990). *Z. Physik B* **80**, 87.

Johnson, N. M., Burnham, R. D., Street, R. A., and Thornton, R. L. (1986). *Phys. Rev. B* **33**, 1102.

Johnson, N. M., Herring, C., and Van de Walle, C. G. (1994). *Phys. Rev. Lett.* **73**, 130.
Kadono, R., Kiefl, R. F., Brewer, J. H., Luke, G. M., Pfiz, T., Riseman, T. M., and Sternlib, B. J. (1990a). *Hyperfine Int.* **64**, 635.
Kadono, R. (1990b). *Hyperfine Int.* **64**, 615–634.
Kadono, R., Matsushita, A., Nagamine, K., Nishiyama, K., Chow, K. H., Kiefl, R. K., MacFarlane, A., Schumann, D., Fujii, S., and Tanigawa, S. (1994a). *Phys. Rev. B* **50**, 1999.
Kadono, R., Matsushita, A., Macrae, R. M., Nishiyama, K., and Nagamine, K. (1994b). *Phys. Rev. Lett.* **73**, 2724.
Kagan, Y. and Prokof'ev, N. V. (1990). *Phys. Lett. A* **150**, 320.
Kiefl, R. F., Holzschuh, E., Keller, H., Kündig, W., Meier, P. F., Patterson, B. D., Schneider, J. W., Blazey, K. W., Rudaz, S. L., and Denison, A. B. (1984). *Phys. Rev. Lett.* **53**, 90.
Kiefl, R. F., Schneider, J. W., Keller, H., Kündig, W., Odermatt, W., Patterson, B. D., Blazey, K. W., Estle, T. L., and Rudaz, S. L. (1985). *Phys. Rev. B* **32**, 530.
Kiefl, R. F., Celio, M., Estle, T. L., Luke, G. M., Kreitzman, S. R., Brewer, J. H., Noakes, D. R., Ansaldo, E. J., and Nishiyama, K. (1987). *Phys. Rev. Lett.* **58**, 1780.
Kiefl, R. F., Celio, M., Estle, T. L., Kreitzman, S. R., Luke, G. M., Riseman, T. M., and Ansaldo, E. J. (1988). *Phys. Rev. Lett.* **60**, 224.
Kiefl, R. F. and Estle, T. L. (1991). *Hydrogen in Semiconductors*, ed. J. I. Pankove and N. M. Johnson. Academic Press, New York, Vol. 34, pp. 547–584.
Kiefl, R. F. and Kreitzman, S. R. (1992). *Perspectives in Muon Science*, ed. T. Yamazaki, K. Nakai, and K. Nagamine. Elsevier, New York, p. 265.
Kreitzman, S. R., Hitti, B., Lichti, R. L., Estle, T. L., and Chow, K. H. (1995). *Phys. Rev. B* **51**, 13117.
Leitch, A. W. R., Prescha, T., and Weber, J. (1991). *Phys. Rev. B* **44**, 1375.
Leon, M. (1992). *Phys. Rev. B* **46**, 6603.
Lichti, R. L., Cox, S. F. J., Schwab, C., Estle, T. L., Hitti, B., and Chow, K. H. (1996). *Hyperfine Int.* **105**, 333.
MacFarlane, W. A., Kiefl, R. F., Schneider, J. W., Chow, K. H., Morris, G. D., Estle, T. L., and Hitti, B. (1994). *Hyperfine Int.* **85**, 23.
Maric, D. M., Vogel, S., Meier, P. F., and Estreicher, S. K. (1989). *Phys. Rev. B* **40**, 8545.
Maric, D. M., Meier, P. F., Vogel, S., Cox, S. F. J., Davis, E. A., and Schneider, J. W. (1991). *J. Phys.: Condens. Matter* **3**, 9675.
Meier, P. F. (1994). *Hyperfine Int.* **86**, 723.
Morrow, Richard A. (1993). *J. Appl. Phys.* **74**, 6174.
Morton, J. R. and Preston, K. F. (1978). *J. Magn. Res.* **30**, 577.
Myers, S. M., Baskes, M. I., Birnbaum, H. K., Corbett, J. W., DeLeo, G. G., Estreicher, S. K., Haller, E. E., Jena, P., Johnson, N. M., Kirchheim, R., Pearton, S. J., and Stavola, M. J. (1992). *Rev. Mod. Phys.* **64**, 559.
Nosov, V. G. and Yakovleva, I. V. (1963). *Sov. Phys. JETP* **16**, 1236.
Odermatt, W. (1988). *Helv. Phys. Acta* **61**, 1087.
Pankove, J. I. and Johnson, N. M. (1991). *Hydrogen in Semiconductors*, ed. J. I. Pankove and N. M. Johnson. Academic Press, New York, Vol. 34, pp. 1–15.
Patterson, B. D. (1988). *Rev. Mod. Phys.* **60**, 69.
Pavesi, L. and Giannozzi, P. (1992). *Phys. Rev. B* **46**, 4621.
Pearton, J., Corbett, W., and Stavola, M. (1992). *Hydrogen in Crystalline Semiconductors*. Springer-Verlag, Berlin, Heidelberg.
Ramirez, R. and Herrero, C. P. (1994). *Phys. Rev. Lett.* **73**, 126.
Roduner, E. and Fischer, H. (1981). *Chem. Phys.* **54**, 261.
Roos, G., Johnson, N. M., Herring, C., and Harris, J. S. (1991). *Appl. Phys. Lett.* **59**, 461.
Schenck, A. (1985). *Muon Spin Rotation Spectroscopy*. Adam Hilger, Bristol.

Scheuermann, R., Schimmele, L., Seeger, A., Stammler, T., Grund, T., Hampele, M., Herlach, D., Iwanowski, M., Major, J., Notter, M., and Pfiz, T. (1995). *Phil. Mag. B* **72**, 161.
Schneider, J. W., Celio, M., Keller, H., Kündig, W., Odermatt, W., Pümpin, B., Savić, I. M., Simmler, H., Estle, T. L., Schwab, C., Kiefl, R. F., and Renker, D. (1990a). *Phys. Rev. B* **41**, 7254.
Schneider, J. W., Keller, H., Odermatt, W., Pümpin, B., Savić, I. M., Simmler, H., Dodds, S. A., Estle, T. L., DuVarney, R. C., Chow, K., Kadono, R., Kiefl, R. F., Li, Q., Riseman, T. M., Zhou, H., Lichti, R. L., and Schwab, C. (1990b). *Hyperfine Int.* **64**, 543.
Schneider, J. W., Kiefl, R. F., Chow, K., Cox, S. F. J., Dodds, S. A., DuVarney, R. C., Estle, T. L., Kadono, R., Kreitzman, S. R., Lichti, R. L., and Schwab, C. (1992a). *Phys. Rev. Lett.* **68**, 3196.
Schneider, J. W., Kiefl, R. F., Ansaldo, E. J., Brewer, J. H., Chow, K., Cox, S. F. J., Dodds, S. A., DuVarney, R. C., Estle, T. L., Haller, E. E., Kadono, R., Kreitzman, S. R., Lichti, R. L., Niedermayer, V. C., Pfiz, T., Riseman, T. M., and Schwab, C. (1992b). *Mat. Sci. Forum* **83–87**, 569.
Schneider, J. W., Chow, K., Kiefl, R., Kreitzman, S. R., MacFarlane, A., DuVarney, R. C., Estle, T. L., Lichti, R. L., and Schwab, C. (1993a). *Phys. Rev. B* **47**, 10193.
Schneider, J. W., Kiefl, R. F., Chow, K. H., Johnston, J., Sonier, J., Estle, T. L., Hitti, B., Lichti, R. L., Connell, S. H., Sellschop, J. P. F., Smallman, C. G., Anthony, T. R., and Banholzer, W. F. (1993b). *Phys. Rev. Lett.* **71**, 557.
Senba, M. (1990a). *J. Phys. B* **23**, 1545.
Senba, M. (1990b). *J. Phys. B* **23**, 4051.
Senba, M. (1991). *J. Phys. B* **24**, 3531.
Slichter, C. P. (1989). *Principles of Magnetic Resonance.* Springer-Verlag, Berlin, Heidelberg, New York.
Stavola, M., Bergman, K., Pearton, S. J., and Lopata, J. (1988). *Phys. Rev. Lett.* **61**, 2786.
Van de Walle, C. G. (1991). *Hydrogen in Semiconductors*, ed. J. I. Pankove and N. M. Johnson. Academic Press, New York, Vol. 34, pp. 585–622.
Van de Walle, C. G. and Pavesi, L. (1993). *Phys. Rev. B* **47**, 4256.
Weissbluth, M. (1978). *Atoms and Molecules.* Academic Press, New York, p. 374.
Yen, H. K. (1988). M.Sc. Thesis, University of British Columbia, 1988.
Yuan, M. H., Wang, L. P., Jin, S. X., Chen, J. J., and G. G. Qin (1991). *Appl. Phys. Lett.* **58**, 925.

CHAPTER 5

# Positron Annihilation Spectroscopy of Defects in Semiconductors

### Kimmo Saarinen and Pekka Hautojärvi

LABORATORY OF PHYSICS
HELSINKI UNIVERSITY OF TECHNOLOGY
ESPOO, FINLAND

### Catherine Corbel

INSTITUT NATIONAL DES SCIENCES ET TECHNIQUES NUCLÉAIRES
CENTRE D'ETUDES NUCLÉAIRES DE SACLAY
GIF-SUR-YVETTE CEDEX, FRANCE

|  |  |
|---|---|
| I. INTRODUCTION | 210 |
| II. POSITRONS IN SOLIDS | 211 |
|     1. *Positrons for Bulk Studies* | 212 |
|     2. *Positrons for Layer Studies* | 212 |
|     3. *Positron Diffusion and Mobility* | 213 |
|     4. *Positron Wave Function and Positron States* | 215 |
|     5. *Annihilation Characteristics* | 216 |
| III. POSITRON TRAPPING | 218 |
|     1. *Trapping Rate and Trapping Coefficient* | 218 |
|     2. *Kinetic Trapping Model* | 220 |
| IV. EXPERIMENTAL TECHNIQUES | 222 |
|     1. *Positron Lifetime Spectroscopy* | 222 |
|     2. *Angular Correlation Spectroscopy* | 225 |
|     3. *Doppler Broadening Spectroscopy* | 227 |
| V. IDENTIFICATION OF VACANCIES AND THEIR CHARGE STATES | 232 |
|     1. *Open Volume and Positron Lifetime* | 232 |
|     2. *Vacancy Charge State: Ga Vacancy in GaAs* | 235 |
|     3. *Identification of Vacancy Sublattice and Impurity Surroundings* | 239 |
| VI. NEGATIVE IONS AS SHALLOW POSITRON TRAPS | 243 |
|     1. *Native Defects in As-grown GaAs as Negative Ions* | 244 |
|     2. *Negative Vacancies and Negative Ions in Electron-Irradiated GaAs* | 246 |
| VII. DEFECTS IN LAYERS STUDIED BY A LOW-ENERGY POSITRON BEAM | 250 |
|     1. *Compensating Defects in Highly Si-doped GaAs Layers* | 250 |
|     2. *Analysis of Depth Profiles of Vacancies in Ion-Implanted GaAs* | 254 |
| VIII. INVESTIGATION OF VACANCY IONIZATION LEVELS | 258 |
|     1. *Arsenic Vacancy in n-Type GaAs: Thermal Ionization* | 258 |
|     2. *Arsenic Vacancy in SI GaAs: Optical Transitions* | 262 |

IX. INVESTIGATION OF THE ATOMIC STRUCTURES OF METASTABLE DEFECTS . . . . . 268
  1. *As-grown SI GaAs: The Midgap Donor EL2* . . . . . . . . . . . . . . . 268
  2. *Metastability of Defects in SI GaAs After Electron Irradiation* . . . . . . . 273
  3. *As-grown n-Type $Al_x Ga_{1-x} As$ Layers: The Deep Donor Level DX* . . . . . 275
  4. *The Atomic Structure of EL2 and DX* . . . . . . . . . . . . . . . . . . 279
X. SUMMARY . . . . . . . . . . . . . . . . . . . . . . . . . . . . . . . . . 282
  *References* . . . . . . . . . . . . . . . . . . . . . . . . . . . . . . . 282

# I. Introduction

There are many techniques to identify defects in semiconductors on an atomic scale. The role of positron annihilation is in its ability to selectively detect vacancy-type defects. This is based on two special properties of the positron: it has a positive charge and it annihilates with electrons.

An energetic positron that has penetrated into a solid rapidly looses its energy and then lives a few hundred picoseconds in thermal equilibrium with the environment. During its thermal motion, the positron interacts with defects, which may lead to trapping into a localized state. Thus, the final positron annihilation with an electron can happen from various states.

The energy and momentum are conserved in the annihilation process. Two photons of 511 keV are emitted into opposite directions and convey the information of the positron annihilation state. The positron lifetime is inversely proportional to the electron density encountered by the positron. The momentum of the annihilated electron causes an angular deviation from 180° between the two 511-keV photons and creates a Doppler shift in their energy. Thus, the observation of positron annihilation radiation gives experimental information on the electronic and defect structures of solids. Three techniques are in use: positron lifetime, angular correlation of the annihilation radiation, and Doppler broadening of the 511-keV line.

The sensitivity of positron annihilation spectroscopy to vacancy-type defects is easy to understand. The free positron in a crystal lattice feels strong repulsion from positive ion cores. An open-volume defect like a vacant lattice cell is therefore an attractive center, where the positron gets trapped. The reduced electron density in the vacant cell increases the positron lifetime. In addition, the missing valence and core electrons cause big changes in the momentum distribution of the annihilated electrons.

The main advantages of the positron spectroscopy can be listed as follows. First, the identification of vacancy-type defects is straightforward. Second, the technique is strongly supported by theory, since the annihilation characteristics can be calculated from first principles. Third, positron annihilation can be applied to bulk crystals and thin layers of any electrical conduction type.

The experimental and theoretical basis of the positron spectroscopy of vacancies in metals and alloys were developed in the 1970s. The applications started gradually to widen to semiconductors in the beginning of the 1980s. At that time, the low-energy positron beam was also developed and it opened an avenue for defect studies of epitaxial layers and surface regions. The impact of positron annihilation can be summarized as follows. It has introduced an experimental technique for the unambiguous identification of vacancies. Native vacancies have been observed at high concentrations in many compound semiconductors, and their role in doping and compensation can now be quantitatively discussed. The vacancy character of the famous metastable defects, EL2 in GaAs and DX in AlGaAs, has been verified.

The aim of this chapter is twofold. We first want to introduce the principles of the positron annihilation technique and then to discuss the physics of some interesting observations in semiconductors. For the sake of coherence, the choice of illustrative examples is much biased towards our own studies in GaAs materials. To cover all the published works of positron annihilation in semiconductors, we refer to earlier review articles (West, 1973; Schultz and Lynn, 1988; Puska and Nieminen, 1994; Asoka-Kumar *et al.*, 1994), books (Hautojärvi, 1979; Brandt and Dupasquier, 1983; Dupasquier and Mills, 1995), conference proceedings (Dorikens-Vanpraet *et al.*, 1989; Kajcsos and Szeles, 1992; He *et al.*, 1995), and to other references therein.

The overview of positron annihilation spectroscopy is presented in Sections II–IV. The identification of vacancies and their charge states is described in Section V. The positron has shallow Rydberg states around negative ions, and Section VI discusses how ion-type acceptors can be detected. Section VII demonstrates the use of the positron beam to characterize thin layers in GaAs. Investigations on the thermal and optical properties of native As vacancies in GaAs are reviewed in Section VIII. Finally, Section IX presents the results obtained on metastable EL2 and DX centers by positron annihilation. Section X gives a brief summary.

## II. Positrons in Solids

In this section, we discuss how to estimate the positron distribution in a crystalline solid. We want to emphasize that after rapidly slowing down, the thermalized positron behaves like a free carrier. Its transport can be described by diffusion, and its density distribution in the lattice is governed by quantum mechanics. The positron also has deep and shallow localized states at lattice defects. The localized positron states lead to annihilation characteristics, which serve as fingerprints for the underlying defects.

1. POSITRONS FOR BULK STUDIES

Positrons are obtained from $\beta^+$ active isotopes like $^{22}$Na, $^{58}$Co, $^{64}$Cu, and $^{68}$Ge. The most commonly used isotope is $^{22}$Na, where the positron emission is accompanied by a 1.28-MeV photon. This photon is used as the time signal of the positron birth in positron lifetime experiments.

The stopping profile of positrons from $\beta^+$ emission is exponential (Paulin, 1983; Valkealahti and Nieminen, 1984):

$$p(x) = \alpha e^{-\alpha x}; \quad \alpha \approx 16 \frac{\rho[\text{g/cm}^3]}{E_{\max}^{1.4}[\text{MeV}]} \text{ cm}^{-1}, \quad (1)$$

where $x$ [cm] is the distance penetrated into the medium, $\rho$ is the density of the solid, and $E_{\max}$ is the maximum energy of the emitted positrons. For the $^{22}$Na source ($E_{\max} = 0.54$ MeV), the positron mean stopping depth $1/\alpha$ is 110 μm in Si and 50 μm in GaAs. The positrons emitted directly from a radioactive source thus probe the bulk of a solid.

2. POSITRONS FOR LAYER STUDIES

Low-energy positrons are needed for studies of thin layers and near-surface regions. Positrons from $\beta^+$ emission or pair production are first slowed down and thermalized in a moderator. This is usually a thin film placed in front of the positron source and made of a material (e.g., Cu or W) that has a negative affinity for positrons. Thermalized positrons close to the moderator surface are emitted into vacuum with an energy of the order of 1 eV, and a beam is formed using electric and magnetic fields. The positron beam is accelerated to a variable energy of 0 to 40 keV, and in this way the positron stopping depth in the sample is controlled. The moderator efficiency is very low, $\varepsilon = 10^{-4}$, and a typical positron beam intensity is $10^4$–$10^6$ e$^+$ s$^{-1}$.

For monoenergetic positrons, the stopping profile can be described by a derivative of a Gaussian function (Valkealahti and Nieminen, 1984; Asoka-

Kumar et al., 1994):

$$p(x) = -\frac{d}{dx} e^{-(x/x_0)^2}. \tag{2}$$

The mean stopping depth is

$$\bar{x} = 0.886 x_0 = AE^n [\text{keV}], \tag{3}$$

where $E$ is the positron energy and

$$A = \frac{4}{\rho} \mu\text{g/cm}^2; \quad n \approx 1.6. \tag{4}$$

The mean stopping depth varies with energy from 1 nm up to a few micrometers. A 20-keV energy corresponds to 2 $\mu$m in Si and 1 $\mu$m in GaAs. The width of the stopping profile is rather broad, and the positron energy must be carefully chosen so that, for example, the signal from an overlayer is not contaminated by that from the substrate or surface. More details on low-energy positrons are given in Section VII.

### 3. Positron Diffusion and Mobility

In a solid, the fast positron rapidly looses its energy via ionization and core electron excitations. Finally, the positron momentum distribution relaxes to a Maxwell-Boltzmann one via electron-hole excitations and phonon emissions. The thermalization time at 300 K is 1 to 3 ps (i.e., much less than a typical positron lifetime of 200 ps). Even at 10 K, theoretical thermalization times are much less than the positron lifetime (Nieminen and Oliva, 1980; Jensen and Walker, 1990). Experiments have shown that down to 25 K, the positron behaves as a fully thermalized particle in semiconductors.

The transport of thermalized positrons in solids is described by diffusion theory. Using the Nernst-Einstein relation and relaxation-time approximation, the positron diffusion coefficient is written as

$$D_+ = \frac{\mu_+}{e} k_B T = \frac{k_B T}{m^*} \tau_{\text{rel}}, \tag{5}$$

where $\mu_+$ is the mobility, $k_B$ the Boltzmann factor, $T$ the temperature, $e$ the charge, $m^*$ the positron effective mass, and $\tau_{rel}$ the relaxation time. Calculations (Bergensen et al., 1974; Boev et al., 1987) have shown that scattering off acoustic phonons is the dominant process in metals and elemental semiconductors leading to the $T^{-1/2}$ dependence of the diffusion coefficient. The scattering off impurities contributes only at high ($>10^{20}$ cm$^{-3}$) impurity concentrations.

The positron diffusion coefficient has been measured in several semiconductors by implanting low-energy positrons at various depths and observing the fraction that diffuses back to the entrance surface (Soininen et al., 1992). The diffusion coefficient at 300 K is in the range of 1.5 to 3 cm$^2$ s$^{-1}$. The total diffusion length during the finite positron lifetime $\tau$ is

$$L_+ = (6D_+\tau)^{1/2} = 5000 \text{ Å}. \tag{6}$$

If defects are present, the positron may get trapped before annihilation, and this naturally reduces the effective diffusion length (see Section VII).

Figure 1 shows the diffusion coefficients in Si (Soininen et al., 1992). $D_+(T)$ varies as $T^{-1/2}$ from 30 to 500 K, consistent with positron scattering

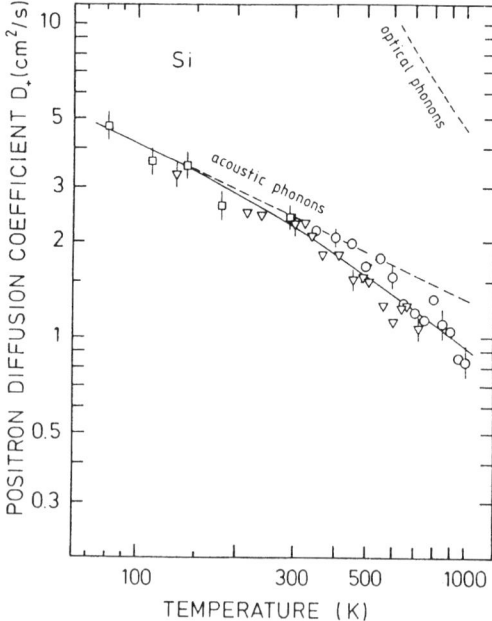

FIG. 1. Positron diffusion coefficient in pure Si. Positron diffusion is limited by scattering off acoustic phonons. The contribution of optical phonons becomes visible at elevated temperatures. (From Soininen et al., 1992, with permission.)

off longitudinal acoustic phonons. At $T > 500$ K, the diffusion coefficient is lower than that extrapolated from the $T^{-1/2}$ dependence, indicating the onset of optical-phonon scattering. In GaAs, the diffusion coefficient is only weakly dependent on temperature between 300 and 800 K, which is attributed to the strong scattering off polar-optical phonons.

The positron mobility in Si has been measured from 30 to 300 K by implanting positrons into the space-charge region of an Au-Si Schottky diode (Mäkinen et al., 1991). The mobility is consistent with the experimental diffusion coefficient according to the Nernst-Einstein relation in Eq. (5). It is 110 cm$^2$ V$^{-1}$ s$^{-1}$ at 300 K and varies as $T^{-3/2}$ in accordance with Eq. (5). The positron mobility in GaAs is around 70 cm$^2$ V$^{-1}$ s$^{-1}$ (Shan et al., 1996). The mobility of the positron is an order of magnitude less than that of free carriers, mainly because the positron always has a "heavy" effective mass of $m^* \approx 1.5 m_0$.

## 4. Positron Wave Function and Positron States

In the spirit of the density functional theory, the positron wave function can be calculated from the one-particle Schrödinger equation

$$-\frac{\hbar^2}{2m}\nabla^2 \Psi_+(\mathbf{r}) + V(\mathbf{r})\Psi_+(\mathbf{r}) = E_+ \Psi_+(\mathbf{r}). \qquad (7)$$

The positron potential consists of two parts:

$$V(\mathbf{r}) = V_{\text{Coul}}(\mathbf{r}) + V_{\text{corr}}(\mathbf{r}), \qquad (8)$$

where the first term is the electrostatic Coulomb potential and the second term takes into account the electron–positron correlation effects in the local density approximation. For various practical schemes to solve the Schrödinger equation, see the review of Puska and Nieminen (1994).

In a perfect lattice, the positron is delocalized in a Bloch state with $\mathbf{k}_+ = 0$. Due to the Coulomb repulsion from positive ion cores, the positron wave function has its maximum at the interstitial space between the atoms. The positron energy band $E_+(\mathbf{k})$ is parabolic and free particle-like with an effective band mass 1.0 to $1.1 m_0$ (Boev et al., 1987). The increase of the positron effective mass to $m^* \approx 1.5 m_0$ is due to phonons and the screening cloud of electrons (Mikeska, 1970; Bergensen and Pajanne, 1969).

In analogy to free carriers, the positron also has localized states at lattice imperfections. At vacancy-type defects where ions are missing, the repulsion sensed by the positron is lowered and the positron sees these kinds of defects

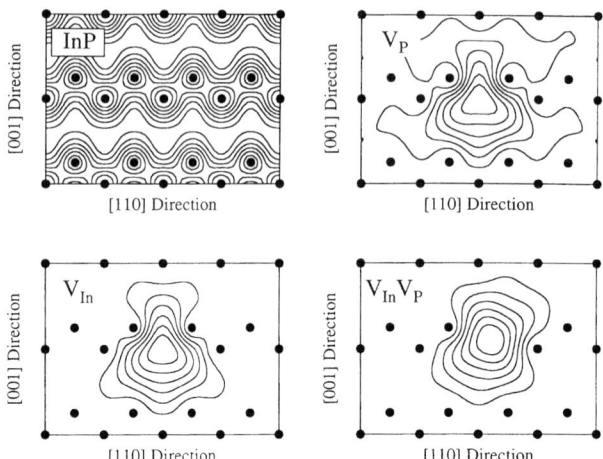

FIG. 2. Positron wave functions in InP for the free state in the lattice and for the trapped states at the In and P vacancies as well as at the divacancy. (From Hakala and Puska, with permission.)

as potential wells. As a result, localized positron states at open-volume defects are formed. The positron ground state at a vacancy-type defect is generally *deep*; the binding energy is about 1 eV or more. Figure 2 shows the free and localized positron wave functions in InP (Hakala and Puska, 1996; Puska et al., 1989). The free positron wave function has its maximum in the interstitial regions between the ions. The positron at the P vacancy is only weakly localized because of the small size of the P atom. At the In vacancy, the positron wave function still leaks into the interstitial region, but at the divacancy, the positron state becomes well localized and isotropic.

Negative ions can trap the positron into Rydberg states with the binding energies of 0.01 to 0.1 eV. These states are *shallow*, and detrapping from them occurs already below 300 K.

The positron can have two-dimensional states at surfaces. At a clean surface, the positron is trapped by its own image potential (Puska and Nieminen, 1994). An interface is often imperfect and full of defects capable of trapping positrons (Peng et al., 1996).

## 5. Annihilation Characteristics

In solids like metals and semiconductors, where positronium is not formed, the positron annihilates with an electron into two 511-keV photons. The probability of three-photon annihilation is only 0.3% and can be neglected.

A positron state can be characterized by its annihilation properties (i.e., the positron lifetime and the momentum distribution of the annihilation radiation). These experimental quantities can be easily calculated once the corresponding electronic structure is known. The direct comparison between experiment and theory allows one to identify the defects generating the localized positron states with a high degree of confidence.

The positron annihilation rate $\lambda$, the inverse of the positron lifetime $\tau$, is proportional to the overlap of the electron and positron densities:

$$1/\tau = \lambda = \pi r_0^2 c \int d\mathbf{r} |\Psi_+(\mathbf{r})|^2 n(\mathbf{r}) \gamma[n(\mathbf{r})], \tag{9}$$

where $r_0$ is the classical radius of the electron, $c$ the velocity of light, $n(\mathbf{r})$ the electron density, and $\gamma[n]$ the enhancement factor of the electron density at the positron. There are various interpolation formulas for $\gamma[n]$ based on many body calculations of electron–positron correlations (Puska and Nieminen, 1994; Boronski and Nieminen, 1986; Barbiellini et al., 1995, 1996; Alatalo et al., 1996). In each positron state, the positrons probe a different electron density, leading to different lifetime values. In a vacancy-type defect, where the positron wave function is confined, the electron density is locally reduced and thus the lifetime of a trapped positron is longer than that of free positrons in the lattice. The bigger the open volume of the vacancy defect, the longer the trapped positron lifetime.

The momentum of an annihilating electron–positron pair is transferred to the annihilation photons. The momentum distribution $\rho(\mathbf{p})$ of the annihilation radiation is that of the annihilated electrons, because the momentum of the thermalized positron is negligible. The momentum distribution is a nonlocal quantity and requires knowledge of all the electron wave functions $\psi_i$ overlapping with the positron. It can be written in the form

$$\rho(\mathbf{p}) = \frac{\pi r_0 c}{V} \sum_i \left| \int d\mathbf{r} e^{-i\mathbf{p}\cdot\mathbf{r}} \Psi_+(\mathbf{r}) \Psi_i(\mathbf{r}) \sqrt{\gamma(\mathbf{r})} \right|^2, \tag{10}$$

where $V$ is the normalization volume.

Positron annihilation at a vacancy-type defect leads to changes in $\rho(\mathbf{p})$. The momentum distribution arising from valence electron annihilation becomes narrower due to a lower eletron density. In addition, the localized positron at a vacancy has a reduced overlap with ion cores, leading to a considerable decrease in annihilation with high-momentum core electrons.

## III. Positron Trapping

1. TRAPPING RATE AND TRAPPING COEFFICIENT

The transition from a free Bloch state to a localized state at a defect is called positron trapping. The trapping is analogous to carrier capture. However, it must be fast enough to compete with annihilation. This determines the type and concentration of defects observable by the positron technique.

The positron trapping rate $\kappa_D$ onto defect $D$ is proportional to its concentration $c_D$:

$$\kappa_D = \mu_D c_D. \tag{11}$$

The trapping coefficient $\mu_D$, sometimes called the "specific trapping rate," depends on the defect and the host lattice. Next we briefly summarize the present view of the trapping mechanisms in semiconductors (Puska et al., 1990).

The transition rate from a free to a localized state is limited by the energy and momentum transfer from the positron to the host lattice. For deep-vacancy-type defects, only electron processes (i.e., electron-hole excitation and Auger processes), are fast enough. For shallow Rydberg states, phonon processes are relevant.

*Positive vacancies:* Due to the Coulombic repulsion, the trapping coefficient is so small that the trapping does not occur during the short positron lifetime. Therefore, the positron technique does not detect vacancies in their positive charge states.

*Neutral vacancies:* The trapping coefficient is independent of temperature. Its value depends on the vacancy and the host lattice, especially via the density of states in the valence and conduction bands. Typical values are $\mu_{V^0} \approx 10^{14} - 10^{15}$ at s$^{-1}$. This value means that neutral vacancies are observed at the concentrations $\geqslant 10^{16}$ cm$^{-3}$.

*Negative vacancies:* The fingerprint of a negative vacancy is the increase of $\mu_{V^-}$ with decreasing temperature. There are two reasons for this. First, the amplitude of the free-positron Coulombic wave increases as the thermal velocity of the positron decreases. This leads to a $T^{-1/2}$ dependence of the direct trapping into the positron ground state at the vacancy. Second, at low temperature, the shallow Rydberg states of the negative vacancy act as precursor states to the final localization into the deep ground state. The trapping to and detrapping from the Rydberg states lead to an exponential temperature dependence of $\mu_{V^-}$ (see Section V.2). At 300 K, the trapping coefficient is typically $\mu_{V^-} \approx 10^{15}–10^{16}$ at s$^{-1}$, meaning that negative vacancies are observed at concentrations $\geqslant 10^{15}$ cm$^{-3}$.

*Negative ions:* The binding energies, 0.01 to 0.1 eV, of the positron Rydberg states around isolated negative ions are of the same order as phonon energies. At low temperature, below 50 K, the trapping coefficient $\mu_{Ry}$ varies as $T^{-1/2}$ because the free positron is a Coulomb wave. The trapping coefficient increases with the quantum number $n$ due to the larger extent of the states with higher $n$. The detrapping $\delta_{Ry}$ from a Rydberg state is exponentially activated (Puska *et al.*, 1990)

$$\delta_{Ry} = \mu_{Ry} \left(\frac{2\pi m^* k_B T}{h^2}\right)^{3/2} \exp\left(-\frac{E_{b,Ry}}{k_B T}\right), \tag{12}$$

and leads to a rapid decrease of the positron trapping at negative ions above 100 K. Typically, ion concentrations above $10^{15}$ cm$^{-3}$ influence positron annihilation at low temperatures.

*Defect complexes:* Multiple vacancies or vacancy clusters are even more effective positron traps than are single vacancies. The charge-state dependence is the same as just described for vacancies. Whenever open volume exists in a defect complex, it appears for the positron as a vacancy-type defect. For example, if a negative ion is bound to a vacancy,

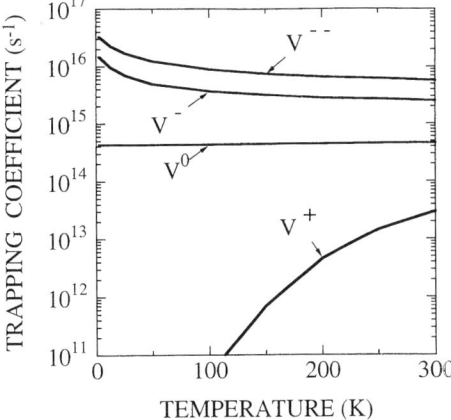

FIG. 3. Theoretical temperature dependence of the positron trapping coefficient for vacancies in various charge states. (From Puska *et al.*, 1990, with permission.)

the defect appears as a vacancy for the positron. It is the total charge of the complex that matters: a neutral pair consisting of a negative ion and a positive vacancy is a positron trap, where the positron is localized into the vacancy.

Figure 3 shows the positron-trapping coefficients as functions of temperature calculated for vacancies in various charge states (Puska et al., 1990). Note that only the direct trapping from the free state into the ground state at the negative vacancies has been taken into account.

## 2. Kinetic Trapping Model

Positron trapping and annihilation are described by kinetic equations. Because the slowing down and thermalization processes are fast compared to trapping and annihilation, the conventional trapping model (West, 1979; Hautojärvi and Corbel, 1995) is based on the following assumptions:

1. At time $t = 0$ the positron is free in the Bloch state.
2. The positron trapping rate $\kappa_D$ is proportional to the defect concentration $c_D$: $\kappa_D = \mu_D c_D$, where $\mu_D$ is the defect-specific trapping coefficient.
3. The trapped positron may escape from the trap. The detrapping rate is denoted by $\delta_D$.

We further assume that there are $N$ different defects. The probability of the positron to be in the free state is $n_B$ and to be trapped in a defect is $n_{Dj}$. The annihilation rates in these states are $\lambda_B$ and $\lambda_{Dj}$, respectively. We get a set of linear differential equations ($j = 1, \ldots, N$)

$$\frac{dn_B}{dt} = -\left(\lambda_B + \sum \kappa_{Dj}\right)n_B + \sum \delta_{Dj} n_{Dj}, \qquad (13)$$

$$\frac{dn_{Dj}}{dt} = \kappa_{Dj} n_B - (\lambda_{Dj} + \delta_{Dj}) n_{Dj}, \qquad (14)$$

where $\Sigma = \Sigma_{j=1}^{N}$ is the sum over all defects $Dj$. The boundary condition from the assumption (1) is

$$n_B(0) = 1, \quad n_{Dj}(0) = 0. \qquad (15)$$

The probability for the positron to be alive at time $t$ is of the exponential form

$$n(t) = n_B(t) + \sum_{j=1}^{N} n_{Dj}(t) = \sum_{i=1}^{N+1} I_i \exp[-\lambda_i t], \qquad (16)$$

where the decay constants $\lambda_i$ and intensities $I_i$ are found by solving Eqs. (13) and (14). The lifetime spectrum is the probability for the annihilation at time $t$

$$-\frac{dn(t)}{dt} = \sum_{i=1}^{N+1} I_i \lambda_i \exp[-\lambda_i t], \qquad (17)$$

with $\sum_{i=1}^{N+1} I_i = 1$. The following sum rule is immediately deduced from Eqs. (13) through (15):

$$-\frac{dn}{dt}\bigg|_{t=0} = \lambda_B = \sum_{i=1}^{N+1} I_i \lambda_i. \qquad (18)$$

Equation (18) is often used to test whether an experimental decomposition $(I_i, \lambda_i)$ is compatible with the trapping model.

The center of mass of the lifetime spectrum is called the average positron lifetime $\tau_{av}$:

$$\tau_{av} = \int_0^\infty dt\, t\left(-\frac{dn}{dt}\right) = \int_0^\infty dt\, n(t) = \sum_{i=1}^{N+1} I_i \tau_i. \qquad (19)$$

The average lifetime is an important quantity because it can be deduced from the experimental lifetime spectrum without any knowledge of the decomposition (see Section IV.1).

The annihilation fraction $\eta_j$ is the probability for the positron to annihilate in the state $j$. We get

$$\eta_B = \int_0^\infty dt\, \lambda_B n_B(t) = \lambda_B n_B^*, \qquad (20)$$

$$\eta_{Dj} = \int_0^\infty dt\, \lambda_{Dj} n_{Dj}(t) = \lambda_{Dj} n_{Dj}^*. \qquad (21)$$

The time-integrated probabilities $n_i^* = \int dt\, n_i(t)$ can be easily calculated from the so-called stationary trapping equations obtained from Eqs. (13) and (14) by integrating them from $t = 0$ to infinity and using the boundary conditions [Eq. (15)]:

$$\left(\lambda_B + \sum \kappa_j\right) n_B^* + \sum \delta_j n_{Dj}^* - 1 = 0, \tag{22}$$

$$\kappa_j n_B^* - (\lambda_{Dj} + \delta_j) n_{Dj}^* = 0. \tag{23}$$

The general solution of Eqs. (22) and (23) is ($j = 1, \ldots, N$; $j' = 1, \ldots, N$)

$$\eta_B = \frac{\lambda_B}{\lambda_B + \sum_{j'} \dfrac{\kappa_{Dj'}}{1 + \varepsilon_{Dj'}}} \tag{24}$$

$$\eta_{Dj} = \frac{\kappa_{Dj}}{(1 + \varepsilon_{Dj})\left(\lambda_B + \sum_{j'} \dfrac{\kappa_{Dj'}}{1 + \varepsilon_{Dj'}}\right)}, \tag{25}$$

where $\varepsilon_{Dj} = \delta_{Dj}/\lambda_{Dj}$ is the "escape ratio" of positrons from trap $j$.

The experimental quantities, the positron lifetime $\tau$, and momentum distribution $\rho(\mathbf{p})$ are superpositions of their characteristic values in the various positron states. For example, for the average lifetime, we get

$$\tau_{av} = \eta_B \tau_B + \sum_j \eta_{Dj} \tau_{Dj}. \tag{26}$$

Notice that Eq. (26) is indeed compatible with Eq. (19), but the annihilation fractions $\eta_j$ are different from the lifetime intensities $I_i$.

## IV. Experimental Techniques

### 1. Positron Lifetime Spectroscopy

Lifetime spectroscopy is a powerful technique in defect studies because the various positron states appear as different exponential decay components [see Eq. (17)]. Thus the number of positron states, their annihilation rates, and relative intensities are the object of a lifetime experiment.

In a lifetime measurement, one needs to detect the start and stop signals corresponding to the positron entrance and annihilation times in the sample, respectively. A suitable start signal is the 1.28-MeV photon accompanying the positron emission from the $^{22}$Na isotope. The 511-keV annihilation photon serves as the stop signal. The positron source is prepared by sealing about 10 μCi of radioactive isotope between two thin foils. The source is then sandwiched between two identical pieces (e.g., $5 \times 5 \times 0.2$ mm$^3$) of the sample material. This technique is standard for bulk crystal studies. Pulsed positron beams have been constructed for lifetime spectroscopy of thin layers (Schödlbauer et al., 1988; Suzuki et al., 1991), but they are not yet in routine use.

The standard lifetime spectrometer consists of start and stop detectors, each of them made by coupling a fast scintillator to a photomultiplier. The timing pulses are obtained by differential constant-fraction discrimination. The time delays between the start and stop signals are converted into amplitude pulses, the heights of which are stored in a multichannel analyzer. About $10^6$ lifetime events are recorded in one hour.

There are limitations for the technique. The experimental spectrum is the convolution of the theoretical spectrum [Eq. (17)] and a Gaussian resolution function that has a width of 200 to 250 ps (full width at half maximum). About 5 to 10% of positrons annihilate in the source material, and proper "source corrections" must be made. Due to the finite time resolution, annihilations in the source materials, and random background, typically only 1 to 3 lifetime components can be resolved in the analysis of the experimental spectra. The separation of two lifetimes is successful only if the ratio $\lambda_1/\lambda_2$ is $>1.5$.

A good and statistically accurate parameter is the average lifetime $\tau_{av}$, defined in Eq. (19). This is because the center of mass of the experimental lifetime spectrum, $\tau_{av}$, can be correctly calculated from the intensity and lifetime values even if the decomposition represented only a good fit to the experimental data without any physical meaning. Changes below 1 ps are reliably observed.

The lifetime corresponding to free positron annihilation in the lattice can be determined in several ways. The positron lifetime spectrum in a "defect-free" reference sample has only the single component $\tau_B$. The temperature dependence of $\tau_B$ is very weak (only $<1$ ps in 100 K) and results mainly from the thermal expansion of the lattice. A p-type semiconductor crystal is often a good reference sample, because all vacancy defects are in their most positive charge state, thus minimizing the effect of positron trapping. Even in the presence of vacancies, the free positron lifetime $\tau_B$ can be estimated from the decomposition of the lifetime spectra [Eq. (18)]. Moreover, theoretical calculations are nowadays able to predict the free positron lifetime with high accuracy (Barbiellini et al., 1995).

The average lifetime is the superposition of the characteristic lifetimes of various states, and from Eq. (26) we get

$$\tau_{av} - \tau_B = \sum_j \eta_{Dj}(\tau_{Dj} - \tau_B). \qquad (27)$$

since for open-volume defects $\tau_{Dj} > \tau_B$, we see that an experimental value of $\tau_{av} > \tau_B$ is direct evidence of the presence of vacancy-type defects. Thus, the first task is always to check whether vacancy-type defects are detected or not (i.e., whether $\tau_{av} > \tau_B$ is valid). The next task is to try to decompose the spectrum in order to determine the characteristic lifetimes $\tau_{Dj}$ and the trapping rates $\kappa_{Dj}$. These quantities are used to identify the defects and estimate their concentrations.

Examples of positron lifetime spectra are shown in Fig. 4. All the spectra have been measured at 300 K. The GaAs(Zn:$10^{18}$ cm$^{-3}$) crystal has only one lifetime component, which is assigned to free positrons with $\tau_B = 231$ ps. The 2-MeV electron-irradiated undoped GaAs has the average lifetime

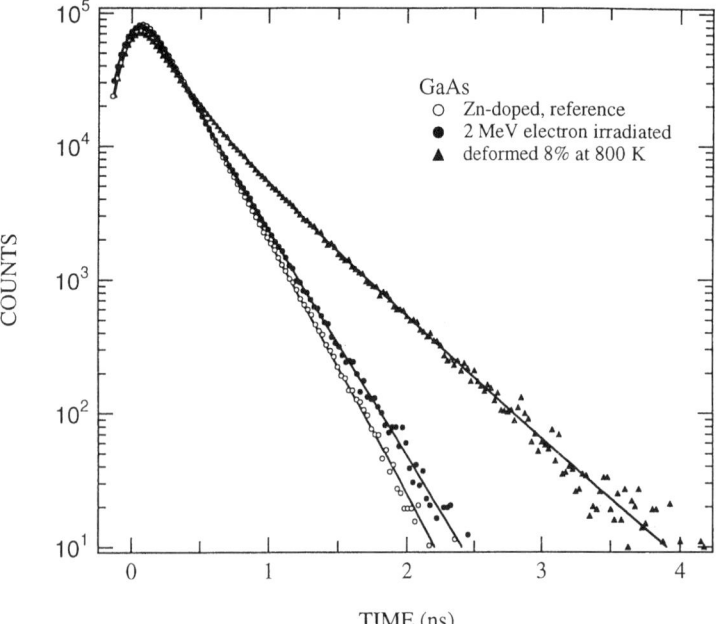

FIG. 4. Positron lifetime spectra in various GaAs crystals. Vacancy-type defects are seen as long lifetime components.

$\tau_{av} = 241$ ps and thus contains vacancy-type defects. The decomposition gives $\tau_1 = 186 \pm 15$ ps, $I_1 = 29 \pm 11\%$, and $\tau_2 = 263 \pm 6$ ps, $I_2 = 71 \pm 11\%$. The second lifetime corresponds to irradiation-induced Ga vacancies (see Section VI.2). The first one represents the inverse of the total disappearance rate of free positrons (annihilation and trapping) and is therefore below the free positron lifetime $\tau_B$ (see Section V.2). The deformed GaAs(Si:$10^{18}$ cm$^{-3}$) has a very high $\tau_{av} = 342$ ps with the decomposition $\tau_1 = 106 \pm 14$ ps, $I_1 = 13 \pm 3\%$; $\tau_2 = 273 \pm 18$ ps, $I_2 = 45 \pm 2\%$; and $\tau_3 = 487 \pm 10$ ps, $I_3 = 42 \pm 4\%$. The very long value of $\tau_3$ indicates that positrons annihilate at big open volumes (i.e., vacancy clusters or voids) (Saarinen et al., 1990).

## 2. Angular Correlation Spectroscopy

Because of the momentum of the annihilating positron–electron pair, the two 511-keV photons are not emitted strictly into opposite directions. The small angular deviation $\theta$ is proportional to the transverse momentum of the pair, $p_T = \theta m_0 c$. In an angular correlation experiment of annihilation radiation (ACAR), the two photons are detected in coincidence. Using big position-sensitive detectors, such as Anger cameras or multiwire proportional chambers, the two-dimensional momentum distribution (2D-ACAR) can be measured as a projection:

$$N(p_x, p_y) = \int \rho(\mathbf{p}) dp_z. \tag{28}$$

If also a second integration is performed over $p_y$, one speaks of 1D-ACAR.

The ACAR experiments give detailed information on electronic structures. The resolution is high, $\Delta p \approx 0.2 \times 10^{-3} m_0 c = 0.03$ a.u., but the measurement time is several days long. The main applications have been in bulk electronic structure and Fermi surface studies.

The basic problem of momentum spectroscopies in defect studies is that the experimental quantity is always a superposition of all positron annihilation states in the sample:

$$\rho_{tot}(\mathbf{p}) = \eta_B \rho_B(\mathbf{p}) + \sum_j \eta_{Dj} \rho_{Dj}(\mathbf{p}). \tag{29}$$

A defect-specific quantity, $\rho_{Dj}(\mathbf{p})$, can be obtained only if all positrons annihilate in the same state $D_j$ ($\eta_{Dj} = 1$) or if the annihilation fractions $\eta_{Dj}$ have been determined for the same sample by positron lifetime measurements.

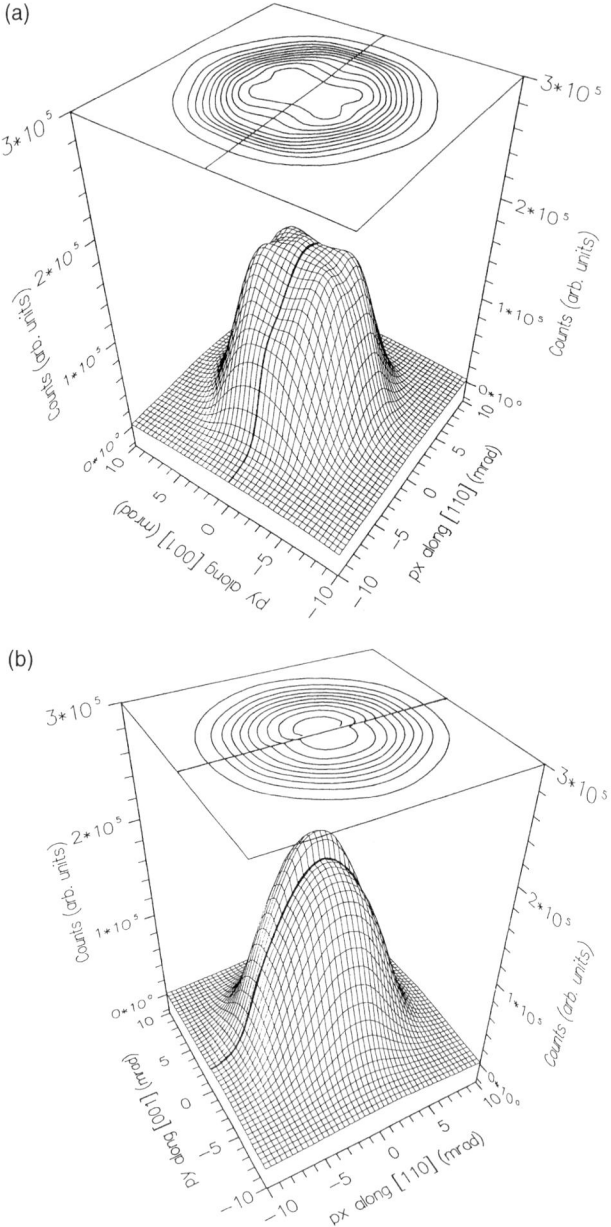

FIG. 5. Combined perspective and contour plots of 2D-ACAR results for (a) the GaAs lattice at 90 K ($p_y < 0$) and at 300 K ($p_y > 0$) and (b) $V_{As}^-(p_y < 0)$ and $V_{As}^0(p_y > 0)$. (From Ampigapathy et al., 1994, with permission.)

Figure 5 shows the 2D-ACAR curves for the negative and neutral state of the As vacancy in GaAs. The curves have been obtained by combining lifetime and 2D-ACAR experiments (Ambigapathy et al., 1994). The data show that the momentum distribution at a vacancy is narrower and more isotropic than that in the GaAs lattice. The momentum distribution in the neutral state is even narrower than that in the negative state, reflecting the larger open volume due to a strong outward relaxation of the neighboring atoms (see Section VIII.1). The 2D-ACAR, combined with an intense low-energy positron beam, can produce unique information on the electronic and defect structures in layers and interfaces (Peng et al., 1996).

3. DOPPLER BROADENING SPECTROSCOPY

The motion of the annihilating pair causes a Doppler shift in the annihilation radiation:

$$\Delta E_\gamma = \tfrac{1}{2} c p_L, \tag{30}$$

where $p_L$ is the longitudinal momentum component of the pair in the direction of the annihilation photon emission. This causes the broadening of the 511-keV annihilation line. The shape of the 511-keV peak gives the one-dimensional momentum distribution of the annihilation radiation

$$L(E_\gamma) \propto \int_{-\infty}^{\infty} \int_{-\infty}^{\infty} dp_x dp_y \rho(\mathbf{p}); \qquad p_z = \frac{2}{c}(E_\gamma - m_0 c^2). \tag{31}$$

A Doppler shift of 1 keV corresponds to a momentum value of $p_z = 3.91 \times 10^{-3} m_0 c$.

*a. Experimental Aspects*

The Doppler broadening is experimentally measured using a Ge gamma detector. Because the broadening shows only small changes due to defects, the stabilization against electronic drifts is needed. For measurements of bulk samples, the same source-sample sandwich is used as in the lifetime experiments. For layer studies, the positron beam hits the sample, and the Doppler broadening is often monitored as a function of the beam energy. The typical resolution of a detector is around 1 keV at 500 keV. This is considerable compared to the total width of 2 to 3 keV of the annihilation peak, meaning that the experimental lineshape is strongly influenced by the detector resolution. Therefore, various shape parameters are used to characterize the 511-keV line.

The high-momentum part of the Doppler broadening spectrum arises from annihilations with core electrons, which contain information on the chemical identity of the atoms. Thus, the detailed investigation of core electron annihilation can reveal the nature of the atoms in the regions where positrons annihilate. To study the high-momentum part, the background needs to be reduced. A NaI or BGO detector is added opposite to the Ge detector, and the only events that are accepted are those for which both 511-keV photons are detected. This coincidence technique enables the measurement of electron momenta up to $p \approx 40 \times 10^{-3} m_0 c \approx 5.5$ a.u.

### b. Momentum Parameters

As indicated, the Doppler-broadened 511-keV line is the one-dimensional momentum distribution of electron–positron pairs convoluted by a poor detector resolution. To easily compare different lineshapes (momentum distributions), it is customary to use parameters called $S$ and $W$.

The low electron-momentum parameter $S$ is defined as the ratio of the counts in the central region of the annihilation line to the total number of the counts in the line. In the same way, the high electron-momentum parameter $W$ is the fraction of the counts in the wing regions of the line (Fig. 6). Due to their low momenta, mainly valence electrons contribute to the region of the $S$ parameter. On the other hand, only core electrons have momentum values high enough to contribute to the $W$ parameter. Therefore, $S$ and $W$ are called the valence and core annihilation parameters, respectively.

To get maximum information out of the parameters, the proper choice of the energy windows is important. The central window for $S$ is chosen so that it covers about 50% of the total peak ($S \sim 0.5$). The $W$ window is far enough from the peak to ensure that only core annihilations are accepted. Typically for GaAs and related materials, we have used the window $|E_\gamma - 511\,\text{keV}| < 0.7\,\text{keV}$ for $S$ and $2.5\,\text{keV} < |E_\gamma - 511\,\text{keV}| < 4.1\,\text{keV}$ for $W$.

The absolute values of the parameters are meaningless, because they depend on the widths and positions of the windows. Only the changes of the parameters are important. To facilitate comparisons between various experiments, it is customary to report relative values like $S/S_{\text{ref}}$ and $W/W_{\text{ref}}$, where a good reference is provided by free annihilation in the lattice. These relative values are rather independent of the energy windows and of the small variations in the energy resolutions of different Ge detectors.

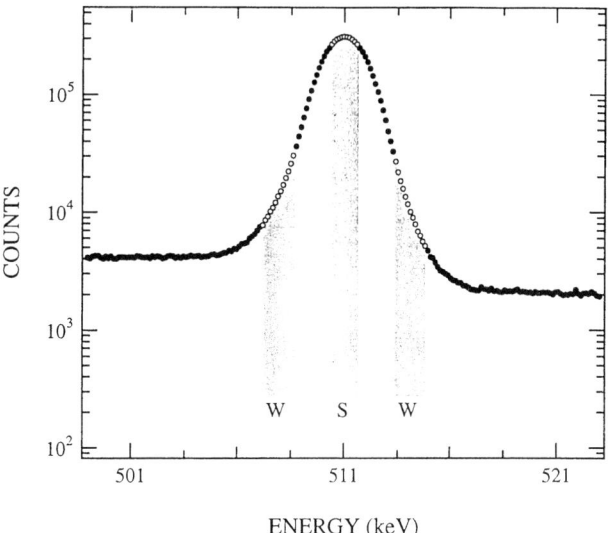

FIG. 6. The Doppler broadened line of the 511-keV annihilation radiation in InP. The low and high electron-momentum parameters $S$ and $W$, respectively, are calculated as the relative numbers of counts in the marked areas. The 1-keV deviation from 511 keV corresponds to a momentum value of $3.91 \times 10^{-3} m_0 c$.

The lineshape parameters have characteristic values for free positrons in each material, depending on the material's electron-momentum distribution. When positrons are trapped, the lineshape is characteristic of the trapping defect. For a vacancy-type defect, the density of valence electrons is reduced. This leads to a narrowing of the momentum distribution, which is seen as an increase in $S$. On the other hand, the localized positron in a vacancy-type defect has less overlap with core electrons than a free positron, leading to a decrease in the core annihilation parameter $W$.

If the fractions $\eta_D$ of positrons are trapped and annihilate at a defect $D$, then the parameters can be expressed as superpositions

$$S = (1 - \eta_D)S_B + \eta_D S_D, \tag{32}$$

$$W = (1 - \eta_D)W_B + \eta_D W_D, \tag{33}$$

where the subscript $B$ refers to the free state in the lattice. The changes $S-S_B$ and $W-W_B$ are proportional to the trapping fraction $\eta_D$. If the ratio of the changes are taken, a new parameter (Mantl and Triftshäuser, 1978; Saarinen et al., 1991b) is obtained:

$$R_D = \left|\frac{\Delta S}{\Delta W}\right| = \left|\frac{S - S_B}{W - W_B}\right| = \left|\frac{S_D - S_B}{W_D - W_B}\right|, \quad (34)$$

which is now independent of $\eta_D$ and thus characteristic of the defect $D$.

A further illustration of the usefulness of the parameters is given in an $(S, W)$ plot. Equations (32) and (33) define a segment of a straight line in the $(S, W)$ plane. The slope of the line is $R_D$. If there is no trapping ($\eta_D = 0$), the point $(S_B, W_B)$ represents the free positron state. With saturation trapping ($\eta_D = 1$), we get the point $(S_D, W_D)$ corresponding to the trapped state. A partial trapping is just an intermediate point between these two points. Thus, any sample that contains an unknown concentration of the defect $D$ should have its $(S, W)$ value on the line from $(S_B, W_B)$ to $(S_D, W_D)$, provided that no other positron traps are present. Because the average lifetime also fulfills the superposition in Eqs. (32) and (33), the $(S, \tau_{av})$ and $(W, \tau_{av})$ plots can be constructed in the same way.

Figure 7 shows how the $(S, W)$ plot can be used to identify vacancies in wide-gap $ZnS_xSe_{1-x}$ layers (Saarinen et al., 1996). The Doppler broadening

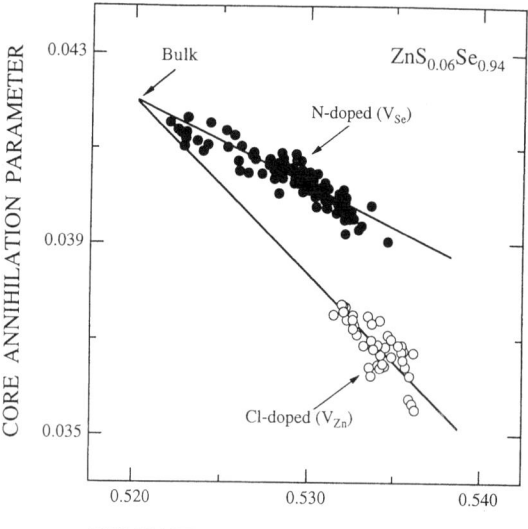

FIG. 7. The $(S, W)$ plot of the Doppler broadening is used to identify the Se vacancy in N-doped and the Zn vacancy in Cl-doped $ZnS_{0.06}Se_{0.94}$. (From Saarinen et al., 1996, with permission.)

was measured in six N-doped and two Cl-doped layers at various temperatures using a positron beam. The values of the $S$ and $W$ parameters have been marked on the $(S, W)$ plot. All the points from the N-doped layers fall on a same line with the slope $|\Delta S/\Delta W| \approx 5$, suggesting that a single type of vacancy defect explains the results for the N-doped layers. Only the fraction of annihilations at the vacancy varies from point to point. Similarly, the points from the Cl-doped layers form a straight line, but with a clearly different slope ($|\Delta S/\Delta W| \approx 2.5$), meaning that, again, a single type of vacancy defect, but different from the one in N-doped layers, explains the results. The crossing point of the two straight lines corresponds to the free positron state common to both types of layers. Based on core electron annihilation, the defects have been identified as the Se vacancy in N-doped layers and the Zn vacancy in Cl-doped layers. The Se vacancy is surrounded by Zn atoms, which have bigger 3d shells than the Se atoms. This leads to a higher core annihilation intensity and indeed, all the points from N-doped layers have higher $W$ values than the Cl-doped layers. Evidently, the Se and Zn vacancies are compensating centers for the n-type (Cl) and p-type (N) dopants, respectively. (For more discussion, see Section VI.3.)

*c. Core-Electron Momentum Spectroscopy*

Figure 8 shows how vacancies can be identified via core electron annihilation. Electron-irradiated InP crystals reveal two different vacancies with

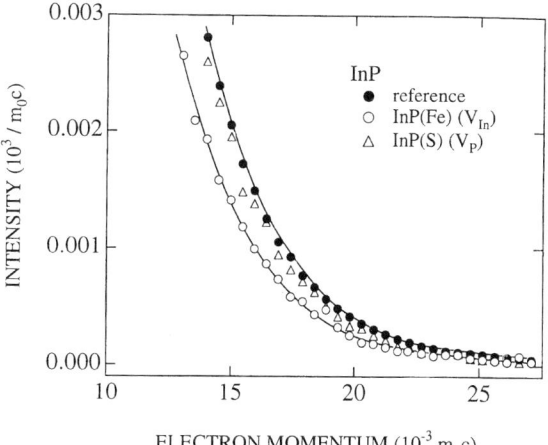

FIG. 8. The high-momentum parts of the Doppler curves for the InP lattice, the phosphorous vacancy, and the indium vacancy. (From Alatalo et al., 1995, with permission.)

lifetimes of 263 and 283 ps in S and Fe doped samples, respectively (Törnqvist et al., 1994). The high-momentum parts of the Doppler curves corresponding to these two different vacancies are presented in the upper part of Fig. 8, together with the one corresponding to the free positron state (reference) (Alatalo et al., 1995). The 263-ps vacancy in InP(S) has an almost equal core annihilation distribution compared to that of the reference material, whereas the 283-ps vacancy in InP(Fe) has a much lower distribution. The conclusion is evident: the 263-ps vacancy is on the P sublattice and the 283-ps vacancy is on the In sublattice. The wave functions presented in Fig. 2 explain well the observed differences in the core electron annihilations (see Fig. 8). The P vacancy is small and the positron is not well localized, but has a considerable overlap with the neighboring In core electrons. This leads to almost as strong a core annihilation component as in the bulk InP. On the other hand, the positron is well localized in the In vacancy surrounded by P atoms with small cores, and thus a big reduction in the core electron annihilations is observd.

## V. Identification of Vacancies and Their Charge States

### 1. Open Volume and Positron Lifetime

The annihilation characteristics of trapped positrons serve as fingerprints in defect identification. The positron lifetime at a defect is a basic quantity for two reasons. First, it reflects the open volume of the defect, and second, it can be predicted by calculations.

Electron irradiation at 1- to 2-MeV energies is the experimental way to produce a controlled concentration of vacancies. By changing the fluence, one can vary the fraction of positrons annihilating at vacancies. We have often used electron irradiation to find out the annihilation characteristics representing monovacancies. Table I lists the lifetime values for monovacancies in Si, Ge, GaAs, and InP. The identification of the sublattices will be discussed later (see Section V.3). The theoretical results (Puska et al., 1989) are also given. Table I shows that the ratio of the positron lifetime in a monovacancy to that in the lattice is the same in different semiconductors, $\tau_V/\tau_B = 1.1-1.2$. Table I also demonstrates that theory can predict positron lifetimes at various defects.

An interesting question is the effect of the vacancy charge state $q$ on the trapped positron lifetime. Different lifetime values, depending on the charge state, have been measured for, for example, the vacancy–phosphorous pair in Si and the As vacancy in GaAs (see Section VIII.1). Puska et al. (1986)

TABLE I

Lifetime Values for Free Positrons in the Lattice (Bulk) and for Trapped Positrons in Monovacancies Introduced by MeV Electron Irradiations. (The theoretical values have been calculated for unrelaxed vacancies.)

| Material/Defect | $\tau_{exp}$ | $\tau_{theory}$[a] |
|---|---|---|
| Si bulk | $218 \pm 1$[b,c] | 219 |
| $V_{Si}^-$ | $273 \pm 3$[b] | 256 |
| $(V_{Si}\text{-}P)^0$ | $268 \pm 3$[c] | |
| $(V_{Si}\text{-}P)^-$ | $250 \pm 1$[c] | |
| Ge bulk | $230 \pm 3$[d] | 226 |
| $V_{Ge}$ | $290 \pm 4$[d] | 263 |
| GaAs bulk | $229 \pm 1$[e] | 229 |
| $V_{Ga}^{3-}$ | $260 \pm 2$[f] | 265 |
| $V_{As}^0$ | $295 \pm 2$[e,g] | 268 |
| $V_{As}^-$ | $257 \pm 3$[e,g] | |
| InP bulk | $244 \pm 1$[h,i] | 246 |
| $V_{In}$ | $283 \pm 2$[h,i] | 295 |
| $V_P$ | $263 \pm 2$[h,i] | 273 |

[a]Puska et al. (1989).
[b]Mäkinen et al. (1989).
[c]Mäkinen et al. (1992).
[d]Corbel et al. (1985).
[e]Saarinen et al. (1991a).
[f]Corbel et al. (1992).
[g]Corbel et al. (1988).
[h]Törnqvist et al. (1994).
[i]Alatalo et al. (1995).

have studied this problem by ab initio calculations. When the charge state has been changed without any relaxation effects, keeping the atoms at fixed positions, the results indicate a very small effect, $\Delta\tau_V/\Delta q \approx 1\text{--}2 \text{ ps/e}$. This can be understood as a competition between the changes in the electron density and positron localization. When a vacancy is more negative, the positron is better localized inside the vacant cell. On the other hand, as the electron density in the vacant cell increases, the total electron–positron overlap remains almost unchanged. Thus, we can conclude that the trapped positron lifetime reflects the open volume of the defect independent of its charge state. The direct consequence is that different lifetimes for different charge states are due to changes in the open volume caused by lattice relaxation.

The question arises whether the trapped positron itself could affect the lattice relaxation because of the positron–ion repulsion. The first Car-

Parrinello calculations (Gilgien et al., 1994) seemed to point in this direction. However, it has been realized that much of this effect may come from insufficient knowledge of the positron–electron correlation (Puska et al., 1995). The numerical artefacts due to the finite size of the computational supercell may also affect the theoretical ionic relaxation and the positron state (Korhonen et al., 1996).

Calculations also have been performed for vacancy clusters. They indicate that the positron lifetime increases with the open volume (Puska et al., 1989; Puska and Corbel, 1988; Saito and Oshiyama, 1996). Experimentally, it is difficult to prepare samples with controlled populations of vacancy clusters. The only well-defined cluster where the positron lifetime has been measured is the divacancy in Si. The result is $\tau_V/\tau_B \approx 1.4$ (Mascher et al., 1989), in good agreement with theory.

In defect studies of layers, the only characteristic that can usually be measured is the Doppler broadening (i.e., the one-dimensional electron–positron momentum distribution). The momentum distribution is much more complicated to interpret and calculate than the positron lifetime, because the individual electron wave functions around the defect are needed [see Eq. (10)].

The low electron-momentum parameter $S$ arises predominantly from annihilation with valence electrons, the wave functions of which are strongly relaxed in the vicinity of a defect. Therefore, heavy computations are needed to calculate the $S$ parameter, and only recently have such calculations been initiated (Hakala et al., 1996). It is, however, straightforward to predict that a defect with a bigger open volume gives rise to a higher $S$ value than a defect with a smaller open volume. There exist some experimental rules of thumb based on comparative lifetime and Doppler measurements in bulk crystals: $S_D/S_B \approx 1.01–1.02$ for monovacancies, $S_D/S_B \approx 1.04$ for divacancies, and $S_D/S_B > 1.05$ for vacancy clusters.

The high electron-momentum parameter $W$ arises from annihilation with rigid core electrons. Their momentum distribution and the value of the $W$ parameter can be calculated once the overlap of the positron and core electron wave functions are known (Alatalo et al., 1995, 1996). The partial core annihilation rates, $\lambda_{core}$, sometimes reported in the positron lifetime calculations (Puska et al., 1989), can also be used to estimate the relative changes of $W$. If the atoms surrounding defects are the same, then a smaller $W$ value corresponds to a bigger open volume. However, the $W$ parameter depends strongly on the chemical identities of the atoms surrounding the annihilation site. The values of $W$ are therefore not comparable for defects in different materials nor on different sublattices. Successful applications of the $W$ parameter to identify vacancies, divacancies, and vacancy clusters in CdTe and CdHgTe have been reported (Liszkay et al., 1994, 1995).

In summary, the open volume of a defect can be estimated from the trapped positron lifetime. The lifetime does not tell whether a vacancy is isolated or complexed with an substitutional impurity. The $S$ parameter, when arising mainly from valence electrons, reflects also the open volume. The $W$ parameter and core electron annihilation are sensitive to the chemical nature of the defect surroundings, and this will be further discussed in Section V.3.

## 2. Vacancy Charge State: Ga Vacancy in GaAs

To illustrate how positron lifetime experiments and the positron trapping model can be used to study vacancies in semiconductors, we review here results on the observation of native Ga vacancies in semi-insulating (SI) GaAs crystals (Saarinen et al., 1993; Le Berre et al., 1995a). Figure 9 shows the average positron lifetime as a function of measurement temperature in

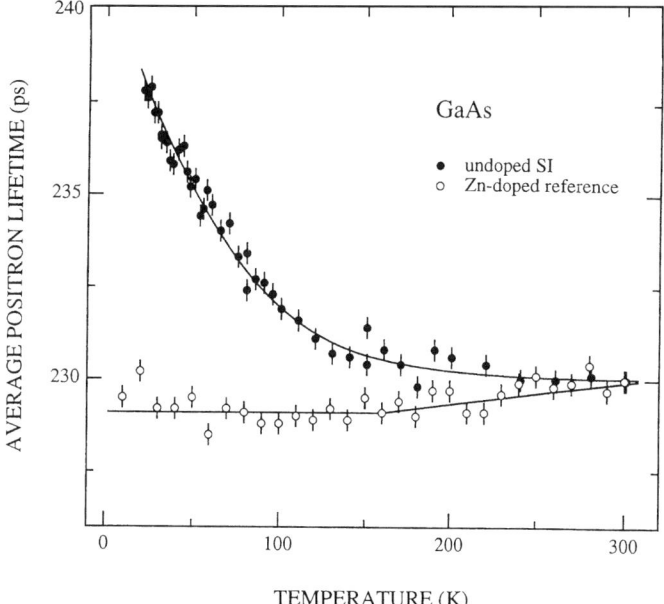

Fig. 9. Average positron lifetime in Zn-doped and undoped SI GaAs as functions of temperature. The curve in GaAs(Zn) corresponds to free positron annihilation in the lattice. The higher values in SI GaAs and the strong decrease with temperature are due to a negative vacancy identified as the Ga vacancy.

SI GaAs. The reference lifetime $\tau_B$ in the perfect GaAs lattice was obtained for heavily Zn-doped material, where the positron lifetime varies only weakly as a function of temperature, mainly due to lattice expansion (Corbel et al., 1988; Saarinen et al., 1991a). The average positron lifetime at 300 K in SI GaAs is very near the free positron lifetime (namely, between 232 and 234 ps in all the samples). When the measurement temperature decreases, the average lifetime increases, and at temperatures $T < 100$ K, it is clearly longer than the free positron lifetime in the GaAs lattice. In several samples, the lifetime spectra measured at temperatures below 80 K can be decomposed into two components. The shorter lifetime is $\tau_1 = 150–200$ ps, and the second component is $\tau_2 = 250–260$ ps, with an intensity around 50%.

The results of the measurements show that the average positron lifetime in darkness is longer than the free positron lifetime. This indicates that there are vacancy-type defects present in as-grown SI GaAs samples. When a single type of vacancy defect exists in a material and no positron detrapping from the vacancy occurs, the positron lifetime spectrum has two components, and the lifetime and intensities can be obtained from the solution of Eqs. (13) and (14) with $N = 1$ and $\delta_D = 0$:

$$\tau_1^{-1} = \tau_B^{-1} + \kappa, \tag{35}$$

$$\tau_2 = \tau_D, \tag{36}$$

$$I_2 = 1 - I_1 = \frac{\kappa}{\kappa + \lambda_B - \lambda_D}. \tag{37}$$

The first lifetime, $\tau_1$, represents the effective lifetime in the lattice in the presence of positron trapping at defects. Since $\kappa > 0$ in Eq. (35) and $I_2 > 0$ in Eq. (37), $\tau_1$ is less than $\tau_B$. The second lifetime component, $\tau_2$, characterizes positrons trapped at monovacancies [(Eq. (36)]. In SI GaAs, we obtain $\tau_2 = 255 \pm 5$ ps, which is a typical value for monovacancies in GaAs and clearly less than that expected for a divacancy (Corbel et al., 1985, 1988; Saarinen et al., 1991a; Puska and Corbel, 1988; Laasonen et al., 1991). The positron lifetime experiments thus show that SI GaAs contains native monovacancy defects.

The positron trapping rate at the vacancies in SI GaAs can be calculated accurately from the average positron lifetime (Section IV.1). With $N = 1$ and $\delta_D = 0$, Eqs. (24) and (26) yield

$$\kappa = \lambda_B \frac{\tau_{av} - \tau_B}{\tau_D - \tau_{av}}. \tag{38}$$

The Fermi level in SI GaAs is pinned to the EL2 level, and the charge states

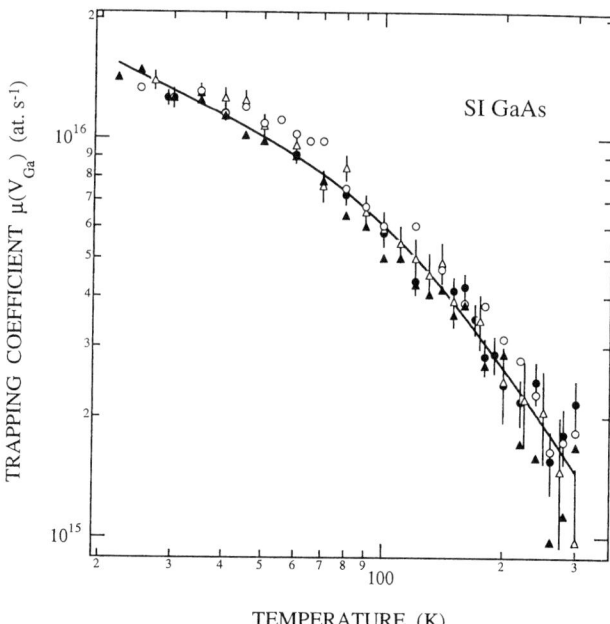

FIG. 10. The temperature dependence of the positron trapping coefficient at Ga vacancies in four SI GaAs samples, indicated by different markers.

of defects other than EL2 do not change as a function of temperature. The increase of the positron lifetime (see Fig. 9) and the consequent increase in the positron trapping rate as the temperature decreases are thus due to the temperature dependence of the positron trapping coefficient at the vacancy ($\kappa(T) = \mu(T)c$, where $c$ is the concentration of vacancies). The temperature dependence $\mu(T) = \kappa(T)/c$ can be calculated from Eq. (38) and is presented in Fig. 10 for several SI GaAs samples. The absolute value of the trapping coefficient has been fixed at $\mu(300\,\text{K}) = 1.4 \times 10^{15}\,\text{s}^{-1}$ on the basis of electron irradiation data (Corbel et al., 1992). Since positron trapping at neutral defects is independent of temperature, the increase of the trapping coefficient at low temperatures (see Fig. 10) indicates that the observed vacancies are negatively charged. According to theoretical calculations, the As vacancies are positive and the Ga vacancies are negative in SI GaAs (Baraff and Schlüter, 1985; Puska, 1989; Xu and Lindefelt, 1990). Positive As vacancies repel positrons, but the negative Ga vacancies are able to trap positrons. We thus associate the vacancies with negative Ga vacancies. The positron experiments are unable to identify, however, whether the Ga vacancies are isolated or bound to a defect complex.

As shown in Fig. 10, two different regions can be observed in the temperature dependence of the positron trapping coefficient at the Ga vacancy. First, the trapping coefficient decreases with temperature as $\mu \propto T^{-0.5}$ at 30 to 80 K. This dependence is a direct consequence of the Coulomb wave nature of the positron wave function far away from the defect (see Section III.1). Second, the decrease of the trapping coefficient is much stronger, $\mu \propto T^{-1.2}$ at 100 to 300 K. This behavior demonstrates that the dominant mechanism for positron trapping into the ground state of the Ga vacancy takes place through shallow Rydberg-like precursor states (see Section III.1), from which the positron can also escape back to the free state. The temperature dependence of the positron-trapping coefficient through the Rydberg state into the ground state of the negative vacancy can be expressed as (Puska et al., 1990; Mäkinen et al., 1992; LeBerre et al., 1995a)

$$\mu_V(T) = \frac{\mu_R}{1 + \frac{\mu_R/N_{at}}{\eta_R}\left(\frac{m_+ k_B T}{2\pi \hbar^2}\right) \exp(-E_b/k_B T)}. \quad (39)$$

In Eq. (39), $\mu_R = \mu_{R0} T^{-1/2}$ is the positron trapping coefficient from the free state to the Rydberg precursor state, and $\eta_R$ is the transition rate from the precursor state to the positron ground state at the vacancy. $E_b$ is the binding energy of the positron to the Rydberg precursor state, and $N_{at}$ is the atomic density of the host material.

The temperature dependence of the positron trapping coefficient can be modeled with Eq. (39), and the quantities $E_b$, $\mu_R$, and $\eta_R$ can be determined as fitting parameters. The fit is shown by the solid line in Fig. 10 and it is in excellent agreement with the data in all samples in the temperature range $T = 30$–$300$ K. The recent estimate for the positron binding energy at the Rydberg precursor state is $E_b = 15$ meV. The positron trapping coefficient into the Rydberg state is $\mu_R = 1.1 \times 10^{16}$ s$^{-1}$ at 30 K, and the transition rate to the ground state is $\eta_R = 3.0 \times 10^{11}$ s$^{-1}$. These values are reasonable compared with the theoretical and experimental results for positron trapping at negative vacancies in Si (Puska et al., 1990; Mäkinen et al., 1989; LeBerre et al., 1995a).

The gallium vacancy concentrations can be estimated from the experimental trapping rates. The concentrations are between 1 and $12 \times 10^{15}$ cm$^{-3}$ in the samples that have been studied. These concentrations are less than typically found for the EL2 defect, but they are equal to or larger than the usual impurity concentrations in LEC GaAs. Since the Ga vacancies are negatively charged, they also play a role in the electrical compensation of undoped SI GaAs.

5 POSITRON ANNIHILATION SPECTROSCOPY OF DEFECTS IN SEMICONDUCTORS    239

3. IDENTIFICATION OF VACANCY SUBLATTICE AND IMPURITY
   SURROUNDINGS

As discussed in Section V.1, the lifetime of the trapped positron reflects the open volume of the defect, but it is insensitive to its chemical surrounding. Hence, this type of experiment cannot be used to directly identify the sublattice of the vacancy or to determine whether the vacancy is isolated or complexed with impurity atoms.

The Doppler broadening experiments provide information on the momentum distribution of the annihilating electrons. By the coincidence technique, one can reveal the core electron momentum distribution, which carries information about the type of atoms in the region of annihilation. In the case of a vacancy, the positron wave function is localized and overlaps predominantly with the core electrons of the neighboring atoms. Therefore, vacancies on different sublattices can be distinguished in compound semiconductors, and impurities associated with vacancies may be identified. Next, we discuss this subject systematically.

*a. Elemental Sensitivity of the Core Annihilation Spectrum*

The sensitivity of the Doppler broadening experiments to the atomic number of the elements is clearly visible in the core annihilation spectra of free positrons in various semiconductors (Fig. 11). The spectrum in GaAs has a high intensity and wide shape, whereas that recorded in GaSb is much narrower. In InP, the core annihilation spectrum decreases even faster than in GaSb. In Si, the intensity of the spectrum is clearly less than that in the other materials shown in Fig. 11. It is thus evident that different elements give observable changes in the shape of the Doppler broadening spectrum of the positron annihilation radiation.

The core annihilation spectra can be qualitatively understood by taking into account only the momentum distribution of the outermost core electron shells of the material (Alatalo *et al.*, 1995). In GaAs, most of the core annihilation occurs with Ga and As 3d electrons, which are relatively tightly bound to the nuclei and quite localized in $r$ space. The corresponding momentum distribution is broad and extends to large momentum values in $p$ space. In InP, the dominant contribution to the core annihilation comes from In 4d electrons. They are less tightly bound and extend to a wider region in $r$ space and thus give a narrower and more rapidly descending contribution in $p$ space. In GaSb, the core annihilation spectrum has contributions from both Ga 3d and Sb 4d electrons, which results in a shape between those observed in InP and GaAs. Since there are no d electrons in

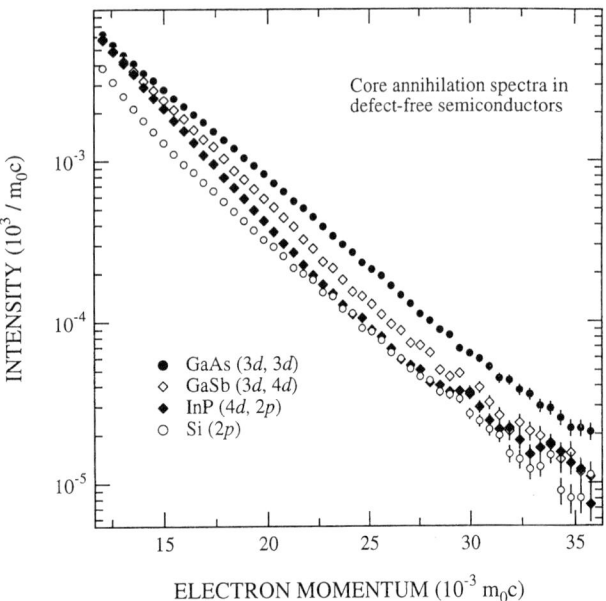

FIG. 11. The high-momentum parts of the Doppler curves due to free positron annihilation with core electrons in various semiconductors.

Si, and the positron overlap with the Si 2p electrons is small due to the nuclear Coulomb repulsion, the core region of the momentum distribution has lower intensity than for GaAs or InP.

### b. *Identification of the Sublattice of a Vacancy Defect*

Due to its sensitivity to the chemical nature of atoms, the core electron-momentum distribution can be used to identify the sublattice of a vacancy in a binary compound. As an example, Fig. 12 shows the high-momentum part of the Doppler broadening spectrum, which was recorded at 300 K for the $[N] = 1.7 \times 10^{19}$ cm$^{-3}$ and $[Cl] = 1.5 \times 10^{18}$ cm$^{-3}$ doped $ZnS_xSe_{1-x}$ overlayers using a low-energy positron beam (Saarinen et al., 1996). Both layers contain vacancies. The data in the N-doped material are above those in Cl-doped $ZnS_xSe_{1-x}$ in the momentum range of $p_L = (10-25) \times 10^{-3} m_0 c$, indicating that the magnitude of the core electron annihilation is about 20% larger at the vacancy in N-doped than in Cl-doped $ZnS_xSe_{1-x}$. However, at the larger momenta of $p_L = (25 - 40) \times 10^{-3} m_0 c$, the data obtained in

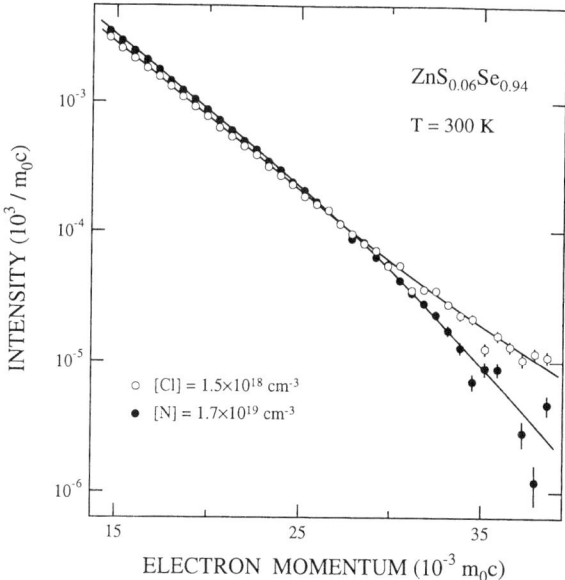

FIG. 12. The core electron-momentum distributions at vacancy defects in Cl-doped and N-doped $ZnS_{0.06}Se_{0.94}$ layers. The vacancies have been associated with the Zn and Se sublattices, respectively. (From Saarinen et al., 1996, with permission.)

the N-doped material fall below those recorded in Cl-doped $ZnS_xSe_{1-x}$. The two curves in Fig. 12 have different shapes: in Cl-doped $ZnS_xSe_{1-x}$, the core electron-momentum distribution is clearly broader than in N-doped $ZnS_xSe_{1-x}$.

The comparison of the momentum distributions indicates that the core electrons around the vacancy have higher momenta in Cl-doped than in N-doped $ZnS_xSe_{1-x}$. This implies that the chemical nature of the atoms surrounding the vacancy are different in the two cases. The 3d electrons of Zn atoms are less localized than those of Se atoms, which leads accordingly to a narrower momentum distribution. Hence, the data of Fig. 12 indicate that positrons trapped at vacancies annihilate preferentially with the core electrons of Zn atoms in N-doped $ZnS_xSe_{1-x}$ and with those of Se atoms in Cl-doped $ZnS_xSe_{1-x}$. The core electron-momentum distribution thus gives a direct identification of the Se vacancy in N-doped $ZnS_xSe_{1-x}$ and the Zn vacancy in Cl-doped material.

The detection of positron trapping at Se vacancies is itself interesting, and the following conclusions can be made (Saarinen et al., 1996). According to theoretical calculations, the Se vacancy is positive and repulsive to positrons in p-type $ZnS_xSe_{1-x}$ (Garcia and Northrup, 1995). The Se vacancy detected in the positron experiments is perhaps part of a complex, where the positive Se vacancy is closely associated with a negatively charged acceptor, possibly with the nitrogen dopant. The concentration of Se vacancy complexes is of the same order of magnitude as the nitrogen doping concentration ($\geqslant 10^{18}\,\text{cm}^{-3}$), suggesting that the role of $V_{Se}$ is important in the electrical compensation of p-type $ZnS_xSe_{1-x}$.

### c. Identification of Vacancy–Impurity Complexes

An interesting application of core annihilation spectroscopy is the identification of vacancy–impurity complexes. The decoration of a vacancy by an impurity normally does not change the positron lifetime. However, if the electronic structure of the impurity differs significantly from that of the host atoms, the impurity in the vicinity of the vacancy may be distinguished simply from the shape of the core electron-momentum distribution.

Heavily Zn-doped InP has been found to contain native vacancy defects with a positron lifetime of about 325 ps (Dlubek et al., 1985). By analyzing the positron lifetime experiments in terms of the positron trapping model (see Section III.2), the fraction of annihilations at vacancy defects can be estimated. With this information, the Doppler broadening curve can be decomposd into bulk and vacancy components. The core annihilation spectrum due to positrons trapped at the 325-ps vacancy in $[Zn] = 6.4 \times 10^{18}\,\text{cm}^{-3}$ doped InP is shown in Fig. 13 (Alatalo et al., 1995). The curve for an electron-irradiated InP sample containing P vacancies is shown for comparison.

The shape of the curve for the vacancy in the Zn-doped sample is broader than those for the In and P vacancies. This indicates that the native vacancy is different from both isolated $V_{In}$ and $V_P$. As explained in the case of free positron annihilation (see Fig. 11), 3d electrons give a broader distribution in **p** space than do 4d electrons. The broad high-momentum part can thus be associated with the 3d electrons of the Zn atoms surrounding the vacancy. Although the lifetime at the vacancy is large (325 ps), the core electron spectrum in InP(Zn) lies high at all momenta, indicating a large number of core electron annihilations. This effect is the fingerprint of the phosphorus vacancy (see Section IV.3 and Fig. 8): the positron in the In vacancy gives only a small core electron annihilation signal from the neighboring P atoms, whereas the positron in the P vacancy yields a strong core annihilation component due to the surrounding In atoms. We can thus identify the native vacancy observed in InP(Zn) as the P vacancy decorated

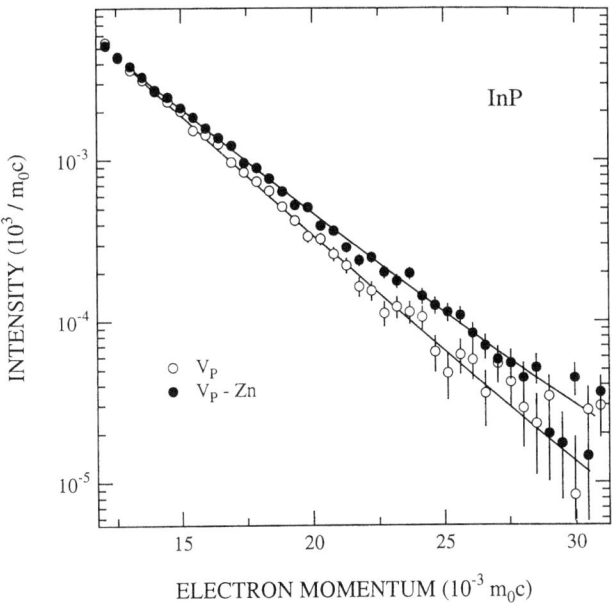

FIG. 13. Core electron-momentum distributions for the phosphorous vacancy and for a native defect in highly Zn-doped InP identified as a $V_P$–Zn complex. (From Alatalo et al., 1995, with permission.)

by Zn impurities. The formation of this vacancy complex may play an important role in the electrical compensation of heavily doped p-type InP.

## VI. Negative Ions as Shallow Positron Traps

In this section, we show that positrons are trapped by negative ions with no open volume (e.g., antisites, interstitials, or impurities) in GaAs materials. The negative ions have been observed by positrons to be native defects in n-type bulk GaAs (Saarinen et al., 1989), undoped SI GaAs (LeBerre et al., 1995a), and in n-type layers grown by molecular beam epitaxy (Laine et al., 1996a). In addition, we have observed positron trapping at negative ions in GaAs after electron irradiation (Corbel et al., 1990, 1992; Saarinen et al., 1995b) and deformation (Saarinen et al., 1990), which shows that some intrinsic defects with no open volume in their atomic structure are acceptors in GaAs. We have attributed these intrinsic defects to Ga antisites.

### 1. NATIVE DEFECTS IN As-GROWN GaAs AS NEGATIVE IONS

In n-type Si-doped liquid encapsulated Czochralski (LEC)-grown GaAs, the positron average lifetime is longer than in the lattice ($\tau_{av} > \tau_B$), and the lifetime spectra can be resolved into two components. These observations indicate that positrons detect native vacancy defects, which are generally attributed to As vacancies (for details, see Section VIII.1). Depending on the position of the Fermi level, these vacancies are either neutral or negatively charged. The positron lifetime is 295 ps at $V_{As}^0$ and 257 ps at $V_{As}^-$ (see Section VIII.1). As a function of temperature (Fig. 14), the configuration $\tau_2 = 257$ ps ($V_{As}^-$) is replaced at $T = 100-200$ K by the configuration $\tau_2 = 295$ ps ($V_{As}^0$). Interestingly, the intensity $I_2$ of the component $\tau_2 = 257$ ps increases with

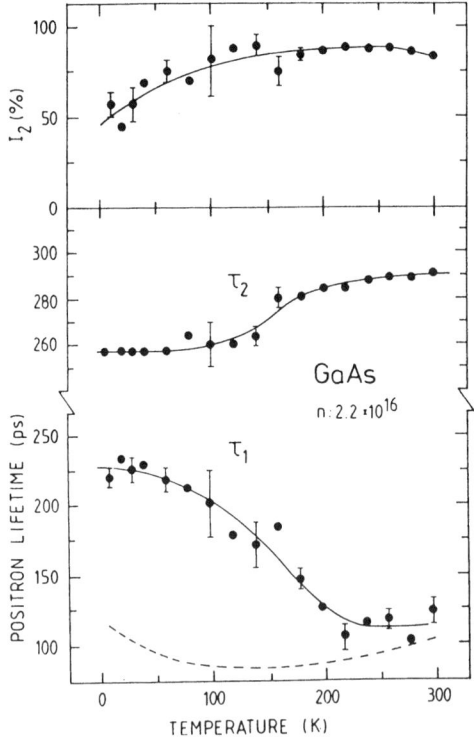

FIG. 14. Decomposition of the positron lifetime spectra as a function of temperature in n-type GaAs. The dashed line indicates the values for $\tau_1$ calculated from the one-defect trapping model [Eq. (40)]. It corresponds to the situation in which only vacancy defects are present. (From Saarinen et al., 1989, with permission.)

temperature, although enhanced positron trapping at negative vacancies can be expected at low temperatures (see Section V.2). This behavior suggests that another defect competes with the As vacancy as a positron trap at low temperatures.

Although more than one type of positron trap is present in the lattice, the positron lifetime analysis with more than two lifetimes is often not possible due to the high number of free parameters in Eq. (17). However, in a two-component fit, the longest lifetime $\tau_2$ is usually clearly separated, while the shorter lifetime $\tau_1$ becomes a superposition of the other decay components. In this case, either Eq. (18) or Eqs. (35) through (38) of the simple positron trapping model can be used to check whether the analysis is compatible with one defect type only. If only one type of trap is present and no detrapping occurs, the test lifetime $\tau_{1,\text{test}}$ calculated from Eqs. (35) and (38) using the experimental values of $\tau_{\text{av}}$ and $\tau_2$,

$$\tau_{1,\text{test}}^{-1} = \tau_B^{-1} + \kappa = \tau_B^{-1}\left(1 + \frac{\tau_{\text{av}} - \tau_B}{\tau_2 - \tau_{\text{av}}}\right), \quad (40)$$

should coincide with the experimental values of $\tau_1$. The values of $\tau_{1,\text{test}}$ calculated from Eq. (40) are shown in Fig. 14 by the dashed line.

The difference between the values of $\tau_1$ and $\tau_{1,\text{test}}$ is evident in Fig. 14: below 200 K, the experimental values are systematically 50 to 100 ps larger than those expected from Eq. (40). This indicates that, in addition to the vacancy defects with $\tau_2 = 257–295$ ps, other positron traps are active at low temperatures. The experimental shortest lifetime component $\tau_1$ is a superposition of annihilations in the lattice $[\tau_1' = \tau_B/(1 + \kappa \tau_B)$ from Eq. (35)] and at the additional traps ($\tau_1'' = \tau_{\text{ST}}$), which are effective only at low temperatures. Below 100 K, where $\tau_2$ is constant, $\tau_2 = 257$ ps, $\tau_1$ stops increasing and tends to reach the constant value of 230 ps. At the saturation $\tau_1 = \tau_{\text{ST}}$, annihilation in the lattice is completely suppressed and all positrons annihilate either at the vacancies with $\tau_2 = 257$ ps, or at the defects characterized by the lifetime $\tau_{\text{ST}} = 230$ ps. The saturation level $\tau_1 = \tau_{\text{ST}} = 230$ ps coincides with the lifetime of free positrons in GaAs, indicating that the additional traps have no open volume. Furthermore, these defects are shallow positron traps in the sense that they are able to capture positrons only below 200 K, as seen in Fig. 14.

The shallow positron state observed in n-type GaAs can be induced by a negative ion, which is a negatively charged impurity atom or an intrinsic point defect without any open volume. Being a positive particle, the positron can be localized in a Rydberg state of the Coulomb field around a negatively charged center, and the situation is analogous to the binding of a hole to an acceptor atom. The positron binding energy at the negative ion can be estimated from the simple effective-mass theory:

$$E_{b,n} = \frac{13.6\,\text{eV}}{\varepsilon^2}\left(\frac{m^*}{m_e}\right)\frac{Z^2}{n^2} \approx 10\text{--}100\,\text{meV}, \qquad (41)$$

where $\varepsilon$ is the dielectric constant, $m^* \approx m_e$ is the effective mass of the positron, $Z$ is the charge of the negative ion, and $n$ is the quantum number. With $Z = 1\text{--}3$ and $n = 1\text{--}4$, Eq. (41) yields typically $E_b = 10\text{--}100\,\text{meV}$, indicating that the detrapping of positrons from Rydberg states takes place at 100 to 200 K. Experimentally, this is detected in Fig. 14 as the strong decrease of the lifetime component $\tau_1$ towards the value $\tau_{1,\text{test}}$ predicted by the simple trapping model. Alternatively, it can be seen as an increase in the average positron lifetime, as will be discussed in Section VI.2.

The negative ions in Si-doped n-type GaAs can be residual impurities like $C_{As}^-$, compensating centers like $Si_{As}^-$, or intrinsic acceptors like $Ga_{As}^{2-}$. Native negative ions have been observed by positrons in undoped bulk SI GaAs (LeBerre et al., 1995a) and in molecular beam epitaxy (MBE)-grown Si-doped layers (see Section VII.1) (Laine et al., 1996a). Almost the same positron binding energy, $E_b \approx 50\,\text{meV}$, and the trapping coefficient, $\mu_{ST}$ (30 K) $\approx 10^{16}\,\text{s}^{-1}$, have been estimated for the negative ions present in all these systems, thus suggesting that the nature of the ions is the same. Interestingly, the same values for $E_b$ and $\mu_{ST}$ (30 K) have been determined also for the intrinsic negative ion, which is formed in the electron irradiation of GaAs and attributed to the Ga antisite defect (see Section VI.2). These observations point out that the negative ion detected in as-grown GaAs could be an intrinsic defect like the Ga antisite. However, no direct experimental information on the identity of the negative ion has been obtained in the positron annihilation measurements.

2. NEGATIVE VACANCIES AND NEGATIVE IONS IN ELECTRON-IRRADIATED GaAs

Negative ions give a lifetime that is indistinguishable from the lifetime in the lattice and can be detected by positron lifetime experiments only when they are competing with vacancy-type defects. We present in this section a quantitative analysis of the concentration of negative ions that positrons detect after the electron irradiation of GaAs.

As mentioned in Section V.1, we have investigated electron-irradiated GaAs crystals to better understand the positron interactions with intrinsic defects in GaAs. Undoped SI GaAs crystals were irradiated with 1 to 3-MeV electrons in liquid hydrogen at 20 K. After irradiation, the samples were transferred in liquid nitrogen into the positron lifetime cryostat and then measured as a function of temperature after *in situ* annealing in the range 77 to 450 K. Before irradiation, practically no trapping at vacancies was found in the undoped SI GaAs material.

After 1.5 to 3-MeV electron irradiation and annealing at 77 K, positrons are trapped at irradiation-induced defects in SI GaAs and annihilate with a lifetime of $260 \pm 3$ ps. This lifetime is typical for positrons trapped at monovacancies (Corbel *et al.*, 1988; Saarinen *et al.*, 1991a; Puska and Corbel, 1988; Laasonen *et al.*, 1991). The annealing of these vacancies occurs in a very broad range, from 80 to 500 K, with a sharp main stage around 280 K.

Figure 15 shows the positron lifetime as a function of temperature for three crystals, which have been irradiated to the fluences $(1 \text{ to } 13) \times$

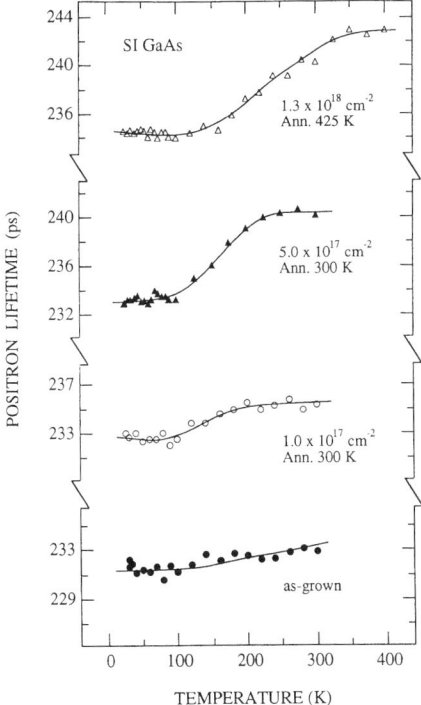

FIG. 15. Fluence effects on the temperature dependence of the average positron lifetime in SI GaAs after 1.5-MeV electron irradiation at 20 K. After irradiation, the samples have been annealed at either 300 or 425 K. The solid lines correspond to the fits of the positron trapping model [Eq. (42)]. (From Saarinen *et al.*, 1995a, with permission.)

$10^{17}$ e$^-$ cm$^{-2}$ at 20 K and thereafter annealed at $T \geqslant 300$ K. Positron trapping at the irradiation-induced vacancies is clearly seen at room temperature, where the average positron lifetime $\tau_{av}$ is well above the lattice value in all three crystals. The average lifetime, however, decreases with temperature and can even reach the value corresponding to the GaAs lattice at $T \leqslant 100$ K ($\tau_B = 230$ ps).

The lifetime values found at 300 and 77 K vary independently, and the temperature range where the lifetime decreases is crystal-dependent. The decrease of the average lifetime is due to the decrease of the intensity $I_2$, and the second lifetime component $\tau_2$ remains constant. As explained in Section VI.1, this behavior reflects competition in positron trapping between vacancies and other defects. The other defects become efficient positron traps when the temperature decreases and the positron lifetime at them is the same as in the lattice. These properties are characteristic for positrons annihilating at Rydberg states of negative ions.

In Fig. 15, the thermal detrapping of positrons from the Rydberg states is marked by the increase of $\tau_{av}$ at $T \geqslant 100$ K. At high temperature, the detrapping is so efficient that no positrons annihilate at the Rydberg states. Consequently, only vacancies trap positrons at the high-temperature plateau of Fig. 15. The onset temperature, where the plateau starts, becomes higher as the irradiation fluence increases. This indicates that the concentration of ions increases with irradiation fluence.

The average lifetime curves in Fig. 15 can be quantitatively analyzed even without knowledge of the intensities $I_1$ and $I_2$. Including the trapping at the vacancies $\kappa_V(T)$, at the negative ions $\kappa_{ST}(T)$, and the thermal detrapping from the ions $\delta_{ST}$ in the positron trapping model [Eqs. (24) and (25)], the average positron lifetime $\tau_{av}$ can be written as:

$$\tau_{av} = \eta_B \tau_B + \eta_{ST} \tau_{ST} + \eta_V \tau_V,$$

$$\eta_B = 1 - \eta_{ST} - \eta_V,$$

$$\eta_{ST} = \frac{\kappa_{ST}}{(1 + \delta_{ST}/\lambda_{ST})\left(\lambda_B + \kappa_V + \dfrac{\kappa_{ST}}{1 + \delta_{ST}/\lambda_{ST}}\right)}, \qquad (42)$$

$$\eta_V = \frac{\kappa_V}{\lambda_B + \kappa_V + \dfrac{\kappa_{ST}}{1 + \delta_{ST}/\lambda_{ST}}},$$

where the detrapping rate from the ions is given by Eq. (12). The expression for $\tau_{av}$ in Eq. (42) can be fitted to the experimental data in Fig. 15 by adjusting the values of the ion concentration $c_{ST} = \kappa_{ST}/\mu_{ST}$, the positron binding energy $E_b$, the vacancy trapping rate $\kappa_V$, and the ion trapping coefficient $\mu_{ST}$, and by postulating the temperature dependencies of the trapping coefficients $\mu_V(T)$ and $\mu_{ST}(T)$. Reasonable estimates of the values of $\kappa_V(300\text{ K})$ and $\kappa_{ST}(20\text{ K})$ can be calculated in the positron trapping model from the experimental data in the high-temperature region, where $\delta_{ST}$ is high ($\delta_{ST} \to \infty$), and in the low-temperature region, where $\delta_{ST}$ is low ($\delta_{ST} \approx 0$), respectively.

After electron irradiation, vacancies can compete at low temperature with the negative ions as positron traps. This indicates that they are also negative. The curves in Fig. 15 could be fitted, postulating that, as predicted by theory, $\mu_{ST}(T)$ varies as $T^{-0.5}$ and that $\mu_V(T)$ follows the dependence [Eq. (39)] determined for the native Ga vacancy. The solid lines in Fig. 15

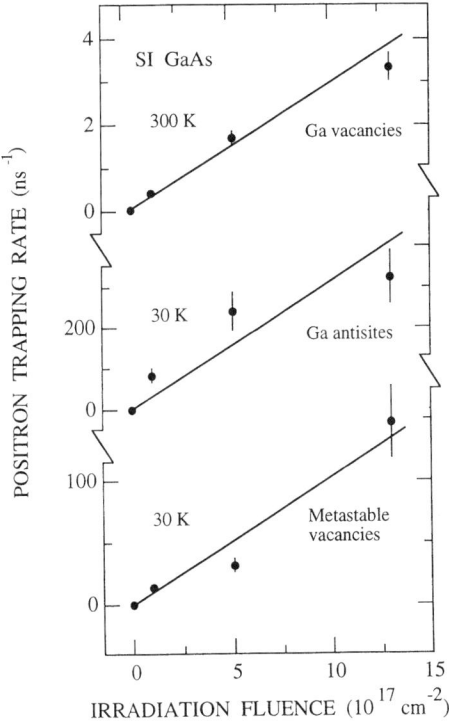

FIG. 16. Positron trapping rate at Ga vacancies, Ga antisite defects, and at the metastable vacancies $V_{\text{irr}}^*$ as a function of the electron irradiation fluence. The trapping rates at $\text{Ga}_{\text{As}}$ and $V_{\text{irr}}^*$ have been determined at 30 K, and the trapping rate at $V_{\text{Ga}}$ at 300 K. After irradiation with 1.5-MeV electrons at 20 K, the samples have been annealed at 300 K. For a detailed discussion of the nature of the metastable vacancies $V_{\text{irr}}^*$, see Section IX.2. (From Saarinen et al., 1995a, with permission.)

correspond to fits with the values of $E_b = 52 \pm 5$ meV for the positron binding energy, $\mu_{ST}(T) = [(5.0 \pm 1.5) \times 10^{16}$ at. s$^{-1}$ K$^{1/2}$]$T^{-0.5}$ for the trapping coefficient, and the positron trapping rates at vacancies and ions shown in Fig. 16.

The positron trapping rates at ions are proportional to the fluences in Fig. 16. This shows that the negative ions are produced by irradiation and that they are likely to be primary defects. The introduction rate of the negative ions is as high as $2.0 \pm 0.5$ cm$^{-1}$. The negative ions are thermally more stable than the vacancies: they recover at about 520 K, whereas the main stage for vacancies is at 250 K.

Positron data give evidence that irradiation introduces two types of acceptor defects: one is vacancy-like and the other ion-like. The simple intrinsic defects that are believed to be negatively charged in SI GaAs (see Section 5.2) are the gallium vacancies and the gallium antisites. We assign the vacancies to the gallium vacancies and the ions to the gallium antisites. Monte Carlo simulations show that antisites are directly formed during irradiation by replacement collisions (Mattila and Nieminen, 1995). The annealing stages in the positron experiments, 250 K for Ga vacancies and 520 K for Ga antisites, correlate well with the recovery stages of the resistivity in irradiated GaAs (Thommen, 1970; Lang, 1977). Therefore, the acceptors $V_{Ga}$ and $Ga_{As}$ contribute strongly to the compensation of n-type GaAs by electron irradiation (Saarinen et al., 1995b).

## VII. Defects in Layers Studied by a Low-Energy Positron Beam

In this section, we describe how vacancies and negative ions in thin overlayers can be studied using a low-energy positron beam. As an example, we first present results on the identification of the compensating vacancies and negative ions formed during the MBE growth of very highly Si-doped GaAs (Laine et al., 1996a). In the second example, we demonstrate how detailed information can be obtained on the depth distribution of vacancy defects formed by the ion implantation of GaAs (Pfeiffer et al., 1994, 1996).

### 1. Compensating Defects in Highly Si-doped GaAs Layers

N-type doping of GaAs with Si impurities is easy until the doping level is very high. At Si concentrations $>10^{19}$ cm$^{-3}$, strong deactivation of the dopant atoms is seen. This phenomenon is often explained with the autocompensation mechanism: at high concentration, the Si impurities occupy both the group III and the group V lattice sites. The acceptors $Si_{As}^-$ compensate the donors $Si_{Ga}^+$, resulting in a strong reduction of the efficiency of the n-type doping. However, there is increasing evidence that the autocompensation mechanism alone cannot explain the compensation (Schuppler *et al.*, 1993, 1995). There exist both theoretical (Zhang and Northrup, 1991; Northrup and Zhang, 1993) and experimental (Maguire *et al.*, 1987; Lee *et al.*, 1990; Chicihibu *et al.*, 1993) arguments that propose that Ga vacancy-related defects may exist at high concentrations in n-type GaAs. Positron annihilation is an ideal method to clarify the role of vacancies in the compensation.

Figure 17 shows the valence annihilation parameter $S$ as a function of the incident positron energy in a 2-$\mu$m-thick GaAs ([Si] = 5 × 10$^{19}$ cm$^{-3}$) overlayer on top of a SI GaAs substrate. For a given positron implantation

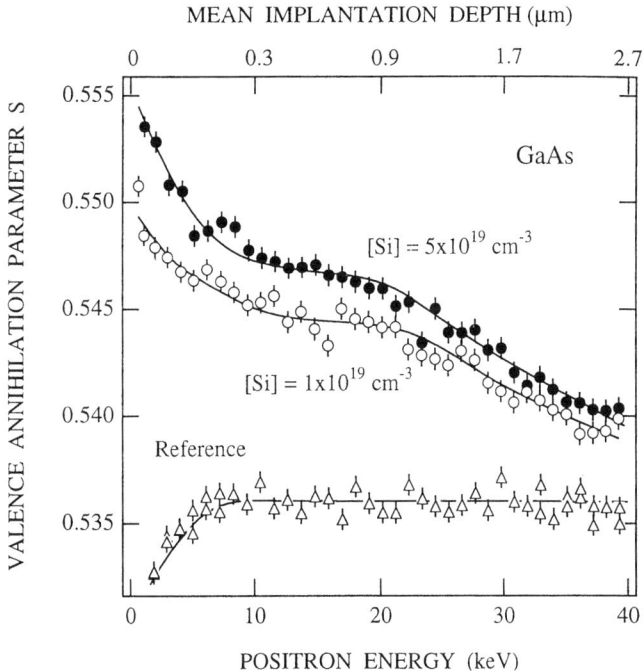

FIG. 17. The low electron-momentum parameter $S$ as a function of positron implantation energy in a defect-free referene GaAs sample and in two, 2-$\mu$m-thick Si-doped GaAs layers that have been grown on SI substrates. The mean implantation depth corresponding to the positron energy scale is indicated on the top axis.

energy $E$, the lineshape parameter $P(E)$ [$S(E)$ or $W(E)$] is a linear superposition of the parameter values for the different positron annihilation states, weighted with the annihilation fraction $\eta(E)$ for the corresponding state:

$$P(E) = \eta_S(E)P_S + \sum_i \eta_{Di}(E)P_{Di} + \eta_B(E)P_B. \tag{43}$$

In Eq. (43) $P_S$, $P_B$, and $P_{Di}$ represent the values of $P$ ($S$ or $W$) at the surface, in the lattice, and in the defect $i$, respectively. In the absence of positron trapping at defects ($\eta_{Di} = 0$), the surface annihilation fraction $\eta_S(E)$ depends only on the positron diffusion length and implantation depth. In the presence of defects that trap positrons, the fractions $\eta(E)$ and thus $S(E)$ and $W(E)$ depend also on the defect concentrations. The $S$ parameter in Fig. 17 decreases rapidly with increasing energy from the surface value $S_S$, whereafter it reaches a plateau. This plateau is clearly higher than the value of $S$ in the defect-free reference GaAs (Fig. 17), indicating that the Si-doped layer contains a large concentration of open volume defects. At $E > 25$ keV, the $S$ parameter decreases again towards the reference level $S_B$. At these incident energies, positrons penetrate and annihilate in the substrate with a low defect concentration, leading to a decrease in $\eta_D(E)$ in Eq. (43).

The experiment as a function of the positron beam energy thus reveals a high concentration of vacancy defects in the GaAs ([Si] = $5 \times 10^{19}$ cm$^{-3}$) layer. The core annihilation spectrum indicates further that the vacancies are on the Ga sublattice (Laine et al., 1996a). The positron data do not tell whether the detected Ga vacancy is isolated or a part of a complex. However, the results of localized vibrational mode (LVM) spectroscopy (Maguire et al., 1987) and theoretical calculations (Northrup and Zhang, 1993) have provided evidence that the Ga vacancy could exist in a complex with the Si atom in the configuration $V_{Ga} - Si_{Ga}$.

The defects present in the highly Si-doped layers were further studied by recording the valence annihilation parameter $S$ as a function of temperature. This experiment was performed at a fixed positron energy of 20 keV, because at this energy the contributions of the annihilation events at the surface and in the substrate are negligible (see Fig. 17). As seen in Fig. 18, the $S$ parameter decreases when temperature is lowered. This behavior is typical when both negative ion-type defects and negative vacancies are present as competitive positron traps (see Section VI). The positron trapping model presented in Section VI.2 can be used to analyze the concentrations of both ions and vacancies. The fits are shown by solid lines in Fig. 18, and the

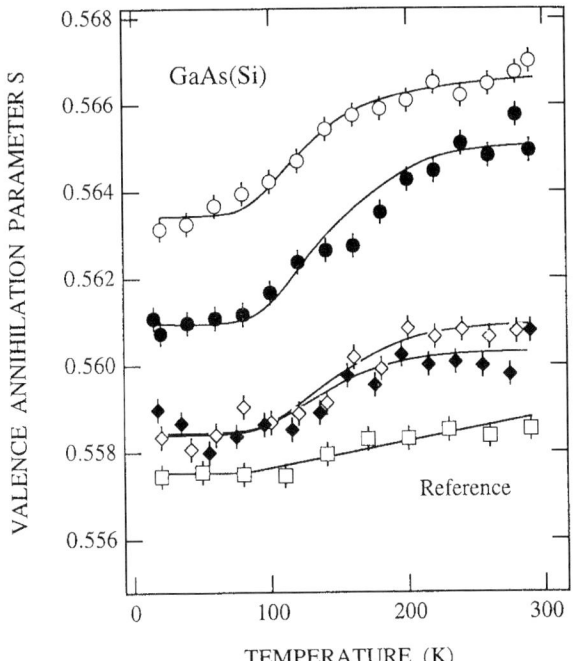

FIG. 18. The low electron-momentum parameter $S$ in the defect-free reference GaAs sample ($\square$) and in Si-doped layers with [Si] = $1 \times 10^{18}$ cm$^{-3}$ ($\blacklozenge$), $3 \times 10^{18}$ cm$^{-3}$ ($\lozenge$), $9 \times 10^{18}$ cm$^{-3}$ ($\bullet$), and $5 \times 10^{19}$ cm$^{-3}$ ($\bigcirc$) as a function of temperature. The solid lines correspond to the fits of the positron trapping model [Eq. (42)]. (From Laine et al., 1996a, with permission.)

adjusted values for the ion and vacancy concentrations are presented in Fig. 19.

The analysis indicates that the concentrations of Ga vacancies and negative ions increase strongly when the Si concentration exceeds $5 \times 10^{18}$ cm$^{-3}$ (see Fig. 19). At this doping concentration, the activation of Si atoms decreases strongly and the material becomes compensated. The concentration of negative ions is almost the same as that of $Si_{As}^-$ measured by NEXAFS experiments (Schuppler et al., 1995), suggesting that the negative ion detected by positrons could be the Si atom acting as an acceptor on the As lattice site. When [Si] = $5 \times 10^{19}$ cm$^{-3}$, the concentration of Ga vacancies is as high as $(1-2) \times 10^{19}$ cm$^{-3}$. This indicates that the Ga vacancies detected in positron experiments play an important role in the compensation of highly Si-doped GaAs.

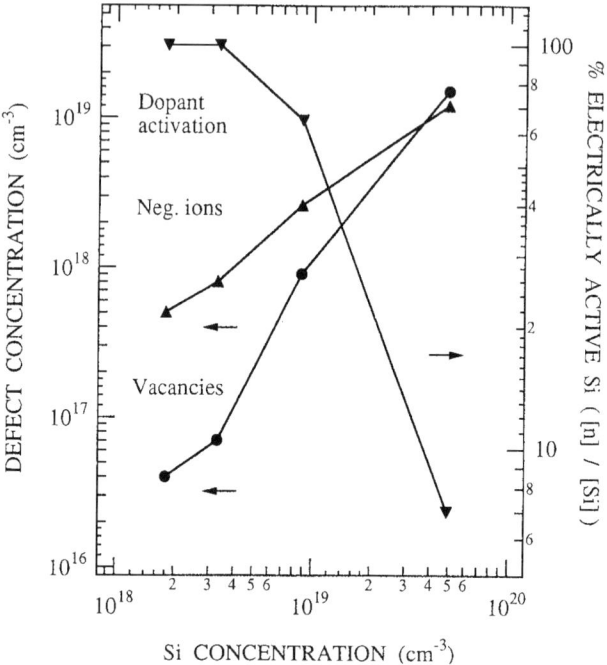

FIG. 19. The dopant activation, expressed as the ratio of free carrier and Si concentrations (▼), the concentration of negative ions (▲), and the concentration of Ga vacancies (●) as functions of Si concentration in the GaAs layer. (From Laine et al., 1996a, with permission.)

### 2. ANALYSIS OF DEPTH PROFILES OF VACANCIES IN ION-IMPLANTED GaAs

The positron beam can be used to analyze also the depth distributions of vacancies close to the sample surface. As an example, Fig. 20 shows the $S$ parameter as a function of positron energy in SI GaAs, which has been implanted with 350-keV In ions at room temperature (Pfeiffer et al., 1994, 1996). The implanted In atoms stop at a depth of 100 nm from the surface. As seen in Fig. 20, the vacancies introduced in the slowing down of the In atoms are clearly visible in the positron experiment: the $S$ parameter has a maximum at 5 keV, which corresponds to a positron mean implantation depth of 100 nm. At higher energies, the $S$ parameter decreases gradually toward the level of the unimplanted crystal.

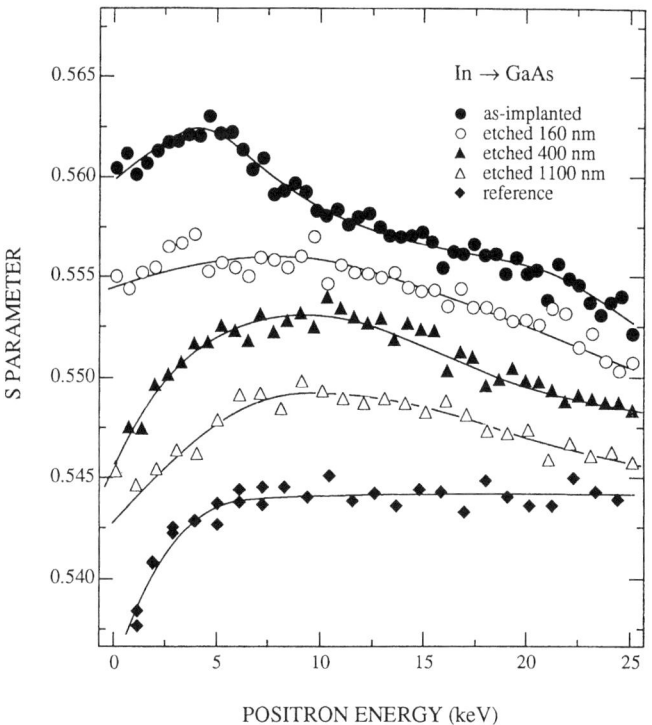

FIG. 20. The low electron-momentum parameter $S$ as a function of positron energy in a reference GaAs sample and in In-implanted GaAs after various etching treatments. The In implantation was performed at 300 K with 350-keV In ions to a fluence of $10^{13}$ cm$^{-2}$. The solid lines are fits to the positron diffusion model [Eqs. (45) through (48)] (Pfeiffer et al., 1996).

Obviously, information on the depth profile of vacancies can be extracted from the analysis of the data in Fig. 20. However, the analysis is not straightforward due to two reasons. First, the stopping profile of positrons has a width comparable to the stopping depth. Second, after stopping, the positron distribution is modified by diffusion. Both of these phenomena can be taken into account by solving for the quasi-stationary positron density $n^*(x, E)$ at the depth $x$ from the diffusion–annihilation equation (Dupasquier and Mills, 1995)

$$D_+\nabla^2 n^*(x, E) - [\lambda_B + \kappa(x)]n^*(x, E) + p(x, E) = 0, \qquad (44)$$

where $D_+$ is the positron diffusion coefficient, $\lambda_B$ the annihilation rate in the

GaAs lattice, $p(x, E)$ the positron stopping profile (see Section II), and $\kappa(x)$ the positron trapping rate. The trapping rate $\kappa(x)$ is directly proportional to the defect concentration $c(x)$ by $\kappa(x) = \mu c(x)$, where $\mu$ is the positron trapping coefficient. From the positron density $n^*(x, E)$, we can calculate the fraction of positrons annihilating at the surface, in the lattice, and at the defect species $i$:

$$\eta_S(E) = D_+ \left.\frac{dn^*(E)}{dx}\right|_{x=0}, \qquad (45)$$

$$\eta_B(E) = \int_0^\infty dx\, \lambda_B n^*(x, E), \qquad (46)$$

$$\eta_{Di}(E) = \int_0^\infty dx\, \kappa_{Di}(x) n^*(x, E), \qquad (47)$$

With Eqs. (45) through (47), a model lineshape curve,

$$\tilde{P}(E) = \sum_j \eta_j(E) P_j, \qquad (48)$$

can be constructed as a function of the incident positron energy $E$. To extract the spatial vacancy-defect distribution, we vary the positron trapping profile $\kappa(x)$ until a satisfactory fit of the model $\tilde{P}(E)$ curve to the experimental lineshape data $P(E)$ is obtained.

Unfortunately, the $S(E)$ or $W(E)$ data are often insensitive to the detailed shape of the trapping profile $\kappa(x)$ because of the stopping and diffusion broadening of the positron density. Therefore, simple model functions like constant trapping profiles with sharp cut-off depths are traditionally used in the analysis [$\kappa(x) = \kappa_0$ at $x \leq d$ and $\kappa(x) = 0$ at $x > d$]. This method then yields an average trapping rate $\kappa_0$ up to an approximate depth $d$, but the sharp cut-off of the vacancy distribution at the depth $d$ is usually nonphysical.

The sensitivity of the vacancy profiling measurement can be considerably improved if the $S(E)$ or $W(E)$ curves are measured after successive etching treatments, which remove a controlled amount of surface layer. Figure 20 shows this type of study for In-implanted GaAs. As the depth of the etch increases, the $S$ parameter decreases, indicating that the vacancy concentration is smaller deeper in the sample ($x > 500$ nm) than at the stopping range of In ions (100 nm). The lowest curve indicates that vacancy defects can be found even at the depth of about 1000 nm from the surface.

The data of Fig. 20 were analyzed with Eqs. (45) through (48) in the following way. A constant trapping profile with a sharp cut-off depth was first fitted to the data measured after the deepest etching treatment (1100 nm). Then, this vacancy profile was kept fixed in the analysis of the data measured after the second deepest etching treatment (700 nm), and an additional trapping rate block was fitted to the surface layer (700–1100 nm) of this sample. Applying this analysis to the $S(E)$ curves after each etching treatment, we consistently obtain the monovacancy profile shown in Fig. 21. In addition, a detailed analysis using the $R$ parameter [Eq. 34)] reveals that divacancies or small vacancy clusters are also present up to a depth of about 200 nm.

Figure 21 shows that the maximum concentration of the monovacancies and divacancies exists roughly at the stopping range of the implanted In. This is natural, since the energy deposited in the lattice during the slowing down is at its maximum in this region. However, there is an exponential tail in the vacancy distribution extending up to 1500 nm (i.e., much below the stopping range of In ions). This tail can result from focused collision cascades or possibly from channeling of the implanted In ions.

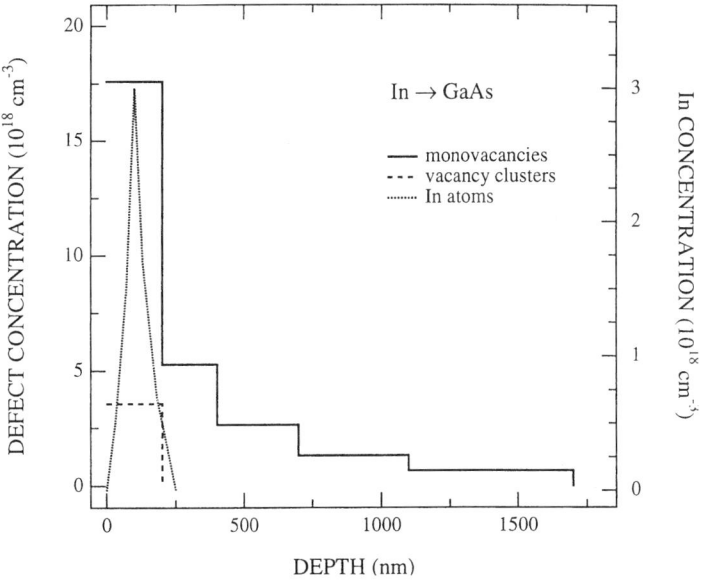

FIG. 21. The depth distributions of In atoms and vacancy-type defects after 350-keV In implantation of GaAs to a fluence of $10^{13}$ cm$^{-2}$ (Pfeiffer et al., 1996).

## VIII. Investigation of Vacancy Ionization Levels

In this section, we show that the sensitivity of the positron lifetime to the open volume and the sensitivity of the positron trapping coefficient to the charge state of a defect give an opportunity to investigate vacancy ionization levels in GaAs crystals. Ionization levels are investigated by varying the temperature of n-type GaAs. The Fermi level moves down in the gap with increasing temperature; a negative vacancy can become neutral or even positive and then no longer be detected (Corbel et al., 1988; Saarinen et al., 1991a). Ionization levels are investigated also under illumination of SI GaAs; vacancies escaping positron detection due to their positive charge can become neutral or negative under illumination (Saarinen et al., 1993; LeBerre et al., 1995b).

### 1. Arsenic Vacancy in n-Type GaAs: Thermal Ionization

In n-type bulk GaAs, native vacancies are generally found in positron experiments. As a function of measurement temperature, the positron trapping at the vacancies exhibits two interesting properties (Corbel et al., 1988; Saarinen et al., 1991a). The first is a lifetime transition. The lifetime component $\tau_2$ of the positrons trapped at the vacancies increases reversibly from $257 \pm 3$ ps to $295 \pm 3$ ps as a function of temperature. The transition is shifted to higher temperature, from 80 to 500 K, as the electron concentration measured at 300 K, $n(300\text{ K})$, increases from $2 \times 10^{15}$ to $2 \times 10^{17}\text{cm}^{-3}$. The width of the transition increases, when the transition occurs at higher temperature. The transition disappears in heavily doped crystals where the lifetime remains at 257 ps from 15 to 600 K. The second property is a transition of the trapping coefficient. The positrons trapped at the 295-ps vacancy have a trapping rate, $\kappa$, which decreases strongly and tends towards zero as temperature increase. The transition $\kappa \to 0$ shifts also to higher temperature, from 260 to 410 K, as the electron concentration $n(300\text{ K})$ increases from $2 \times 10^{15}$ to $6 \times 10^{16} \text{ cm}^{-3}$. Thus, the lifetime transition, $257 \to 295$ ps, and the trapping rate transition, $\kappa \to 0$, depend on the carrier concentration as well as on the temperature. This behavior suggests that both processes are controlled by a single parameter, the Fermi level.

The role played by the Fermi level in the lifetime transition, $257 \to 295$ ps, is clearly seen in Fig. 22. The transition occurs in the same narrow range of Fermi level position even though the electron concentration $n(300\text{ K})$ varies over two orders of magnitude. The native vacancy has two distinct configurations, $V$ and $V'$, in the upper part of the energy gap. The $\tau_V = 257$ ps

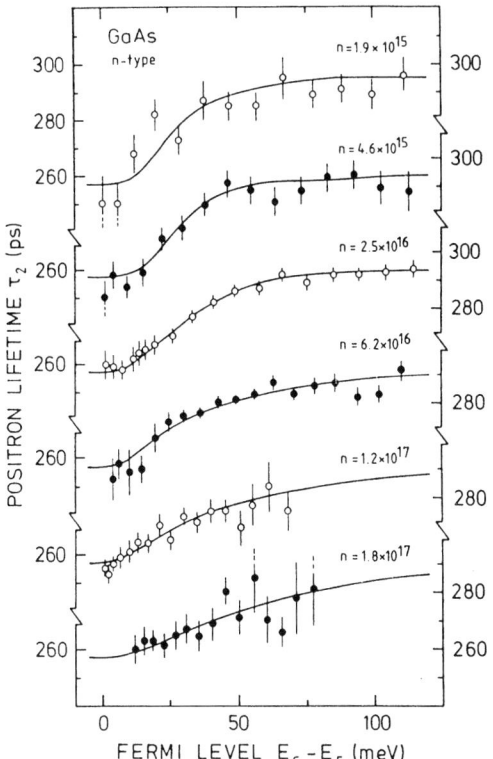

FIG. 22. Positron lifetime $\tau_2$ at the As vacancy in n-type GaAs as a function of the position of the Fermi level $E_F$ in reference to the bottom of the conduction band $E_c$. The solid line is a fit of Eq. (50) to the experimental data. The carrier concentrations $n$ are given in cm$^{-3}$ in the figure. (From Saarinen *et al.*, 1991a, with permission.)

configuration is replaced by the one with $\tau_{V'} = 295$ ps when the Fermi level shifts down from $E_c - 10$ meV to $E_c - 50$ meV. The $\tau_V = 257$ ps configuration corresponds to a smaller open volume than the $\tau_{V'} = 295$ ps configuration. Calculations indicate that the lattice relaxation in the breathing mode should change by about 10% in order to increase the lifetime in the vacancy from 257 to 295 ps (Mäkinen and Puska, 1989).

The occupancy of the vacancy configuration with $\tau_V = 257$ ps is controlled by the relative positions of the Fermi level $E_F$ and the ionization level $E_i$ in the gap. It is given by the Fermi distribution

$$f = \{1 + g \exp[(E_i - E_F)/k_B T]\}^{-1}, \tag{49}$$

where the factor $g$ is the ratio $g = Z'/Z$ of the internal degeneracies $Z$ and $Z'$ of the ground states corresponding to the $\tau_V = 257$ ps and $\tau_{V'} = 295$ ps configurations, respectively. For a large concentration of vacancies, the lifetime $\tau_2$ of the trapped positrons is a linear combination of the lifetimes 257 ps and 295 ps corresponding to the two configurations of the vacancy defect:

$$\tau_2 = f^* \times 257 \text{ ps} + (1 - f^*) \times 295 \text{ ps}. \tag{50}$$

The function $f^*$ is the function $f$ of Eq. (49), where the factor $g$ has been replaced by $(g\mu'/\mu)$ to take into account the different positron trapping coefficients $\mu$ and $\mu'$ for the $\tau_V = 257$ ps and $\tau_{V'} = 295$ ps configurations, respectively. The solid lines in Fig. 22 correspond to the best fits of Eq. (50) to the experimental data $\tau_2$, with the adjusted values of the transition level at $E_c - (30 \pm 3)$ meV. The ratio $\mu/\mu'$ is about $6 \pm 2$ at 300 K (Saarinen et al., 1995b).

Figure 23 shows that the transition $\kappa \to 0$ is also controlled by the Fermi level. The trapping rate in all the different samples decreases in the range of Fermi level positions between $E_c - 50$ meV and $E_c - 150$ meV. In this transition, the vacancy configuration with the positron lifetime of 295 ps disappears. This transition corresponds to a decrease of the trapping coefficient at the $\tau_{V'} = 295$ ps configuration by a factor of more than 10. As no further vacancy states appear, the positron trapping rate $\kappa$ is directly proportional to the occupancy $f(E_i, E_F)$ of the $\tau_{V'} = 295$ ps vacancy configuration. The solid lines in Fig. 23 are obtained by fitting the equation $\kappa(T)/\kappa_0(T_0) = f(E_i, E_F)$ to the experimental trapping rate normalized to the trapping rate $\kappa_0$ at the temperature $T_0$ before the transition takes place. The ionization level $E_i$ in the fits is determined to be at $E_c - (140 \pm 15)$ meV.

Thus, the native vacancy has three configurations in the upper part of the gap. When the Fermi level moves below $E_c - 140$ meV, the positron trapping coefficient decreases dramatically and we identify the transition $\kappa \to 0$ at $E_c - 140$ meV to a charge-state transition $V^0 \to V^+$ of the native vacancy. Then the lifetime transition $257 \to 295$ ps at $E_c - 30$ meV can be simply ascribed to the charge-state transition $V^- \to V^0$ of the same native vacancy. The trapping coefficient $\mu$ is consistently found to be higher at the negative configuration than at the neutral one. To identify the native vacancy, we use the property that it has two ionization levels in the upper part of the gap. These ionization levels are located in the region where the calculations predict ionization levels for $V_{As}$ (Puska, 1989; Xu and Lindefelt, 1990). A natural conclusion is that the detected native vacancy is $V_{As}$, which is either isolated or bound to other defects, such as impurities or antisites. The As vacancy exhibits a large outward volume relaxation when the

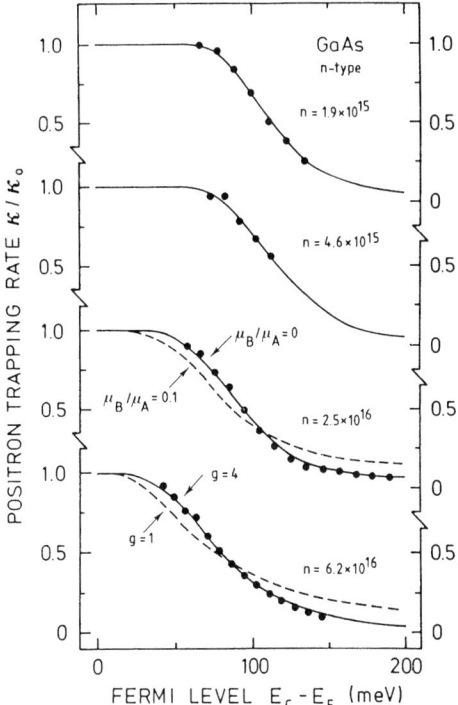

FIG. 23. Positron trapping rate $\kappa$ at the As vacancy in n-type GaAs as a function of the position of the Fermi level $E_F$ in reference to the bottom of the conduction band $E_c$. The solid lines are fits to the Fermi function describing the ionization of the vacancy. The influences of the positron trapping coefficients $\mu'/\mu$ and the degeneracy factor $g$ to the fits are demonstrated. The carrier concentrations $n$ are given in cm$^{-3}$ in the figure. (From Saarinen et al., 1991, with permission.)

electron in the level $E_c - 30$ meV is released (see also Fig. 5 in Section IV.2). Taking $\mu(V_{As}) = \mu(V_{Ga}) \approx 10^{15}$ at s$^{-1}$ at 300 K, the As vacancy concentration is about $10^{16}$–$10^{17}$ cm$^{-3}$ in n-type GaAs.

In summary, positron annihilation shows that a native vacancy exists in as-grown, n-type GaAs. The native vacancy is identified as the As vacancy, which is either isolated or bound to a complex. The As vacancy has two ionization levels, at about $E_c - 30$ meV and $E_c - 140$ meV. Furthermore, the results show that the As vacancy exhibits a large outward volume relaxation when the ionization level $E_c - 30$ meV is depopulated. Discarding the possibility of negative-U effects, we attribute the levels $E_c - 30$ meV and $E_c - 140$ meV to the charge transitions $V^- \to V^0$ and $V^0 \to V^+$ of the As vacancy, respectively.

## 2. ARSENIC VACANCY IN SI GaAs: OPTICAL TRANSITIONS

Under below-bandgap illumination, free carriers are photogenerated in the valence and conduction bands through optical transitions with defects. The occupancy of the vacancies results from the balance between radiative and thermal emission of free carriers and their thermal capture. At low temperature, around 20 K, thermal emission can be neglected for vacancies that generally have deep levels in the bandgap. It follows that a vacancy for which the optical transitions are very weak may still change its charge state at low temperature due to the capture of free carriers photogenerated by optical transitions at other defects.

At 300 K, the positron lifetimes in undoped GaAs vary from 231 to 233 ps in darkness. When a sample is cooled down to low temperatures, the positron lifetime increases as shown in Fig. 24. The increase of $\tau_{av}$ means that the samples contain vacancy defects and further that the charge of those defects is negative. The lifetime spectra at 20 to 80 K can be decomposed into two components. The second component is $\tau_2 = 255 \pm 5$ ps. As explained in Section V.2, these defects detected in darkness are native Ga vacancies in a negative charge state.

Under illumination, the average positron lifetime increases compared with the values in darkness (see Fig. 24) (Saarinen et al., 1993; Kuisma et al., 1996b). The effect of illumination is largest at 20 to 100 K and disappears completely above 180 K. At the lowest temperatures, from 20 to 60 K, part of this increase remains even if the illumination is switched off. The persistent effect is due to the open volume in the atomic structure of the metastable state of the EL2 defect, and it is discussed in detail in Section IX. However, the increase under illumination is clearly larger than the persistent part of it and it is detected up to much higher measurement temperatures. This indicates that most of the effect seen under illumination is not related to the metastable state of the EL2 defect. The lifetime spectra measured under illumination can be decomposed into two components, the second of which is $\tau_2 = 255 \pm 5$ ps (i.e., the same lifetime as that obtained in darkness). Hence, all positron traps present in the samples produce the same lifetime, and the vacancies trapping positrons under illumination can be characterized with $\tau_2 = 255 \pm 5$ ps.

The increase of the average positron lifetime under illumination indicates that some vacancies are converted to more efficient positron traps by capturing electrons. The lifetime component $\tau_2 = 255 \pm 5$ ps indicates that these defects have the open volume of a monovacancy. In an SI sample, the ionization levels of the vacancies located below midgap are occupied by electrons. The levels populated under illumination are thus located above midgap.

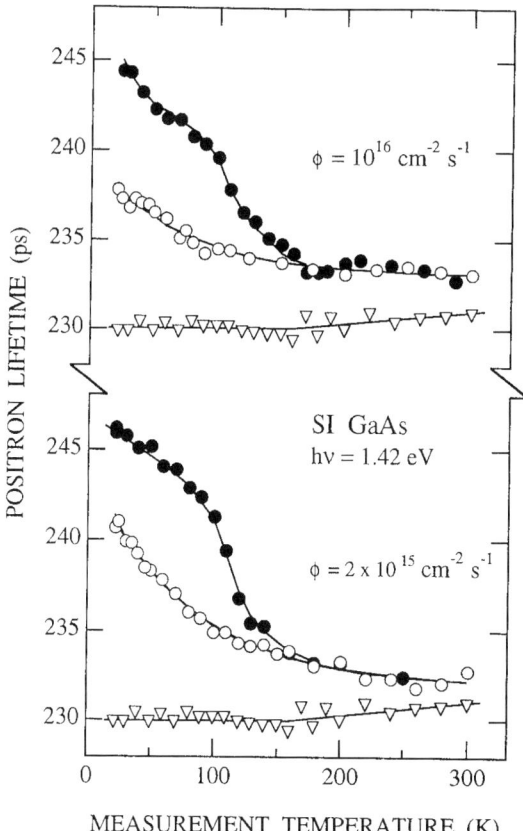

FIG. 24. The average positron lifetime as a function of temperature in two undoped SI GaAs samples. The experiment is performed either in darkness (○) or under $hv = 1.42$-eV illumination with the photon flux $\phi$(●). The reference level measured in Zn-doped GaAs is also shown (▽). (From Kuisma et al., 1996b, with permission.)

According to theory and experiments, ionization levels are expected in the upper part of the gap for As vacancies but not for Ga vacancies. Furthermore, As vacancies are expected to be positive in SI GaAs, whereas Ga vacancies are negative (Baraff and Schlüter, 1985; Puska, 1989). It is thus natural to identify the defects seen under illumination as As vacancies or complexes related to As vacancies. These defects are positive in darkness, but under illumination they are converted to efficient positron traps by electron photoexcitation and capture processes.

For the temperature range of 20 to 70 K, the value of $\tau_{av}$ and, correspondingly, the positron trapping rate at $V_{As}$ decrease when temperature increases (see Fig. 24). This behavior is similar to that observed for the Ga vacancies in darkness, and it indicates that the As vacancies are negatively charged (see Section V.2). Positron lifetimes of 257 and 295 ps have been previously determined for negative and neutral charge states of $V_{As}$, respectively (see Section VIII.1). In this case, both under illumination and in darkness, the positron lifetimes fall in the range $\tau_2 = 255 \pm 5$ ps. No indication of the longer lifetime 295 ps corresponding to neutral $V_{As}$ are found. Thus, we conclude that the As vacancies revealed under illumination are in a negative charge state. From the positron trapping rate we estimate that the concentration of As vacancies is typically $10^{15}$–$10^{16}$ cm$^{-3}$ in undoped SI GaAs.

Figure 25 shows the concentration of the detected As vacancies as a function of the photon energy under illumination with constant photon flux (Kuisma et al., 1996b). When EL2 is in its stable state, the negative charge state of the As vacancy can be generated with photons of $hv > 0.9$ eV, and the spectrum is almost flat for $hv > 1.2$ eV. In Si GaAs, the absorption at $hv < 1.4$ eV is mainly due to the optical exchange of electrons between the EL2 defect and the conduction and valence bands (Martin and Makram-Ebeid, 1986). These processes generate photoelectrons that can be captured by the defect levels in the bandgap. Since the As vacancy becomes negatively charged under these illumination conditions, we associate the generation process with the capture of photoelectrons emitted from the EL2 defect to the conduction band under illumination.

When EL2 is in its metastable state, only photons with $hv > 1.35$ eV are able to excite the As vacancy to the negative charge state (see Fig. 25). The photon flux needed for this excitation is several orders of magnitude higher than that required in the case in which EL2 is in its stable state. These observations indicate that a different optical process leads to the population of the negative As vacancy when EL2 is in the optically inactive metastable state. We attribute this process to the direct excitation of electrons from the valence band to the ionization levels of $V_{As}$.

Above 70 K, the positron trapping rate depends on the light intensity, and its temperature dependence becomes much steeper than for a negative vacancy. To investigate this steep part in more detail, we calculate the concentration of photoinduced negative As vacancies as a function of measurement temperature for different 1.42-eV illumination intensities. We assume that under illumination the positron trapping at the Ga vacancies remains the same as in darkness.

The concentration of $V_{As}^-$ is shown in Fig. 26. $[V_{As}^-]$ is independent of the light intensity at 20 to 70 K and saturates at the value of $1.5 \times 10^{16}$ cm$^{-3}$. Above 70 K, $[V_{As}^-]$ decreases and disappears completely above 150 K.

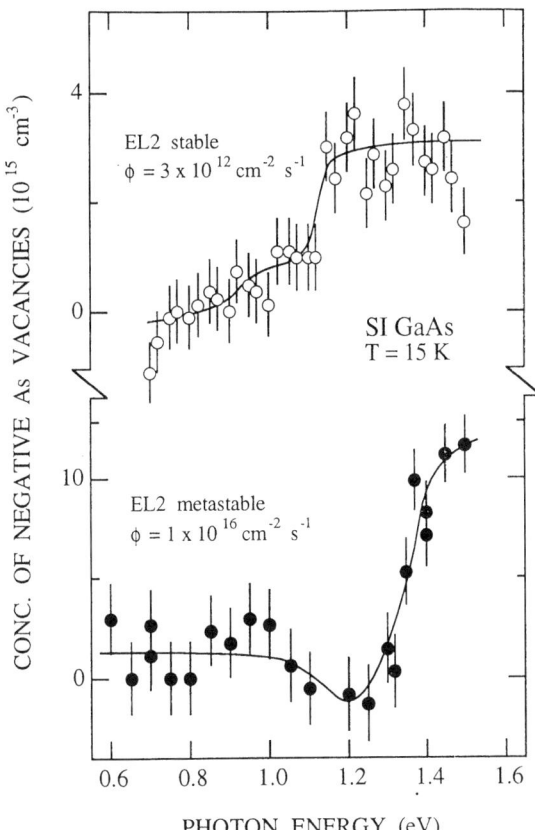

FIG. 25. Concentration of negative As vacancies in SI GaAs as a function of photon energy under illumination with photon flux $\phi$ at 15 K. In the data of the top panel, the EL2 defect remains in the stable state over the experiment. In the data of the bottom panel, EL2 is deliberately converted into the metastable state before the measurements. (From Kuisma et al., 1996b, with permission.)

The relation between the light intensity and the decrease of $[V_{As}^-]$ is clear; the stronger the photon flux, the higher the temperature that is needed to decrease the vacancy concentration from the plateau value. This type of behavior shows that at $T > 70$ K, the thermally activated emission of electrons from the $V_{As}$ levels to the conduction band starts to compete with the optical transitions. The thermal electron emission rate is $g_e = c_e \gamma N_c(T) \exp(-E_d/kT)$, where $c_e = \sigma_e v_e$ is the electron capture coefficient, $v_e$ is the electron thermal velocity, $\gamma$ is the degeneracy factor of the ionization

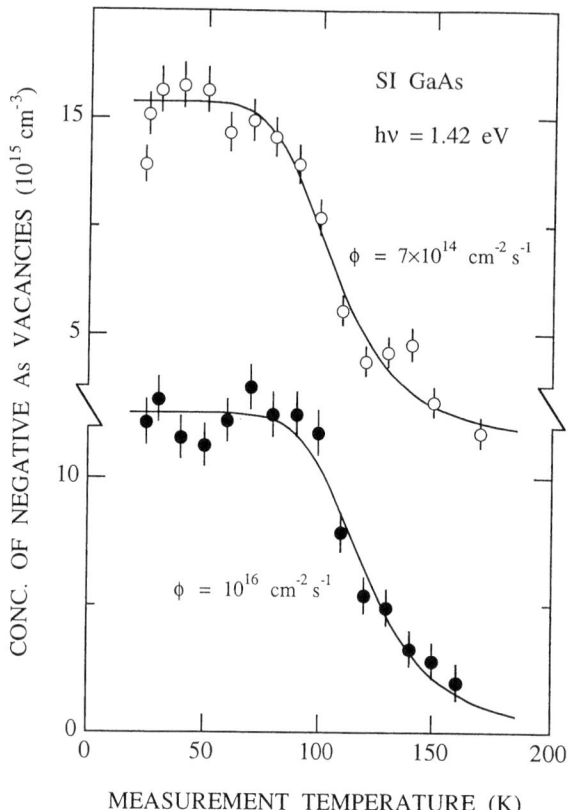

FIG. 26. Concentration of the negative As vacancies in SI GaAs as a function of temperature under $h\nu = 1.42$-eV illumination with the photon flux $\phi$. The solid lines are fits of Eq. (51) to the experimental data. (From Kuisma et al., 1996b, with permission.)

level, $N_c(T)$ is the effective density of states of the conduction band, and $E_d$ is the ionization energy to the conduction band.

We can analyze the data in Fig. 26 by assuming that the concentration of $V_{As}^-$ above 70 K is controlled by the electron emission $g_e$ from a single ionization level and by the optical transitions between the bands and the vacancy level (Saarinen et al., 1993; Kuisma et al., 1996b). Assuming further that the electron capture cross-section $\sigma_e$ is independent of temperature, we can write the fraction of negative vacancies as

$$\frac{[V_{As}^-]}{[V_{As}]} = \frac{A}{(1 + B\Phi^{-1}v_e(T)N_c(T)\exp(-E_d/k_BT))}. \quad (51)$$

Here $\Phi$ is the photon flux, $A = (1 + \sigma_e^0/\sigma_h^0)^{-1}$ and $B = A\gamma\sigma_e/\sigma_h^0$, where $\sigma_e^0$ and $\sigma_h^0$ are the optical electron and hole generation cross-sections of the ionization level, respectively.

The solid lines in Fig. 26 are the fits of Eq. (51) to the experimental data with the ionization energy $E_d$ and the constants $A$ and $B$ as free parameters. The fitted functions reproduce the trends observed in the data. When the intensity increases, the decrease of $[V_{As}^-]/[V_{As}]$ from the saturation level is shifted to higher temperatures and takes place over a wider temperature range. The fitted ionization energy is $E_d = 65 \pm 15$ meV. This value correlates well with the maximum efficiency of the excitation at $h\nu = 1.4$–$1.5$ eV (see Fig. 25), since the GaAs bandgap is 1.5 eV at 30 K. It is also in good agreement with the results in n-type GaAs, where we have found the ionization level $V_{As}^- \to V_{As}^0$ at $E_c - 30$ meV (see Section VIII.1).

In optical experiments on bulk GaAs, a strong absorption of monochromatic light within 50 meV of the conduction band edge is observed below 150 K (Tüzeman and Brozel, 1991). This near-band-edge absorption ["reverse-contrast (RC) absorption"] has been attributed to an unidentified

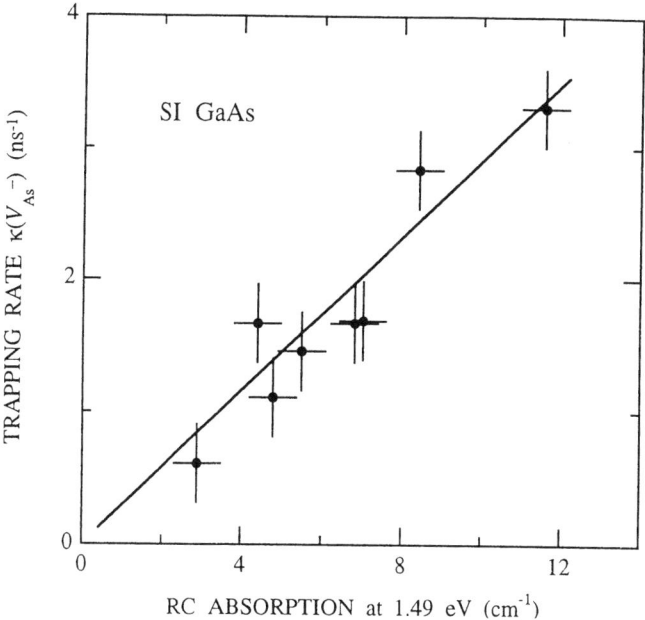

FIG. 27. Positron trapping rate at the negative As vacancies in SI GaAs as a function of the RC absorption coefficient measured at 1.49 eV. (From LeBerre et al., 1995b, with permission.)

point defect, which is not related to EL2. The present positron experiments indicate that the ionization level of the As vacancy is at about $E_c - 50$ meV and that this level can be most efficiently populated using 1.4- to 1.5-eV light. The concentration of the As vacancy correlates quantitatively with the magnitude of the RC absorption (Fig. 27) (LeBerre et al., 1995b). Therefore, our results provide a natural explanation for the near-band-edge absorption in GaAs; it results from the photoinduced electron transition from the valence band to the ionization level of the As vacancy.

In summary, positron data give evidence that an amphoteric vacancy with $-/0/+$ charge states in the upper part of the gap exists in n-type GaAs. Under illumination, this vacancy can trap electrons and holes, depending on its charge state in SI GaAs, which demonstrates its role as an efficient recombination center. The vacancy is also responsible for RC near-band-edge absorption in GaAs. We attribute these properties to the As vacancy, which is either isolated or bound to a complex. The properties of the As vacancy in GaAs are consistent with those of the RC defects, which act as the main nonradiative recombination centers in GaAs.

## IX. Investigation of the Atomic Structures of Metastable Defects

Low-temperature illumination produces metastable changes in electrical and optical properties of undoped GaAs (Martin and Makram-Ebeid, 1986; Kaminska and Weber, 1993) and n-type $Al_{1-x}Ga_xAs$ (Mooney, 1990). The origin of this metastability is attributed to the EL2 defect in GaAs and to the DX center in n-type $Al_{1-x}Ga_xAs$, but the microscopic mechanisms underlying the metastability have been much debated. According to some models, the metastability is related to the rearrangement of atoms induced by optical transitions. This large lattice relaxation creates an open volume, which gives the donor the characteristic of a vacancy-type defect either in the stable (DX) (Chadi and Chang, 1988b) or in the metastable configuration (EL2) (Dabrowski and Scheffler, 1988; Chadi and Chang, 1988a). The only technique able to directly probe the atomic structure of the donor has been positron annihilation (Krause et al., 1990; Saarinen et al., 1994; Mäkinen et al., 1993).

### 1. As-grown SI GaAs: The Midgap Donor EL2

By compensating residual acceptors, the EL2 midgap donor plays a central role in the growth of undoped SI GaAs (Martin and Makram-Ebeid,

1986; Kaminska and Weber, 1993). Due to its technological importance, considerable experimental and theoretical effort has been devoted to the determination of its atomic structure. A key property of EL2 is its optically induced metastability. The defect can be permanently convered to the neutral metastable state EL2* under 0.8- to 1.5-eV illumination at temperatures below 100 K. The photoquenching occurs without generation of any new electrical or optical signals, which could be associated with the metastable state EL2*. The structure of EL2* has consequently been a very

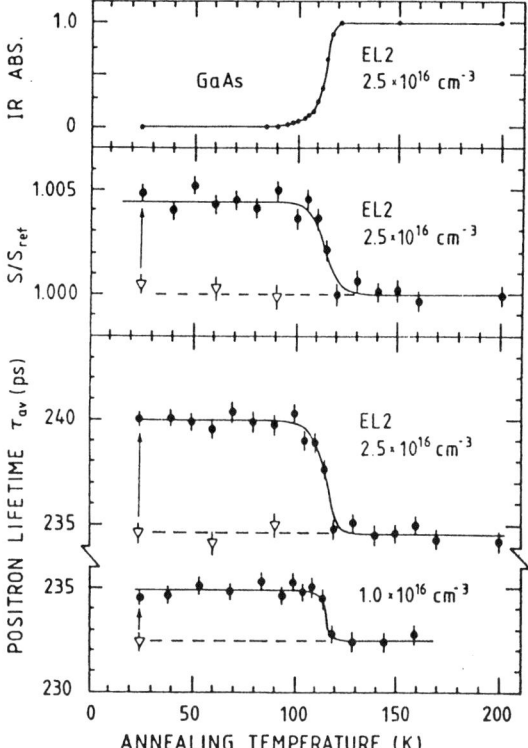

FIG. 28. The average positron lifetime and the low electron-momentum parameter $S$ in SI GaAs as functions of the isochronal annealing temperature after 1.2-eV illumination at 25 K. The illumination transforms EL2 into the metastable state, and the corresponding changes in the positron parameters are indicated by arrows. The open triangles with dashed lines represent the reference levels where EL2 is in the stable state. The normalized infrared absorption coefficient is shown in the top panel. All measurements have been performed in darkness at 25 K. (From Krause et al., 1990, with permission.)

speculative matter, although a consensus has been reached that the stable state involves an As antisite defect.

To investigate the positron behavior when EL2 is photoquenched, we have performed positron lifetime and Doppler broadening experiments in an optical cryostat (Krause et al., 1990; Saarinen et al., 1994). In the experiment of Fig. 28, crystals with EL2 concentrations of $1 \times 10^{16}$ and $2.5 \times 10^{16}$ cm$^{-3}$ were cooled in darkness and illuminated *in situ* with 1.2-eV light. To check that the EL2 defects were photoquenched by the illumination, infrared absorption was also measured. Figure 28 shows that at 25 K there is a clear increase of the positron annihilation parameters, $\tau_{av}$ and $S$, after the photoquenching of the EL2 centers. This property has been reproducibly observed in various other SI GaAs crystals, and it indicates that vacancy-type defects are generated by the photoquenching. Furthermore, the concentration of vacancy defects generated correlates quantitatively with the total EL2 concentration (Saarinen et al., 1994; LeBerre et al., 1994).

The EL2 infrared absorption recovers at 120 K. In perfect agreement, the annihilation characteristics $\tau_{av}$ and $S$ also recover to their reference values after annealing at about 120 K in darkness (see Fig. 28). Similar to the recovery of EL2, the recovery of $\tau_{av}$ also takes place at lower temperature under illumination. For example, the 120-K recovery shifts to 80 K under 0.9-eV illumination. The existence of the metastable vacancy $V^*$ thus appears to correlate with the existence of the metastable state EL2*.

To better correlate the persistent changes in the positron lifetime and 1.15-eV IR absorption, both were investigated at 25 K as functions of the illumination time in the same crystal. The results are shown in Fig. 29 in terms of the absorption coefficient and the positron trapping rate. Both curves follow first-order kinetics for photoquenching, and the metastable vacancy is clearly formed simultaneously with the disappearance of the IR absorption. The experiments were repeated for different photon energies of the photoquenching light, and the curves were analyzed to determine the optical cross-section for the conversion of EL2 to the metastable state. As seen in Fig. 30, there is perfect agreement between both the absolute values and the photon energy dependence of the optical cross-section. The conversion is most efficient at 1.15 eV, in good agreement with earlier infrared absorption, photocapacitance, or photoconductivity measurements in SI GaAs (Kaminska et al., 1985). We can conclude that the metastable vacancy is indeed generated when EL2 is converted to the metastable state.

After illumination, the second lifetime component in SI GaAs is $\tau_2 = 247 \pm 3$ ps. In this state of the sample, two types of vacancies may trap positrons: the Ga vacancies often found as native defects in SI GaAs (see Section V.2) and the metastable vacancies observed after illumination. In

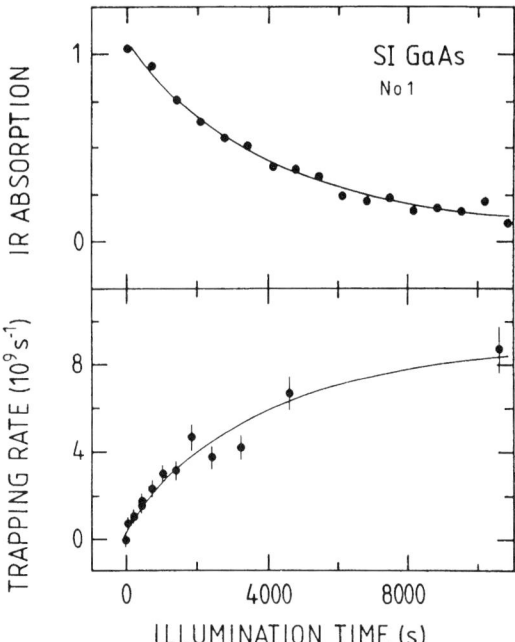

FIG. 29. Normalized absorption coefficient and the positron trapping rate at the metastable vacancy as a function of illumination time in SI GaAs. A photon energy of 1.15 eV and a flux of $10^{14}\,\text{s}^{-1}\,\text{cm}^{-2}$ were used in the illumination. The experiments were performed at 25 K in darkness. (From Saarinen et al., 1994, with permission.)

this case, the second lifetime $\tau_2$ is the superposition of the lifetimes at Ga vacancies $\tau_V = 260$ ps and at the metastable vacancies with the positron lifetime $\tau_{V*}$. Taking into account the positron trapping at native Ga vacancies, one can estimate that the positron lifetime at the metastable vacancy is $\tau_{V*} = 245 \pm 3$ ps (Saarinen et al., 1994). This lifetime is clearly less than the values at the Ga (260 ps) and As vacancies (257–295 ps). The low lifetime $\tau_{V*} = 245 \pm 3$ ps at the metastable vacancy thus indicates that its open volume is less than that of a monovacancy in GaAs.

The positron trapping at the metastable vacancy $V^*$ has been found to be strongly temperature-dependent: between 20 and 70 K, the positron trapping rate at $V^*$ decreases more than an order of magnitude. As explained in Section V.2, this indicates that the metastable vacancy is negatively charged. However, the decrease of the trapping rate at $V^*$ is much stronger than the $T^{-0.5}$ dependence found at $T < 70$ K for the native

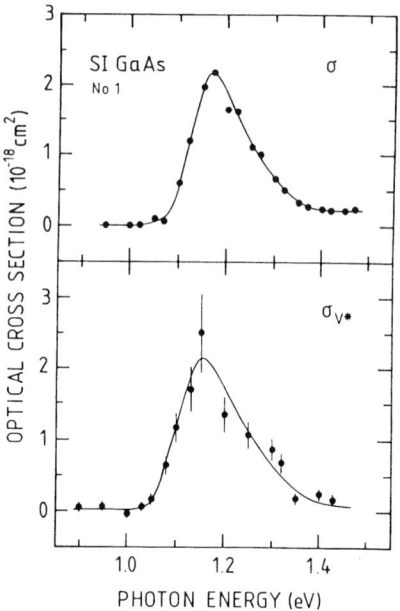

FIG. 30. Optical cross-section for the creation of the metastable vacancy as a function of the photon energy. The data in the upper part of the figure were obtained from IR absorption measurements, and in the lower part of the figure, from positron lifetime measurements. Measurement temperature was 25 K. (From Saarinen et al., 1994, with permission.)

Ga vacancy. Hence, the positron trapping properties at $V^*$ are quantitatively different from those at isolated negative vacancies. A possible explanation for this behavior is that the negatively charged metastable vacancy is a part of a larger defect complex, whose total charge is neutral instead of negative.

In summary, we have observed a metastable vacancy in SI GaAs when the EL2 defect is photoquenched to the metastable state. This vacancy has smaller open volume than Ga or As vacancies, and it possesses a negative charge. The metastable vacancy is generated with 1.1- to 1.3-eV light with exactly the same optical cross-section as the metastable state EL2*. The metastable vacancy is detected as long as EL2 remains in its metastable state, and the vacancy recovers at 120 K with the same annealing kinetics as EL2*. The concentration of the metastable vacancy scales with the total EL2 concentration. These correlations lead us to conclude that the metastable vacancy belongs to the atomic structure of the metastable state of the EL2 defect.

## 2. METASTABILITY OF DEFECTS IN SI GaAs AFTER ELECTRON IRRADIATION

The introduction of As antisites by electron irradiation has been observed by various techniques. Although the irradiation-induced $As_{Ga}$ has properties similar to those of EL2, there has been very little evidence of its metastability. To investigate whether illumination at low temperature generates metastable vacancies in addition to the native ones, we have performed positron lifetime experiments in electron-irradiated GaAs crystals (Saarinen et al., 1995a).

Figure 31 shows that the average positron lifetime at 25 K increases persistently after 1.15-eV illumination of electron-irradiated GaAs. Metastable vacancies $V_{irr}^*$ can thus be generated by illumination. However, the

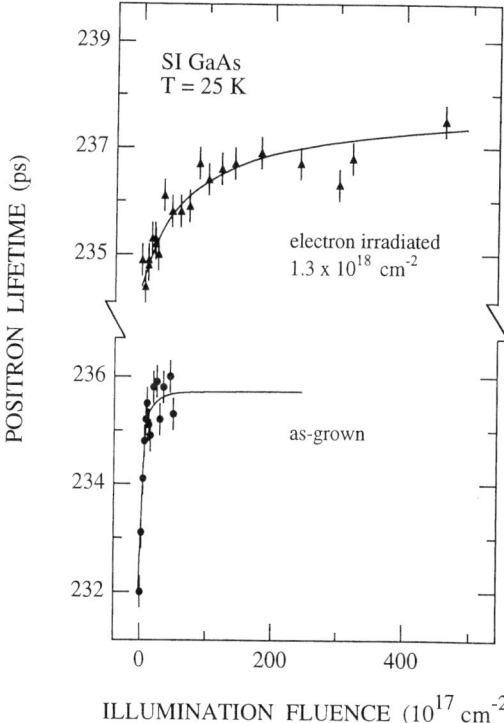

FIG. 31. Positron average lifetime as a function of the illumination fluence in as-grown and electron-irradiated (electron fluence $1.3 \times 10^{18}$ cm$^{-2}$) GaAs samples. A photon energy of 1.15 eV was used in the illumination. All experiments have been performed in darkness at 25 K between each illumination at 25 K. (From Saarinen et al., 1995a, with permission.)

photoquenching process is much slower than before irradiation, as clearly seen in Fig. 31. Taking into account the stronger absorption of the light after electron irradiation, one can estimate the optical cross-section for the generation of $V_{irr}^*$. The cross-section $\sigma(1.15\,\text{eV}) = 3 \times 10^{-19}\,\text{cm}^2$ is an order of magnitude smaller than for the generation of the native $V^*$, $\sigma(1.15\,\text{eV}) = 2 \times 10^{-18}\,\text{cm}^2$. Figure 16 shows that the $V_{irr}^*$ concentration increases with fluence. Consequently, the metastable vacancy $V_{irr}^*$ is related to irradiation-induced defects.

After illumination at 25 K, $V_{irr}^*$ recovers at about 80 to 100 K, thus at somewhat lower temperatures than the native $V^*$ associated with the metastable state of the EL2 defect (120 K). $V_{irr}^*$ is also generated by about

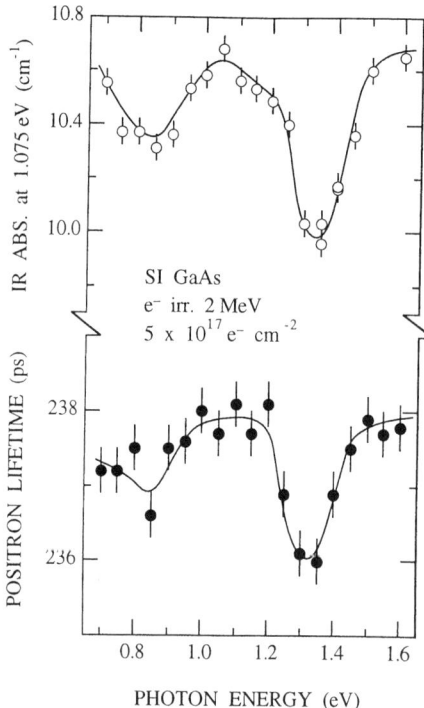

FIG. 32. The optical recovery of the average positron lifetime and the absorption coefficient in electron-irradiated SI GaAs. The sample was first illuminated with 1.075-eV photons at 25 K. Thereafter, the photorecovery was studied by measuring the IR absorption and positron lifetime after an additional illumination at 25 K. The photon energies used in this illumination are indicated on the horizontal axis. For each photon energy, the same illumination fluences of $10^{19}$ and $3 \times 10^{18}\,\text{cm}^{-2}$ in positron and IR absorption experiments, respectively, were used to induce the photorecovery. The measurement temperature was 25 K. (From Kuisma et al., 1996a, with permission.)

0.1-eV lower photon energies than $V^*$. As shown in Fig. 32, it can also be optically recovered at 20 K with 0.9-eV or 1.3-eV light, although for the native EL2, this is not possible. An interesting property of $V_{irr}^*$ is that it absorbs IR light in the range 0.8 to 1.2 eV (Kuisma et al., 1996a). As shown in Fig. 32 at 1.075 eV, this absorption can be optically recovered at 20 K and the efficiency of the recovery has exactly the same energy dependence as for that of $V_{irr}^*$ in positron experiments. The absorption properties of $V_{irr}^*$ indicate that, unlike the metastable state of EL2, the defect containing the metastable vacancy $V_{irr}^*$ has ionization levels in the energy gap.

We can conclude that defects with a vacancy in the metastable state are observed in electron-irradiated GaAs after annealing at 300 K. The metastability of this defect closely resembles that of the native EL2 defect, although some differences are seen in its photoquenching and recovery properties. The most prominent characteristic of the defect is that its metastable state absorbs infrared light, suggesting that its ionization level is in the bandgap. We discuss the atomic structure of the irradiation-induced metastable defect further in Section IX.4.

3. AS-GROWN n-TYPE $Al_xGa_{1-x}As$ LAYERS: THE DEEP DONOR LEVEL DX

A striking property of $Al_xGa_{1-x}As$ alloys with $x = 0.22$ is the persistence of the photoconductivity (PPC) after switching off the illumination. This property has been attributed to the DX deep donor level that is believed to remain photoionized after illumination (Mooney, 1990). To investigate whether the positron annihilation in n-type $Al_xGa_{1-x}As$ exhibits metastable properties under illumination, we have measured the core annihilation parameter $W$ in darkness after illumination at low temperature (Mäkinen et al., 1993, 1995; Laine et al., 1996b).

The core annihilation parameter $W$ is plotted against the AlAs-mole fraction in Fig. 33 in undoped and n-doped AlGaAs(Si, Sn, Te) layers at 25 K (Mäkinen et al., 1993). The annihilation characteristics of undoped AlGaAs vary with the alloy composition and can be attributed to free positron annihilation. When Ga atoms are replaced by Al, the core annihilation parameter decreases, indicating that the fraction of positrons annihilating with the core electrons decreases. This is due the smaller number of core electrons in Al than in Ga. In the n-type AlGaAs layers, the core electron parameters in the dark at 25 K are systematically lower than in the undoped layers. This means that the annihilation probability with the core electrons is smaller in n-type AlGaAs layers, which is typical for samples in which positrons annihilate at vacancy defects.

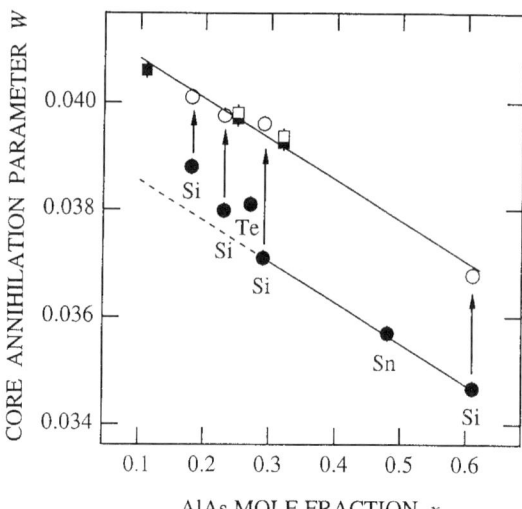

FIG. 33. The high electron-momentum parameter $W$ in $Al_xGa_{1-x}As$. Symbols: undoped AlGaAs before (■) and after (□) illumination, Te-doped and Sn-doped AlGaAs in the dark (●), Si-doped AlGaAs before (●) and after (○) illumination. The experiments were performed at 25 K in the dark. (From Mäkinen et al., 1993, with permission.)

To see whether the vacancies detected in n-type AlGaAs layers exhibit the metastability associated with the DX center, the annihilation spectra were measured after illumination at 25 K. In undoped AlGaAs, the $W$ parameter remains unchanged. In n-type AlGaAs (Si or Te), positron annihilation is sensitive to light. This effect is persistent at 25 K. After illumination, the $W$ parameter values become identical with those measured for undoped AlGaAs, indicating free positron annihilation. This implies that illumination makes the vacancy signal disappear. In Sn-doped $Al_{0.48}Ga_{0.52}As$, illumination at 25 K has no persistent influence on positron annihilation.

The core annihilation parameter $W$ in undoped $Al_{0.25}Ga_{0.75}As$ and Si-doped ($2.5 \times 10^{18}$ cm$^{-3}$) $Al_{0.29}Ga_{0.71}As$ is plotted against temperature in Fig. 34 (Mäkinen et al., 1993). Positron annihilation was measured from 25 to 600 K, keeping the samples in the dark throughout the measurement. In the undoped layer, the increase of temperature causes a small and nearly linear decrease of the $W$ parameter. It can be attributed to the thermal lattice expansion and is commonly observed for free positron annihilation in solids.

In Si-doped $Al_{0.29}Ga_{0.71}As$, the $W$ parameter is independent of temperature up to 300 K, but between 350 and 550 K it increases strongly. The

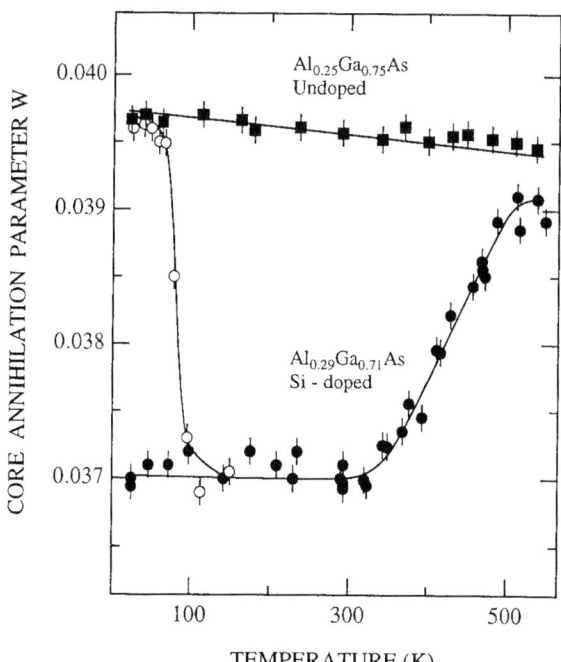

FIG. 34. The high electron-momentum parameter $W$ as a function of temperature in undoped (■) and in Si-doped AlGaAs before (●) and after (○) illumination. (From Mäkinen et al., 1993, with permission.)

change is completely reversible. Above 550 K, the $W$ values in Si-doped and undoped AlGaAs are identical, indicating free positron annihilation in both layers. The temperature behavior of the core annihilation parameter in Te-doped $Al_{0.27}Ga_{0.73}As$ is similar to that of Si-doped $Al_{0.29}Ga_{0.71}As$ between 350 and 550 K.

The effect of illumination on positron annihilation was studied by measuring the core annihilation parameter in the dark as a function of temperature subsequent to the illumination of the sample with IR light at 25 K. Only free positron annihilation is observed below a critical temperature in Si-doped $Al_{0.29}Ga_{0.71}As$. By increasing the temperature, one can restore positron trapping at the vacancies. The temperature at which this occurs depends on the alloy composition and the donor atom. For Si-doped $Al_xGa_{1-x}As$ with $x = 0.18, 0.23, 0.29, 0.33$, and 0.61, the vacancy signal reappears at 120, 110, 80, 60, and 90 K, respectively, and for Te-doped $Al_xGa_{1-x}As$ with $x = 0.27$, it reappears at 60 K.

The findings of the positron annihilation experiments can be compared with the well-known properties of the DX centers. We obtain a cross-section of $\sigma \approx 4 \times 10^{-17}$ cm$^2$ for the removal of positron trapping at a vacancy in Te-doped AlGaAs (Laine et al., 1996b). With this same photon energy of 1.32 eV, Lang et al. measured an optical ionization cross-section of $\sigma_0 = 4 \times 10^{-17}$ cm$^2$ for the transition of the DX center to the metastable state in Te-doped Al$_x$Ga$_{1-x}$As, and they found that this value is independent of the alloy composition (Lang et al., 1979). We have also measured a cross-section of $\sigma \approx 1 \times 10^{-18}$ cm$^2$ for Si-doped AlGaAs with a photon energy of 1.32 eV. The result is markedly different from the DX(Te) center but is in agreement with value ($\sigma_0 = 1 \times 10^{-18}$ cm$^2$) measured for the DX(Si) center using the photocapacitance techniques (Northrop and Mooney, 1991).

Positron trapping at vacancies is lost between 300 and 600 K, both in Si- and Sn-doped layers. The electron binding energies at the DX centers imply that at high donor concentration the thermal ionization occurs above room temperature (Chand et al., 1984; Theis et al., 1991). Positron trapping at vacancies vanishes in the same temperature range. Consistent with the low temperature results, it thus indicates free positron annihilation when the DX center is ionized.

It has been shown that the temperature at which the PPC in n-doped AlGaAs (Si, Sn, Te) is observed varies with the alloy composition (Mooney, 1990) and with the impurity (Lang et al., 1979). We find also that the critical temperature below which the positron trapping is persistently removed depends on the composition in Si-doped AlGaAs. This dependence is similar to that observed for the DX center in PPC experiments.

The electron capture energies $E_b$ of the DX centers related to different donor species (Si, Sn, Te) can be put in the following order: $E_b(\text{Sn}) \leqslant E_b(\text{Te}) \leqslant E_b(\text{Si})$ (Mooney, 1990). The recovery temperature of the metastable state observed in positron annihilation correlates with this result: in Sn-doped Al$_{0.48}$Ga$_{0.52}$As the metastable state is seen only under illumination $T = 12$ K; in Te-doped Al$_{0.27}$Ga$_{0.73}$As, the vacancy signal reappears at 60 K; and in Si-doped Al$_{0.29}$Ga$_{0.71}$As, the vacancies reappear at 80 K.

In conclusion, the cross-sections for optical ionization of the DX center and low-temperature photoquenching of positron trapping at the vacancies are the same. The temperature for the reappearance of positron trapping is in agreement with the barrier for electron capture at the DX center. The temperatures at which annihilation trapping disappears and the DX center is thermally ionized are consistent. These correlations clearly suggest that the DX center acts as a vacancy-like trap for positrons.

In highly Si- or Te-doped Al$_x$Ga$_{1-x}$As layers, we can measure the annihilation characteristics at the vacancy in the DX center. The values are $W_V/W_B = 0.87$ and $S_V/S_B = 1.004$ for both DX(Si) and DX(Sn), whereas for DX(Te), we get the clearly different values of $W_V/W_B = 0.92$ and $S_V/S_B =$

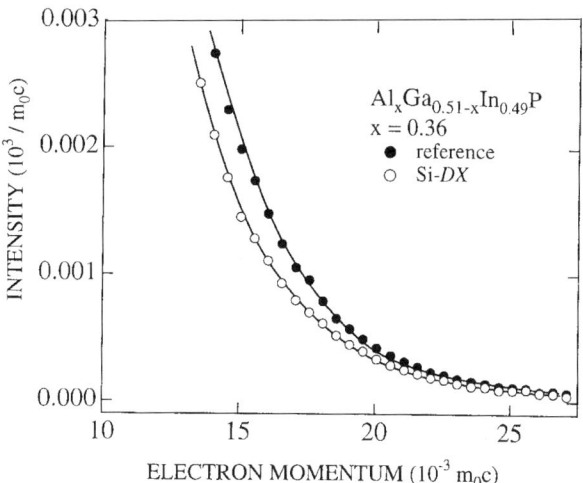

FIG. 35. The core electron-momentum distributions in undoped (●) and heavily Si-doped (○) $Al_xGa_{0.51-x}In_{0.49}P$. In the Si-doped sample, the momentum distribution corresponds to that obtained in the Si-DX center. (From Mäkinen et al., 1996, with permission.)

1.002. The relative $W$ parameters are larger and the $S$ parameters are smaller than those found for Ga or As vacancies in GaAs; $W_V/W_B = 0.73$ and $S_V/S_B = 1.015$ for $V_{Ga}^{3-}$ and $W_V/W_B = 0.72$ and $S_V/S_B = 1.012$ for $V_{As}^-$, respectively. This indicates that the open volume at the DX center is less than that of a monovacancy. Furthermore, we can argue that the open volume is less in DX(Te) than in DX(Si) and DX(Sn) (Laine et al., 1996b).

The DX centers have been observed by positrons also in $Al_xGa_{0.51-x}In_{0.49}P$ layers where the difference in the core electron shell structures for group III (Al, Ga, In) and group V (P) atoms gives a perfect opportunity to determine the lattice site of the vacancy (Mäkinen et al., 1996). Figure 35 shows that the core momentum distribution for the DX(Si) is lower than bulk. The difference from the bulk distribution is much closer to that found in Fig. 8 for the In vacancy than to that observed for the P vacancy. One can therefore infer that the vacancy in the DX(Si) center is on the group III lattice site rather than on the group V site.

### 4. THE ATOMIC STRUCTURE OF EL2 AND DX

The same mechanism has been proposed to explain the metastability of the EL2 defect and the DX center. Both are due to isolated donor atoms that can occupy two different positions in the lattice, depending on their electronic configuration. The theoretical models explain the structural metastability of the donor atoms in III–V compounds simply by two different forms of sp hybridization in covalent crystals. The group III donors like $As_{Ga}$, $Si_{Ga}$, and $Sn_{Ga}$ can either be on their ideal lattice sites in the $sp^3$ configuration or relax away along the [111] direction towards the interstitial position forming $sp^2$ bonds (Dabrowski and Scheffler, 1988; Chadi and Chang, 1988a, 1988b). When the donor occupies a group V site, like $Te_{As}$, it may induce a similar displacement of a nearby group III atom (Chadi and Chang, 1989). Depending on the details of the electronic system, the $sp^3$ or the $sp^2$ configuration can either be the stable or metastable state of the donor atom.

According to the model in which the EL2 defect consists of the isolated As antisite defect, the As antisite relaxes towards the interstitial position in the metastable state. The reaction $As_{Ga} \to V_{Ga} + As_i$ creates the $V_{Ga}$–$As_i$ pair, which is the metastable vacancy $V^*$ with a smaller open volume than that of the isolated Ga vacancy (Dabrowski and Scheffler, 1988; Chadi and Chang, 1988a). This structure for the metastable state of EL2 is consistent with the positron data. The positron experiments show that the metastable state of EL2 contains a vacancy whose open volume is less than that of a monovacancy. There is no experimental evidence that positron trapping occurs at the stable state of EL2. This lack of evidence is consistent with the idea that open volume is not present in the stable state of EL2.

As described in Section IX.2, metastable defects are also found in the positron experiments in electron-irradiated GaAs. This irradiation-induced defect has a structurally similar vacancy, $V_{irr}^*$, in its metastable state as that detected in the metastable state of the native EL2 defect. Therefore, it is natural to conclude that $V_{irr}^*$ is related to the same intrinsic defect as EL2 (i.e., the $As_{Ga}$ antisite), and the metastable vacancy is formed in the similar relaxation: $As_{Ga} \to V_{Ga} + As_i$.

However, some photoquenching, photorecovery, and thermal annealing properties of the irradiation-induced metastable vacancy $V_{irr}^*$ are different from those of the native ones. In contrast to the properties of native EL2, the metastable vacancy $V_{irr}^*$ in electron-irradiated GaAs absorbs infrared light, indicating that it has ionization levels in the bandgap. Our conclusion is that $V_{irr}^*$ is related to a different defect from EL2 but still related to $As_{Ga}$. We infer that it is a defect complex involving the As antisite.

This conclusion is also supported by electron paramagnetic resonance and magnetic circular dichroism of absorption experiments, where the $As_{Ga}$–$Ga_{As}$ pair has been detected with similar photoquenching and recovery properties as those we observe for the metastable vacancy (Krambrock and Spaeth, 1993; Hesse et al., 1994). According to a recent theoretical calculation (Pöykkö et al., 1997), the $As_{Ga}$–$Ga_{As}$ complex can relax to a metastable state $V_{Ga}$–$As_i$–$Ga_{As}$, which contains a vacancy capable of positron trapping. Furthermore, the metastable configuration has ionization levels in the bandgap (Pöykko et al., 1997), thus explaining the infrared absorption observed in the experiments.

In the model attributing the DX center to the isolated donor, the impurity atom $Si_{Ga}$ or $Sn_{Ga}$ relaxes towards the interstitial position, $Si_{Ga} \to V_{Ga} + Si_i$ when the donor is filled with two electrons forming the negative DX state. This relaxation leaves behind an open volume, which is less than that of a Ga vacancy (Chadi and Chang, 1988b). For the donor $Te_{As}$, the nearest-neighbor Ga moves away in a similar relaxation: $Te_{As} + Ga_{Ga} \to Te_{As} + V_{Ga} + Ga_i$ (Chadi and Chang, 1989). Such atomic configurations are consistent with the positron data; positrons reveal the presence of a vacancy whose open volume is less than that of a Ga vacancy. In agreement with the theory, the vacancy found in the positron experiments is on the group III sublattice (i.e., it is the Ga vacncy) (see Fig. 35).

To check whether the positron can be localized in the open volume of the $V_{Ga}$–$As_i$ pair, positron lifetimes have been calculated in a Ga vacancy and in the atomic configuration proposed for the metastable structure of EL2 (Hakala and Puska, 1996; Laasonen et al., 1991). The positron is indeed trapped, and the positron density is illustrated in Fig. 36. The calculated lifetimes are 268 ps for the Ga vacancy and 241 ps for the metastable structure of EL2.

In summary, the positron data obtained for the EL2 and DX donors support the idea that in both cases open volumes is involved in the atomic configuration of the metastable (EL2) or stable (DX) state. The properties of the open volume are consistent with the prediction of the model, which attributes the metastability to atomic relaxations that accompany the changes of electronic charge states.

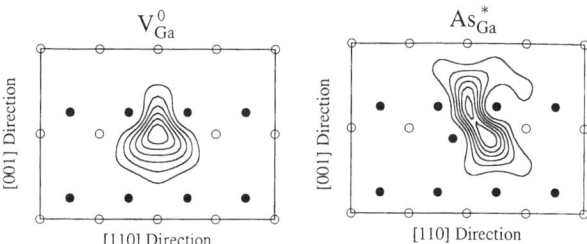

FIG. 36. The positron density in the Ga vacancy in GaAs and in the metastable state of the isolated As antisite consisting of a vacancy–interstitial pair. (From Hakala and Puska, 1996, with permission.)

## X. Summary

Positron annihilation spectroscopy gives microscopic information about the vacancy defects in the concentration range $10^{15}$ to $10^{19}\,\mathrm{cm}^{-3}$. It is applied to study vacancies both in bulk crystals and in epitaxial layers. The positron lifetime is the fingerprint of the open volume associated with a defect. It is used to identify mono- and divacancies and larger vacancy clusters. The Doppler broadening of the annihilation radiation measures the momentum distribution of the annihilating electrons. It can be used to identify the nature of the atoms surrounding the vacancy. Consequently, vacancies on different sublattices of a compound semiconductor can be distinguished and impurities associated with the vacancies can be identified. The charge state of a vacancy defect is determined by the temperature dependence of the positron trapping coefficient. Positron localization into Rydberg states around negative centers yields information about ionic acceptors that have no open volume.

Applications of the technique to GaAs and related materials have been presented. The Ga and As vacancies have been identified and their role as electrically and optically active centers has been discussed. The intrinsic ionic acceptors seen in as-grown, irradiated, or deformed GaAs have been associated with the Ga antisite defect. Positron measurements on metastable defects, EL2 in GaAs and DX in AlGaAs, reveal open volume in the stable (DX) or metastable (EL2) state and thus give experimental support to the vacancy-interstitial model of these defects.

REFERENCES

Alatalo, M., Kauppinen, H., Saarinen, K., Puska, M. J., Mäkinen, J., Hautojärvi, P., and Nieminen, R. M. (1995). *Phys. Rev. B*. **51**, 4176.
Alatalo, M., Barbiellini, B., Hakala, M., Kauppinen, H., Korhonen, T., Puska, M. J., Saarinen, K., Hautojärvi, P., and Nieminen, R. M. (1996). *Phys. Rev. B* **54**, 2397.
Ambigapathy, R., Manuel, A. A., Hautojärvi, P., Saarinen, K., and Corbel, C. (1994). *Phys. Rev. B* **50**, 2188.
Asoka-Kumar, P., Lynn, K. G., and Welch, D. O. (1994). *J. Appl. Phys.* **76**, 4935.
Baraff, G. A. and Schlüter, M. (1985). *Phys Rev. Lett.* **55**, 1327.
Barbiellini, B., Puska, M. J., Torsti, T., and Nieminen, R. M. (1995). *Phys. Rev. B* **51**, 7341.
Barbiellini, B., Puska, M. J., Korhonen, T., Harju, A., Torsti, T., and Nieminen, R. M. (1996). *Phys. Rev. B* **53**, 16201.
Bergensen, B. and Pajanne, E. (1969). *Phys. Rev. B* **186**, 375.
Bergensen, B., Pajanne, E., Kubica, P., Stott, M. J., and Hodges, C. H. (1974). *Solid State Commun.* **15**, 1377.
Boev, O. V., Puska, M. J., and Nieminen, R. M. (1987). *Phys. Rev. B* **36**, 7786.
Boronski, E., and Nieminen, R. M. (1986). *Phys. Rev. B* **34**, 3820.
Brandt, W. and Dupasquier, A. (eds.) (1983). *Positron Solid-State Physics*. North-Holland, Amsterdam.

Chadi, D. J. and Chang, K. J. (1988). *Phys Rev. Lett.* **60**, 2187.
Chadi, D. J. and Chang, K. J. (1988). *Phys Rev. Lett.* **61**, 873.
Chadi, D. J. and Chang, K. J. (1989). *Phys. Rev. B* **39**, 10063.
Chand, N., Henderson, T., Klem, J., Masselink, W. T., Fischer, R., Chang, Y., and Morkoc, H. (1984). *Phys. Rev. B* **30**, 4481.
Chicihibu, S., Iwai, A., Nakahara, Y., Matsumoto, S., Higuchi, H., Wei, L., and Tanigawa, S. (1993). *J. Appl. Phys.* **73**, 3880.
Corbel, C., Moser, P., and Stucky, M. (1985). *Ann. Chim. (Paris)* **8**, 733.
Corbel, C., Stucky, M., Hautojärvi, P., Saarinen, K., and Moser, P. (1988). *Phys. Rev. B* **38**, 8192.
Corbel, C., Pierre, F., Hautojärvi, P., Saarinen, K., and Moser, P. (1990). *Phys. Rev. B* **41**, 10632.
Corbel, C., Pierre, F., Saarinen, K., Hautojärvi, P., and Moser, P. (1992). *Phys. Rev. B* **45**, 3386.
Dabrowski, J. and Scheffler, M. (1988). *Phys Rev. Lett.* **60**, 2183.
Dlubek, G., Brümmer, O., Plazaola, F., Hautojärvi, P., and Naukkarinen, K. (1985). *Appl. Phys. Lett.* **46**, 1136.
Dorikens-Vanpraet, L., Dorikens, M., and Segers D. (eds.) (1989). *Proc. of 8th International Conference on Positron Annihilation.* World Scientific, Singapore.
Dupasquier, A. and Mills, A. P., Jr. (eds.) (1995). *Positron Spectroscopy of Solids, Proc. of the Int. School of Physics Enrico Fermi Course CXXV.* IOS Press, Amsterdam.
Garcia, A. and Northrup, J. E. (1995). *Phys Rev. Lett.* **74**, 1131.
Gilgien, L., Galli, G., Gygi, F., and Car, R. (1994). *Phys Rev. Lett.* **72**, 3214.
Hakala, M. and Puska, M. J. (1996). Unpublished.
Hakala, M. et al. (1996). Unpublished.
Hautojärvi, P. and Corbel, C. (1995). *Positron Spectroscopy of Solids, Proc. of the Int. School of Physics Enrico Fermi Course CXXV.* p. 491 (ed. Dupasquier, A. and Mills, A. P., Jr.). IOS Press, Amsterdam.
Hautojärvi, P. (ed.) (1979). *Positrons in Solids*, Topics in Current Physics, Vol. 12. Springer, Heidelberg.
He, Y.-J., Cao, B.-S., and Jean, Y. C. (eds.) (1995). *Proc. of 10th International Conference on Positron Annihilation.* Mat. Sci. For Vols. 175–178. Trans Tech., Aedermannsdorf.
Hesse, M., Koschnick, F. K., Krambrock, K., and Spaeth, J.-M. (1994). *Solid State Commun.* **92**, 207.
Jensen, K. O. and Walker, A. B. (1990). *J. Phys. Condens. Matter* **2**, 9757.
Kajcsos, Zs. and Szeles, Cs. (eds.) (1992). *Proc. of 9th International Conference on Positron Annihilation.* Mat. Sci. For. Vols. 105–110. Trans Tech, Aedermannsdorf.
Kaminska, M. and Weber, E. R. (1993). *Imperfections in III/V Materials*, p. 59 (ed. Weber, E. R.). Semiconductors and Semimetals, Vol. 38. Academic, New York.
Kaminska, M., Skowronski, M., and Kuszko, W. (1985). *Phys Rev. Lett.* **55**, 2204.
Korhonen, T., Puska, M. J., and Nieminen, R. M. (1996). *Phys. Rev. B* **54**, 15016.
Krambrock, K. and Spaeth, J.-M. (1993). *Phys. Rev. B* **47**, 3987 (1993).
Krause, R., Saarinen, K., Hautojärvi, P., Polity, A., Gärtner, G., and Corbel, C. (1990). *Phys Rev. Lett.* **65**, 3329.
Kuisma, S., Saarinen, K., Hautojärvi, P., and Corbel, C. (1996). *Phys. Rev. B* **53**, R7588.
Kuisma, S., Saarinen, K., Hautojärvi, P., Corbel, C., and LeBerre, C. (1996). *Phys. Rev. B* **53**, 9814.
Laasonen, K., Alatalo, M., Puska, M. J., and Nieminen, R. M. (1991). *J. Phys. Condens. Matter.* **3**, 7217.
Laine, T., Saarinen, K., Mäkinen, J., Hautojärvi, P., Corbel, C., Pfeiffer, L. N., and Citrin, P. (1996). *Phys. Rev. B* **54**, R11030.
Laine, T., Mäkinen, J., Saarinen, K., Hautojärvi, P., Corbel, C., Fille, M. L., and Gibart, P. (1996). *Phys. Rev. B* **53**, 11025.

Lang, D. V., Logan, R. A., and Jaros, M. (1979). *Phys. Rev. B* **19**, 1015.
Lang, D. V. (1977). *Radiation Effects in Semiconductors*. p. 70 (ed. Orli, N. B. and Corbett, J. W.). IOP Conf. Proc. No. 31. Institute of Physics and Physical Society, London.
LeBerre, C., Corbel, C., Brozel, M. R., Kuisma, S., Saarinen, K., and Hautojärvi, P. (1994). *J. Phys.: Condens. Matter* **6**, L759.
LeBerre, C., Corbel, C., Saarinen, K., Kuisma, S., Hautojärvi, P., and Fornari, R. (1995). *Phys. Rev. B* **52**, 8112.
LeBerre, C., Corbel, C., Mih, R., Brozel, M. R., Tüzemen, S., Kuisma, S., Saarinen, K., Hautojärvi, P., and Fornari, R. (1995). *Appl. Phys. Lett.* **66**, 2534.
Lee, J. L., Wei, L., Tanigawa, S., and Kawabe, M. (1990). *J. Appl. Phys.* **68**, 5571.
Liszkay, L., Corbel, C., Baroux, L., Hautojärvi, P., and Tatarenko, S. (1994). *Appl. Phys. Lett.* **64**, 1340.
Liszkay, L., Corbel, C., Baroux, L., Hautojärvi, P., Declamy, A., and Renault, P. O. (1995). *J. Phys.: Condens. Matter* **7**, 8529.
Maguire, J., Murray, R., Newman, R. C., Beall, R. B., and Harris, J. J. (1987). *Appl. Phys. Lett.* **50**, 516.
Mäkinen, S. and Puska, M. J. (1989). *Phys. Rev. B* **40**, 12523.
Mäkinen, J., Corbel, C., Hautojärvi, P., Moser, P., and Pierre, F. (1989). *Phys. Rev. B* **39**, 10162.
Mäkinen, J., Corbel, C., and Hautojärvi, P. (1991). *Phys. Rev. B* **43**, 12114.
Mäkinen, J., Hautojärvi, P., and Corbel, C. (1992). *J. Phys.: Condens. Matter* **4**, 5137.
Mäkinen, J., Laine, T., Saarinen, K., Hautojärvi, P., Corbel, C., Airaksinen, V. M., and Gibart, P. (1993). *Phys Rev. Lett.* **71**, 3154.
Mäkinen, J., Laine, T., Saarinen, K., Hautojärvi, P., Corbel, C., Airaksinen, V. M., and Nagle, J. (1995). *Phys. Rev. B* **52**, 4870.
Mäkinen, J., Laine, T., Partanen, J., Saarinen, K., Hautojärvi, P., Tappura, K., Hakkarainen, T., Asonen, H., Pessa, M., Kauppinen, J. P., Vänttinen, K. A., Paalanen, M., and Likonen, J. (1996). *Phys. Rev. B* **53**, 7851.
Mantl, S. and Triftshäuser, W. (1978). *Phys. Rev. B* **27**, 1645.
Martin, G. M. and Makram-Ebeid, S. (1986). *Deep Centers in Semiconductors*, Chap. 6 (ed. Pantelides, S.). Gordon and Breach, New York.
Mascher, P., Dannefaer, S., and Kerr, D. (1989). *Phys. Rev. B* **40**, 11764.
Mattila, T. and Nieminen, R. M. (1995). *Phys Rev. Lett.* **74**, 2721.
Mikeska, H. (1970). *Z. Phys.* **232**, 159.
Mooney, P. M. (1990). *J. Appl. Phys.* **67**, R1.
Nieminen, R. M. and Oliva, J. (1980). *Phys. Rev. B* **22**, 2226.
Northrop, G. A. and Mooney, P. M. (1991). *J. Electr. Mat.* **20**, 13.
Northrup, J. E. and Zhang, S. B. (1993). *Phys. Rev. B* **47**, 6791.
Paulin, R. (1983). *Positron Solid-State Physics*. p. 565 (ed. Brandt, W. and Dupasquier, A.). North-Holland, Amsterdam.
Peng, J. P., Lynn, K. G., Asoka-Kumar, P., and Becker, D. P. (1996). *Phys Rev. Lett.* **76**, 2157.
Pfeiffer, W., Liszkay, L., Burchard, A., Deicher, M., Magerle, R., Ronning, C., Saarinen, K., and Hautojärvi, P. (1994). *Proc. of the 17th International Conference on Defects in Semiconductors*, p. 305 (ed. Heinrich, H. and Jantsch, W.). Mat. Sci. For. Vols. 143–147.
Pfeiffer, W., Liszkay, L., Saarinen, K., Burchard, A., Deicher, M., Magerle, R., Ronning, C., and Hautojärvi, P. (1996). Unpublished.
Pöykkö, S., Puska, M. J., and Nieminen, R. M. (1997). *Phys. Rev. B* **55**, 6914.
Puska, M. J. and Corbel, C. (1988). *Phys. Rev. B* **38**, 9874.
Puska, M. J. and Nieminen, R. M. (1994). *Rev. Mod. Phys.* **66**, 841.
Puska, M. J., Jepsen, O., Gunnarsson, O., and Nieminen, R. M. (1986). *Phys. Rev. B* **34**, 2695.
Puska, M. J., Mäkinen, S., Manninen, M., and Nieminen, R. M. (1989). *Phys. Rev. B* **39**, 7666.

Puska, M. J., Corbel, C., and Nieminen, R. M. (1990). *Phys. Rev. B* **41**, 9980.
Puska, M. J., Seitsonen, A. P., and Nieminen, R. M. (1995). *Phys. Rev. B* **52**, 10947.
Puska, M. J. (1989). *J. Phys. Condens. Matter* **1**, 7347.
Saarinen, K., Hautojärvi, P., Vehanen, A., Krause, R., and Dlubek, G. (1989). *Phys. Rev. B* **39**, 5287.
Saarinen, K., Corbel, C., Hautojärvi, P., Pierre, F., and Vignaud, D. (1990). *J. Phys.: Condens. Matter* **2**, 2453.
Saarinen, K., Hautojärvi, P., Lanki, P., and Corbel, C. (1991). *Phys. Rev. B* **44**, 10585.
Saarinen, K., Hautojärvi, P., Keinonen, J., Rauhala, E., Räisänen, J., and Corbel, C. (1991). *Phys. Rev. B* **43**, 4249.
Saarinen, K., Kuisma, S., Hautojärvi, P., Corbel, C., and LeBerre, C. (1993). *Phys Rev. Lett.* **70**, 2794.
Saarinen, K., Kuisma, S., Hautojärvi, P., Corbel, C., and LeBerre, C. (1994). *Phys. Rev. B* **49**, 8005.
Saarinen, K., Kuisma, S., Mäkinen, J., Hautojärvi, P., Törnqvist, M., and Corbel, C. (1995). *Phys. Rev. B* **51**, 14152.
Saarinen, K., Seitsonen, A. P., Hautojärvi, P., and Corbel, C. (1995). *Phys. Rev. B* **52**, 10932.
Saarinen, K., Laine, T., Skog, K., Mäkinen, J., Hautojärvi, P., Rakennus, K., Uusimaa, P., Salokatve, A., and Pessa, M. (1996). *Phys. Rev. Lett.* **77**, 3407.
Saito, M. and Oshiyama, A. (1996). *Phys. Rev. B* **53**, 7810.
Schödlbauer, D., Sperr, P., Kögel, G., and Trifthäuser, W. (1988). *Nucl. Instrum. Methods B* **34**, 258.
Schultz, P. J. and Lynn, K. G. (1988). *Rev. Mod. Phys.* **60**, 701.
Schuppler, S., Adler, D. L., Pfeiffer, L. N., West, K. W., Chaban, E. E., and Citrin, P. H. (1993). *Appl. Phys. Lett.* **63**, 2357.
Schuppler, S., Adler, D. L., Pfeiffer, L. N., West, K. W., Chaban, E. E., and Citrin, P. H. (1995). *Phys. Rev. B* **51**, 10527.
Shan, Y. Y., Asoka-Kumar, P., Lynn, K. G., Fung, S., and Beling, C. D. (1996). *Phys. Rev. B* **54**, 1982.
Soininen, E., Mäkinen, J., Beyer, D., and Hautojärvi, P. (1992). *Phys. Rev. B* **46**, 13104.
Suzuki, R., Kobayashi, Y., Mikado, T., Ohgaki, H., Chiwaki, M., Yamazaki, T., and Tomimasu, T. (1991). *Jpn. J. Appl. Phys. B* **30**, L532.
Theis, T. N., Mooney, P. M., and Parker, B. D. (1991). *J. Electron. Mat.* **20**, 35.
Thommen, K. (1970). *Radiat. Eff.* **2**, 201.
Törnqvist, M., Nissilä, J., Kiesling, F., Corbel, C., Saarinen, K., Seitsonen, A. P., and Hautojärvi, P. (1994). *Proc. of the 17th International Conference on Defects in Semiconductors*, p. 347 (ed. Heinrich, H., and Jantsch, W.). Mat. Sci. For. Vols. 143–147.
Tüzemen, S. and Brozel, M. R. (1991). *Appl. Surf. Sci.* **50**, 395.
Valkealahti, S. and Nieminen, R. M. (1984). *Appl. Phys. A* **35**, 51.
West, R. N. (1973). *Adv. Phys.* **22**, 263.
West, R. N. (1979). *Positrons in Solids*, p. 89 (ed. Hautojärvi, P.). Topics in Current Physics, Vol. 12. Spriner, Heidelberg.
Xu, H. and Lindefelt, W. (1990). *Phys. Rev. B* **41**, 5975.
Zhang, S. B. and Northrup, J. E. (1991). *Phys Rev. Lett.* **67**, 2339.

CHAPTER 6

# The *Ab Initio* Cluster Method and the Dynamics of Defects in Semiconductors

## R. Jones

DEPARTMENT OF PHYSICS
UNIVERSITY OF EXETER
EXETER, UNITED KINGDOM

## P. R. Briddon

DEPARTMENT OF PHYSICS
UNIVERSITY OF NEWCASTLE UPON TYNE
NEWCASTLE UPON TYNE, UNITED KINGDOM

|  |  |
|---|---|
| I. INTRODUCTION | 288 |
| II. THE MANY-BODY PROBLEM | 290 |
|     1. The Born-Oppenheimer Approximation | 291 |
|     2. Hartree-Fock Theory | 292 |
|     3. The Homogeneous Electron Gas | 295 |
|     4. The Spin-Polarized Electron Gas | 297 |
|     5. Density Functional Theory | 300 |
| III. PSEUDOPOTENTIAL THEORY | 306 |
| IV. THE REAL-SPACE CLUSTER METHOD | 310 |
|     1. The Hartree Energy | 312 |
|     2. The Exchange-Correlation Energy | 314 |
|     3. Matrix Formulation | 317 |
| V. SELF-CONSISTENCY AND ATOMIC FORCES | 319 |
|     1. Self-Consistency | 319 |
|     2. Evaluation of Forces | 323 |
| VI. STRUCTURAL OPTIMIZATION | 324 |
|     1. Unconstrained Relaxation | 324 |
|     2. Constrained Relaxation | 326 |
| VII. DETERMINATION OF VIBRATIONAL MODES | 327 |
|     1. Energy Second Derivatives and Musgrave-Pople Potentials | 327 |
|     2. Effective Charges | 329 |
|     3. Resonant Modes | 331 |
| VIII. PRACTICAL CONSIDERATIONS | 332 |
|     1. Choice of Basis Sets | 332 |
|     2. The Construction of a Suitable Cluster | 334 |
|     3. Mulliken Populations | 336 |
|     4. Radiative Lifetimes | 337 |

| | | |
|---|---|---|
| IX. APPLICATIONS | . | 337 |
| 1. *General* | . | 337 |
| 2. *Point Defects in Bulk Solids* | . | 338 |
| 3. *Line Defects* | . | 343 |
| X. SUMMARY | . | 345 |
| *References* | . | 346 |

## I. Introduction

First-principles methods of determining the structure and electronic properties of materials have become very popular in the 15 years or so since the pioneering studies of Yin and Cohen (1980). Usually, these calculations are carried out on a supercell employing a basis of plane waves. For many applications, such an approach is not the most efficient. For example, molecules, topological defects like dislocations, kinks, or adsorbates on surfaces are cases in which a cluster approach has definite advantages. Even for point defects in a crystalline environment, there are advantages arising from a cluster method using localized basis set of orbitals. Such methods can give a direct interpretation of defect wave functions in terms of hybridized orbitals on atoms local to the defect and can treat the strict symmetry of a point defect that is lost through interaction between defects in different unit cells. Unlike the supercell approach, the cluster method can easily treat the induced dipole moments of dynamic defects, which is important for determining the integrated absorption intensity of infrared radiation.

In this review, we shall discuss the basis of the two main *ab initio* methods: Hartree-Fock and density functional theories. We then discuss in detail the application to atomic clusters using a localized basis set. We shall almost exclusively deal with the implementation developed by us, and finally we discuss some applications that have been made of the formalism to defects in diamond, silicon, and other semiconductors, although the theory has been used to treat many other systems. However, it seems relevant to begin by discussing our motivation for introducing the technique in the first place.

In 1985, we became interested in the problem of water reacting with the cores of dislocations in quartz. It was believed that water molecules could react with the strained Si–O bonds within the core, or at kinks, breaking them and creating two Si–O–H bonds (Griggs and Blacic, 1965). This mechanism might then explain the very dramatic effect of hydrolytic weakening where the yield stress of dry quartz is more than an order of magnitude higher than wet quartz (Doukhan and Trepied, 1985; Heggie and Jones, 1987). To investigate such a process requires a theoretical method that not only is able to account for the strength of chemical bonds, but also is able to deal with their unusual environment within dislocation cores. The most satisfactory technique would be one that did not rely on empirical information, the applicability of which would be uncertain in this case. There are two *ab initio* schemes that do not rely on empirical methods: the traditional Hartree-Fock method and density functional methods. Hartree-Fock methods have been championed by chemists but are unable to account for the quasi-particle spectrum of metals, and perhaps for this reason density functional methods have been favored by physicists. Moreover, an important consideration is that the latter have a well-developed pseudopotential scheme that makes applications to materials containing, for example, germanium no more difficult than those composed of carbon. In addition, the evaluation of the exchange-correlation energy is simpler but at a cost of the lack of systematic improvements, which progressively reduce the errors in the energies of bonds or multiplets. Another problem area for density functional theory is in the description of Mott insulators such as NiO. Spin-density functional theory predicts these materials to be either metallic or narrow-gap semiconductors, whereas they are observed to be highly insulating. It is usually the case that density functional theory finds bandgaps smaller than the experimental values. Hartree-Fock methods, on the other hand, usually predict bandgaps much larger than experimental values. Both methods could be used to treat clusters or supercells, but the problem of dislocations and kinks is so demanding and requires so many atoms, that it seemed desirable to use a cluster containing a single dislocation—not possible in a supercell—whose surface was terminated by hydrogen. So it seemed sensible at that time to invest a considerable amount of effort in developing an *ab initio* local-density functional cluster method that incorporated pseudo-potentials.

The particular code developed and used by the Exeter, Newcastle, Sussex, and Luleå groups is called AIMPRO, which is an acronym for *Ab Initio Modeling Program*. The code has undergone a great many modifications and improvements since it was first written. These developments have extended the range of applications and most importantly have led to a considerable speed-up, so that nowadays very large clusters of atoms can be considered. At the time of this writing, the largest cluster considered is an 840-atom bucky-onion, which was run, without using any symmetry acceleration options, on a T3D using 256 processors (Heggie *et al.*, 1996a). Typically, about two hours were required to carry out one conjugate gradient iteration, which generates the relaxed structure. This extreme application illustrates the power of the method, but of course most applications to solid state problems use much smaller 70 to 150 atomic clusters. Such clusters can be run on simple RISC workstations, taking several days.

We begin by giving an overview of the problem of determining the equilibrium structure of a multiatom system; then we shall discuss the cluster method in some detail before describing some of the applications that have been made.

## II. The Many-Body Problem

It is desirable to choose a system of units in which the fundamental constants are removed from the equations. We shall use atomic units throughout, except in dealing with applications. In terms of these units, $\hbar, e, m$, and $4\pi\varepsilon_0$ are taken to be unity. The Schrödinger equation for the electron in the hydrogen atom, for example, then becomes:

$$\left\{-\frac{1}{2}\nabla^2 - \frac{1}{r} - E\right\}\psi(\mathbf{r}) = 0.$$

The 1s solution is then $\psi = 1/(\sqrt{\pi})e^{-r}$ and has energy, $E = -\frac{1}{2}$. This establishes the unit of energy to be 1 a.u. = 27.212 eV, and as the radius of the atom is 1 a.u., the unit of length is the Bohr radius of 0.529 Å.

The nonrelativistic many-body Schrödinger equation for the electrons in a fixed field due to ions of charges $Z_a$ at sites $\mathbf{R}_a$ is

$$\left\{-\frac{1}{2}\sum_\mu \nabla_\mu^2 + \frac{1}{2}\sum_{\nu \neq \mu}\frac{1}{|\mathbf{r}_\mu - \mathbf{r}_\nu|} - \sum_{\mu,a}\frac{Z_a}{|\mathbf{r}_\mu - \mathbf{R}_a|} + \frac{1}{2}\sum_{a \neq b}\frac{Z_a Z_b}{|\mathbf{R}_a - \mathbf{R}_b|} - E\right\}\Psi(r) = 0,$$

or, in an obvious notation,

$$(H - E)\Psi(r) = \{T + V_{e-e} + V_{e-i} + V_{i-i} - E\}\Psi(r) = 0. \tag{1}$$

Here $r$ denotes the positions and spins of the electrons: $(\mathbf{r}_1, s_1, \mathbf{r}_2, s_2, \ldots)$. We shall be mainly concerned with the ground-state solution to this equation, and the greater part of this article is devoted to a discussion of the techniques we employ to obtain this.

It should be noted that the Hamiltonian of Eq. (1) does not include the kinetic energy of the ions themselves, if they are also regarded as a quantum mechanical system. The full Hamiltonian, $H_T$, includes this kinetic energy:

$$H_T = H - \sum_a \frac{m}{2M_a}\nabla_a^2.$$

We have written the term $m/M_a$ to remind the reader that this ratio involves the electron and atomic masses. The structure and properties of all atomic clusters are contained in the solution of the Schrödinger equation of the full Hamiltonian. However, to develop a practical method of solving this equation, it is necessary to first decouple the motions of the electrons and ions; then to construct an effective potential acting on each electron due to the other electrons as well as the surrounding nuclei; and finally to calculate the forces acting on each ion so that both the equilibrium structure of the cluster as well as its vibrational modes can be found.

## 1. THE BORN-OPPENHEIMER APPROXIMATION

The first step in which the dynamical equations of the ions are separated from those of the electrons is made using this approximation. We assume that the ions are so much more massive than the electrons that their movement simply modulates the wave function of the electrons. The total wave function can then be written

$$\Psi_T(r, R) = \chi(R)\Psi(r, R),$$

where $\chi(R)$ is an amplitude dependent on the nuclear coordinates alone, and $\Psi(r, R)$ is a solution of Eq. (1). We have denoted the nuclear coordinates by $R = (\mathbf{R}_1, \mathbf{R}_2, \ldots)$. If this is substituted into the Schrödinger equation for the full Hamiltonian, then we find, after multiplying through by $\Psi^*(r, R)$,

$$\left\{-\sum_a \frac{m}{2M_a} \nabla_a^2 + E(R) + W(R) - E_T\right\}\chi(R)$$
$$= \sum_a \int \Psi^*(r, R) \frac{m}{M_a} \nabla_a \Psi(r, R) \nabla_a \chi(R) dr. \qquad (2)$$

The term $dr$ represents the integration over all the electron coordinates, $\mathbf{r}_\mu$, and the summation over all their spins $s_\mu$. The left-hand side represents the Schrödinger equation for the ions moving in a potential $E + W$, where $W$ is a small correction, invariably neglected, due to the electrons moving along with the nuclei:

$$W(R) = -\sum_a \frac{m}{2M_a} \int \Psi^*(r, R) \nabla_a^2 \Psi(rt, R) dr.$$

The term on the right-hand side of Eq. (2) vanishes if $\Psi(r, R)$ is real corresponding to a nondegenerate ground state. Otherwise, it usually represents a small perturbation but can be particularly important for degenerate ground states—as perturbations usually are—for then it can lead to symmetry breaking, as in the Jahn-Teller effect (Stoneham, 1975). If we neglect this term, then the ionic and electron motions are decoupled.

$E(R)$ is an effective potential energy of the ions averaged over the state $\Psi(r, R)$. The minimum value of $E(R)$ is then the ground-state energy of the cluster, and one of the principal objectives of the theory is to deduce this energy and the corresponding ionic positions. If $E(R)$ is expanded about its minimum value, we find:

$$E(R) = E(R_o) + \frac{1}{2} \sum_{la,mb} \left( \frac{\partial^2 E}{\partial R_{la} \partial R_{mb}} \right) \Delta R_{la} \Delta R_{mb} + \cdots \qquad (3)$$

Here, $\Delta R_{la}$ represents the displacement of the ion $a$ in direction $l$ from the equilibrium configuration $R_o$. The harmonic frequencies of vibration $\omega_i$ and their normal coordinates, $u_{la}^i$, are related to the eigenvalues and eigenvectors of the dynamical matrix calculated from the energy derivatives (Born and Huang, 1954); that is,

$$\sum_{mb} E_{la,mb} u_{mb}^i = \omega_i^2 u_{la}^i, \qquad E_{la,mb} = \frac{1}{\sqrt{M_a M_o}} \left( \frac{\partial^2 E}{\partial R_{la} \partial R_{mb}} \right).$$

We shall discuss this further in Section VII. Now that we have separated the motions of the ions and electrons, we are confronted by the problem of the interaction between the electrons implicit in Eq. (1). To deal with this, further approximations are required.

2. Hartree-Fock Theory

The assumption behind this method is that there exists a set of $M$ orthonormal one-electron spin-orbitals $\psi_\lambda(r)$ from which the many-body wave function can be constructed as a single Slater determinant:

$$\Psi(r) = \frac{1}{\sqrt{M!}} \det|\psi_\lambda(r_\mu)|, \qquad \psi_\lambda(r) = \psi_i(\mathbf{r})\chi_\alpha(s),$$

and $\chi_\alpha(s)$ is a spin function satisfying

$$\sum_s \chi_\alpha^*(s)\chi_\beta(s) = \delta_{\alpha\beta}.$$

The sum being over 2-values of $s$ and $\alpha$ being "up" or "down." The orbitals $\psi_i(\mathbf{r})$ satisfy

$$\int \psi_i^*(\mathbf{r})\psi_j(\mathbf{r})d\mathbf{r} = \delta_{ij}.$$

The many-body wave function is clearly antisymmetric with respect to the interchange of two particles, as is required by the Pauli exclusion principle. As a simple example, we may write down the wave function for a two-electron problem such as $H_2$ by expanding the determinant. We obtain the well-known result for a two-particle fermion system:

$$\Psi(r_1, r_2) = \frac{1}{\sqrt{2}}\{\psi_1(r_1)\psi_2(r_2) - \psi_1(r_2)\psi_2(r_1)\}.$$

The average energy of the single normalized determinental wave function is $\langle \Psi|H|\Psi\rangle$ and can be shown to be (Slater, 1960)

$$E = \sum_\lambda \langle\lambda|T + V_{e-i} + V_{i-i}|\lambda\rangle + \frac{1}{2}\sum_{\lambda,\mu}\{\langle\lambda\mu|V_{e-e}|\lambda\mu\rangle - \langle\lambda\mu|V_{e-e}|\mu\lambda\rangle\}. \quad (4)$$

Here, the first term involves the matrix elements of one-particle operators: the kinetic, electron–ion, and ion–ion interactions, respectively, and the sum is over the occupied spin-orbitals $\lambda$. The second and third terms involve four-center integrals of the electron–electron interaction:

$$\langle\lambda\mu|V_{e-e}|\nu\kappa\rangle = \sum_{s_1 s_2} \int \psi_\lambda^*(r_1)\psi_\mu^*(r_2)\frac{1}{|\mathbf{r}_1 - \mathbf{r}_2|}\psi_\nu(r_1)\psi_\kappa(r_2)d\mathbf{r}_1 d\mathbf{r}_2.$$

We can rewrite $E$ in the form

$$E = -\frac{1}{2}\sum_{\lambda s}\int \psi_\lambda^*(\mathbf{r},s)\nabla^2\psi_\lambda(\mathbf{r},s)d\mathbf{r} + \int n(\mathbf{r})V_{e-i}d\mathbf{r} + E_H + E_x + E_{i-i}, \quad (5)$$

$$E_H = \frac{1}{2}\int \frac{n(\mathbf{r}_1)n(\mathbf{r}_2)}{|\mathbf{r}_1 - \mathbf{r}_2|}d\mathbf{r}_1 d\mathbf{r}_2, \quad (6)$$

$$E_x = -\frac{1}{2}\sum_{\lambda\mu}\sum_{s_1 s_2}\int \psi_\lambda^*(r_1)\psi_\mu^*(r_2)\frac{1}{|\mathbf{r}_1 - \mathbf{r}_2|}\psi_\mu(r_1)\psi_\lambda(r_2)d\mathbf{r}_1 d\mathbf{r}_2, \quad (7)$$

$$E_{i-i} = \frac{1}{2}\sum_{a\neq b}\frac{Z_a Z_b}{|\mathbf{R}_a - \mathbf{R}_b|}. \quad (8)$$

Here, we have introduced the electron density

$$n(\mathbf{r}) = \sum_{\lambda s} |\psi_\lambda(\mathbf{r}, s)|^2, \tag{9}$$

and the Hartree energy $E_H$, the exchange energies $E_x$, and the ion–ion energy $E_{i-i}$.

The ground-state orbitals $\psi_\lambda$ are determined by the requirement that $E$ is minimized subject to orthonormal $\psi_\lambda$. This constrained minimization problem can be solved by introducing Lagrange multipliers, $E_{\lambda\mu}$, such that the function

$$E - \sum_{s, \lambda \neq \mu} E_{\lambda\mu} \int \psi_\lambda^* \psi_\mu d\mathbf{r} - \sum_\lambda E_\lambda \left\{ \sum_s \int |\psi_\lambda|^2 d\mathbf{r} - 1 \right\}$$

is minimized with respect to $\psi_\lambda^*$, $E_\lambda$, and $E_{\lambda\mu}$ without constraint. From Eq. (5), we then get the Hartree-Fock equations for each orbital $\lambda$:

$$\left\{ -\frac{1}{2}\nabla^2 + V_{e-i}(\mathbf{r}) + V^H(\mathbf{r}) + V_\lambda^x(\mathbf{r}) - E_\lambda \right\} \psi_\lambda(r) = \sum_{\mu \neq \lambda} E_{\lambda\mu} \psi_\mu(r), \tag{10}$$

$$\sum_s \int \psi_\mu^* \psi_\lambda d\mathbf{r} = \delta_{\lambda\mu},$$

$$V^H(\mathbf{r})\psi_\lambda(r) = \frac{\delta E_H}{\delta \psi_\lambda^*} = \int \frac{n(\mathbf{r}_1)\psi_\lambda(r)}{|\mathbf{r} - \mathbf{r}_1|} d\mathbf{r}_1,$$

$$V_\lambda^x(\mathbf{r})\psi_\lambda(r) = \frac{\delta E_x}{\delta \psi_\lambda^*} = -\sum_{\mu s_1} \int \psi_\mu^*(r_1) \psi_\lambda(r_1) \frac{1}{|\mathbf{r}_1 - \mathbf{r}|} \psi_\mu(r) d\mathbf{r}_1.$$

$V^H$ and $V_\lambda^x$ are the Hartree and exchange potentials. The last involves sum over occupied orbitals $\mu$, whose spin is the same as that of $\lambda$.

Now we can carry out a unitary transformation on the Slater determinant, which diagonalizes $E_{\lambda\mu}$, and then the right-hand side of the differential Eq. (10) vanishes. The exchange potential can be written in terms of the exchange density, $n_\lambda^x$, as

$$V_\lambda^x(\mathbf{r}) = \int \frac{n_\lambda^x(\mathbf{r}, \mathbf{r}_1)}{|\mathbf{r}_1 - \mathbf{r}|} d\mathbf{r}_1$$

$$n_\lambda^x(\mathbf{r}, \mathbf{r}_1) = -\frac{\sum_{\mu s_1} \delta(s_\lambda, s_\mu) \psi_\mu^*(r_1) \psi_\mu(r) \psi_\lambda^*(r) \psi_\lambda(r_1)}{\psi_\lambda(r) \psi_\lambda^*(r)}.$$

The exchange density satisfies

$$\int n_\lambda^x(\mathbf{r}, \mathbf{r}_1) d\mathbf{r}_1 = -1, \quad n_\lambda^x(\mathbf{r}, \mathbf{r}) < 0.$$

These relations show that an exchange hole of total charge unity is introduced around each electron. The exchange integral is a very difficult term to evaluate numerically because it involves the product of orbitals, each of which oscillates in a complicated way. Further, because it depends on $\lambda$, it has to be evaluated many times. This makes practical versions of the theory rather slow, even on the fastest computers.

The total energy $E$ can be found by multiplying the Hartree-Fock equations (10) by $\psi_\lambda^*(r)$ and integrating over $\mathbf{r}$ and summing over $s$ and $\lambda$. This gives

$$E = \sum_\lambda E_\lambda - E_H - E_x + E_{i-i}.$$

The sum is over occupied orbitals only. Notice that the interaction terms must be subtracted from the sum of energy eigenvalues.

The Hartree-Fock equations in Eq. (10) are solved by a self-consistent method. An initial set of orbitals $\psi_\lambda(r)$ are selected, which are usually related to atomic orbitals, and the Hartree and exchange potentials found. Then the Hartree-Fock equations are solved for an output set of orbitals. These are used to construct a new set of input orbitals, and the process is repeated until the output and input sets are equal. This process is called the self-consistent cycle.

The energy-eigenvalues, $E_\lambda$, can be given an interpretation through Koopman's theorem, which states that the difference in energy between two configurations differing by the occupation of an orbital $\lambda$, while all the other orbitals $\psi_\mu$ are unchanged, is $E_\lambda$. Hence $-E_\lambda$ is the ionization energy for the $\lambda$ electron. The correspondence is not exact, as all the orbitals will alter, in general, whenever the configuration is modified.

Hartree-Fock theory usually predicts structures and vibratory modes of small molecules quite accurately. However, bond lenghs are usually underestimated, leading to an overestimate of mode frequencies. Excitation energies are also overestimated.

### 3. The Homogeneous Electron Gas

As an example, we apply the theory to as homogeneous electron gas, sometimes called jellium, in which the ions form a uniform background of density $n$. The total spin $S$ is then a good quantum number, and we begin by looking at nonpolarized states where $S = 0$. A solution of the Hartree-Fock equations consists of orbitals corresponding to plane waves. Then $\lambda$ refers to the wave vector $\mathbf{k}$ and spin state $\alpha$:

$$\psi_\lambda(\mathbf{r}, s) = \frac{1}{\sqrt{\Omega}} e^{i\mathbf{k}\cdot\mathbf{r}} \chi_\alpha(s).$$

Here, $\Omega$ is the volume of the system. The charge density, $n$, is uniform, and hence the Hartree term, $E_H$, and the ion–ion term $E_{i-i}$, which in this case is just the electrostatic energy of the uniform positive background charge, exactly cancel the electron–ion term, $E_{el-i}$, in the total energy $E$.

The energy levels are therefore

$$E_\lambda = E_{\mathbf{k},\alpha} = \tfrac{1}{2}k^2 + V_{\mathbf{k}}^x,$$

where the exchange potential, $V_\lambda^x(\mathbf{r})$ is given by

$$V_{\mathbf{r}}^x(\mathbf{r}) = -\sum_{k_1 < k_f} \int \frac{e^{i(\mathbf{k}-\mathbf{k}_1)\cdot(\mathbf{r}_1 - \mathbf{r})}}{\Omega |\mathbf{r}_1 - \mathbf{r}|} d\mathbf{r}_1.$$

This is in fact independent of $\mathbf{r}$ and spin-state $\alpha$ and depends on the magnitude $k$ of $\mathbf{k}$ alone.

$$V_{\mathbf{k}}^x = -\sum_{k_1 < k_f} \frac{4\pi}{\Omega|\mathbf{k} - \mathbf{k}_1|^2} = -\frac{1}{8\pi^3} \int_{k_1 < k_f} \frac{4\pi}{|\mathbf{k} - \mathbf{k}_1|^2} d\mathbf{k}_1.$$

Here, $k_f$ is the Fermi wave vector related to the electron density by $n = 1/3\pi^2 (k_f^3)$. To carry out the integral, we write $\eta = k/k_f$, and a simple calculation shows

$$V_{\mathbf{r}}^x = -4\left(\frac{3n}{8\pi}\right)^{1/3} F(\eta), \qquad F(\eta) = \frac{1}{2} + \frac{1 - \eta^2}{4\eta} \ln\left(\frac{1 + \eta}{1 - \eta}\right).$$

The function $F(\eta)$ tends to 1 as $\eta \to 0$, and to $\tfrac{1}{2}$ as $\eta \to 1$. Its derivative has a weak singularity as $\eta$ tends to 1, which has catastrophic implications for the applicability of the theory to simple metals. This follows as the density of states, per unit energy range and for each spin, is

$$N(E) = \frac{4\pi k^2}{8\pi^3} \frac{1}{|\nabla E_{\mathbf{k}}|},$$

and tends to zero as $k \to k_f$, thus showing that the density of states is zero at the Fermi level. This is incorrect and is due to absence of correlation in the theory. The form of the Hartree-Fock wave function does not include correlated movement of the electrons. This can be included by constructing wave functions built out of combinations of determinants. These are known as configuration interaction calculations, but the computational demands are so high and the scaling with the number of electrons so poor that such calculations can only be done for a very small number of atoms.

The total energy is

$$E = \frac{1}{2}\sum_{k\alpha} k^2 + E_x,$$

which gives the energy density:

$$\frac{3n}{10}(3\pi^2 n)^{2/3} + n\varepsilon_x(n), \qquad \varepsilon_x(n) = -\frac{3}{2}\left(\frac{3n}{8\pi}\right)^{1/3}.$$

The quantity $\varepsilon_x$ is the exchange energy per electron.

## 4. THE SPIN-POLARIZED ELECTRON GAS

We now consider solutions of the Hartree-Fock equations for nonzero spin values S. This means we have more "up" spins, say, than "down" spins, and each state is specified by the densities of up and down spins, $n_\uparrow$ and $n_\downarrow$, respectively. The orbitals remain plane waves defined by a wave vector **k**, but each spin population has its own Fermi wave vector, and the total energy is then

$$E = \Omega \sum_s \left\{ \frac{3n_s}{10}(6\pi^2 n_s)^{2/3} - \frac{3}{2}\left(\frac{3}{4\pi}\right)^{1/3} n_s^{4/3} \right\}. \tag{11}$$

Improved estimates of the ground-state energy can be found by going beyond Hartree-Fock theory. Ceperley and Alder (1980) used a quantum Monte-Carlo method to find the correlation energy $E_c$—the difference between the ground state and the Hartree-Fock energy—for polarized and nonpolarized electron gases for low-density homogeneous electron gas. This can be combined with results of perturbation theory for the high-density case to produce an energy for a wide range of densities. If the correlation energy-per-electron, $\varepsilon_c$, polarization $\zeta$ and the Wigner-Seitz radius of each

## TABLE I
### PARAMETERIZATION OF THE EXCHANGE-CORRELATION ENERGY

|  | $\gamma$ | $\beta_1$ | $\beta_2$ |  |
|---|---|---|---|---|
| Nonpolarized | −0.1423 | 1.0529 | 0.3334 |  |
| Polarized | −0.0843 | 1.3981 | 0.2611 |  |
|  | A | B | C | D |
| Non polarized | 0.0311 | −0.0480 | 0.0020 | −0.0116 |
| Polarized | 0.0155 | −0.0269 | 0.0007 | −0.0048 |
| $i$ | $A_i$ | $p_i$ | $q_i$ |  |
| 1 | −0.9305 | 0.3333 | 0 |  |
| 2 | −0.0361 | 0 | 0 |  |
| 3 | 0.2327 | 0.4830 | 1 |  |
| 4 | −0.2324 | 0 | 1 |  |
| $i$ | $A'_i$ | $p'_i$ | $q'_i$ |  |
| 1 | −0.9305 | 0.3333 | 0 |  |
| 2 | −0.0375 | 0.1286 | 0 |  |
| 3 | −0.0796 | 0 | 0.1286 |  |

electron $r_s$ are defined by

$$E_c = \Omega n \varepsilon_c(n, \xi), \qquad \xi = \frac{(n_\uparrow - n_\downarrow)}{n}, \qquad r_s = (4\pi n/3)^{-1/3};$$

then $\varepsilon_c$ for the nonpolarized and fully polarized electron gases are given by Perdew and Zunger (1981) as

$$\varepsilon_c = \begin{cases} \gamma\{1 + \beta_1\sqrt{r_s} + \beta_2 r_s\}^{-1}, & \text{for } r_s \geq 1 \\ B + (A + Cr_s)\ln(r_s) + Dr_s, & \text{for } r_s < 1 \end{cases}$$

The values of the coefficients are given for both cases in Table I.

In the case of a partially polarized gas, where $1 > \xi > 0$, the correlation energy is averaged over the polarized and nonpolarized cases using the procedure of von Barth and Hedin (1972):

$$\varepsilon_c(n, \xi) = \varepsilon_c^{np}(n) + f(\xi)(\varepsilon_c^p - \varepsilon_c^{np}), \qquad f(\xi) = \frac{(1 + \xi)^{4/3} + (1 - \xi)^{4/3} - 2}{2^{4/3} - 2}.$$

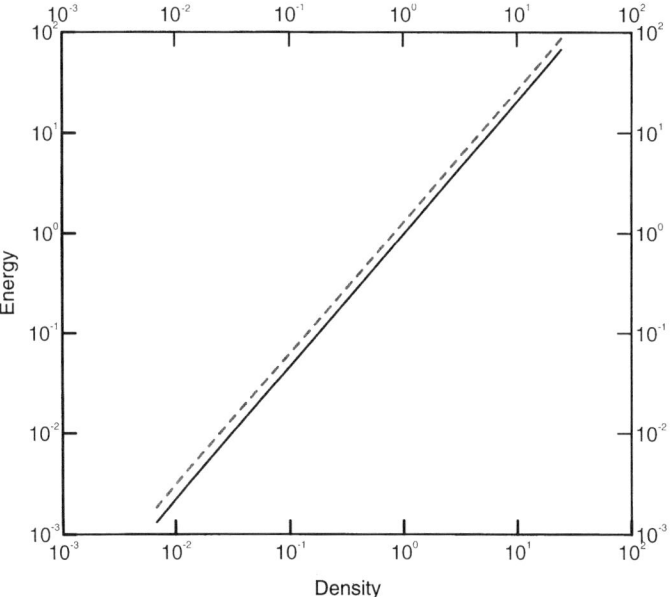

FIG. 1. Variation of polarized (full) and nonpolarized (dashed) exchange-correlation energy, $\times(-1)$ a.u., per unit volume with density.

In Fig. 1 we show the exchange-correlation energy per unit volume for the nonpolarized and fully polarized gases for the same density. It is clear that, to a good approximation, these energies are power series in the densities $n$ and $n_s$, respectively. In developing the theory of clusters in Section IV, it is necessary to simplify the expressions for $E_{xc}$. For the nonpolarized case, we can write

$$E_{xc} = \Omega A n^p, \quad (12)$$

where $p$ is 1.30917. This fit is accurate to within 0.002 a.u. for $n < 1.2$. For larger values, the error increases with $E_{xc}$, but the percentage error is less than 3% for $n$ up to 15.

For polarized gases, we use

$$E_{xc} = \Omega \sum_{i,s} A_i n_s^{p_i+1} n_{1-s}^{q_i}, \quad (13)$$

where $A_i$, $p_i$, and $q_i$ are given in Table I. The error in this expression is less than 0.001 a.u. for $n_\uparrow, n_\downarrow < 1$. For larger density values, it is desirable to use the values of $A'_i p'_i$, and $q'_i$, also given in Table I. The error, then, is less than 3% for large $n$ but is 4% for $n$ around 0.1.

It is necessary, when dealing with the core electrons of heavy elements, to multiply $\varepsilon_x$ by a small factor arising from relativistic corrections.

## 5. Density Functional Theory

The difficulty of evaluating the exchange energy and the need to include correlation has prompted the development of alternative methods. Density functional theory is one such development that has proved to be very successful. There are several ways of deducing the relevant equations. The simplest approach is to argue that the nonlocal exchange energy in Hartree-Fock theory is a very complicated integral that is considerably simplified if we treat the inhomogeneous problem locally as jellium and replace the exchange energy in Eq. (7) with its known electron gas value:

$$\int n(\mathbf{r}) \varepsilon_{xc}(n_\uparrow, n_\downarrow) d\mathbf{r}.$$

The up and down spin densities are defined in terms of the orbitals $\psi_\lambda(\mathbf{r}, s)$ through

$$n_s(\mathbf{r}) = \sum_\lambda \delta(s, s_\lambda) |\psi_\lambda(\mathbf{r}, s)|^2.$$

The theory then proceeds by minimizing the total energy with respect to the orbitals.

This approach, called local spin-density functional (LSDF) theory, has several advantages. Many of the problems with Hartree-Fock theory are solved, and LSDF theory is far more efficient computationally. Nevertheless, there are problems caused by the replacement of the Hartree-Fock exchange energy by its electron gas value. In particular, we have introduced a self-interaction term. In Eq. (4), the diagonal term with $\lambda = \mu$ in the expression for the Hartree energy cancels out the diagonal term in the exchange energy. This is no longer the case in LSDF theory, and there is then a potential due to an electron acting on itself. This has the consequence that when the theory is applied to the H atom, for example, the 1s energy level is found to be $-0.269$ a.u. and the energy of the neutral atom is $-0.479$ a.u. instead of each being $-0.5$ a.u. For this reason, the ionization energies of atoms are not in very good agreement with experimental values. This deficiency can be corrected by incorporating an extra term, which removes the self-interaction (Perdew and Zunger, 1981). This approach has been successful in treating transition metal oxides (Svane and Gunnarsson, 1990; Szotek et al., 1993).

The more usual approach to LSDF theory is based on the work of Hohenberg and Kohn (1964) and Kohn and Sham (1965). These authors showed there is a 1:1 correspondence between a nondegenerate non-polarized ground-state wave function $\Psi(r)$ and the electron density $n(\mathbf{r}_1)$ defined by

$$n(\mathbf{r}_1) = \sum_\mu \int \delta(\mathbf{r}_1 - \mathbf{r}_\mu)|\Psi(r)|^2 dr.$$

The proof rests on the preliminary result that in the Hamiltonian

$$H = T + V_{e-e} + V_{e-i},$$

the ground-state electron density is in 1:1 correspondence with the external potential $V_{e-i}$. Suppose this is false; that is, there exist two external potentials $V_1$ and $V_2$ having the same $n$. Then from the variational principle, if $\Psi_1$ and $\Psi_2$ are the corresponding normalized wave functions, and if $H_i$ is the Hamiltonian with potential $V_i$ and energy $E_i$, then

$$E_1 = \langle \Psi_1|H_1|\Psi_1\rangle < \langle \Psi_2|H_1|\Psi_2\rangle$$
$$= E_2 + \langle \Psi_2|V_1 - V_2|\Psi_2\rangle$$
$$= E_2 + \int (V_2 - V_1)n(\mathbf{r})d\mathbf{r}.$$

But, in a similar way we can show

$$E_2 < E_1 + \int (V_2 - V_1)n(\mathbf{r})d\mathbf{r}.$$

Adding these equations gives us

$$E_2 + E_1 < E_1 + E_2,$$

which is a contradiction unless the states are degenerate.

This shows that $V_{e-i}$ is determined uniquely by $n$, and hence the Schrödinger equation for $\Psi$ can be solved in terms of $n$. Thus $\Psi$ is a unique functional of $n$. Since $E$ is the expectation of $H$ with respect to $\Psi$, it also follows that $E$ is determined uniquely by $n$. This is a remarkable result, as it shows that the many-body wave function dependent on the set of $\mathbf{r}_\mu$ and $s_\mu$ (i.e., 4-$M$ variables where $M$ is the number of electrons) is determined uniquely by a function of three variables [i.e., $n(\mathbf{r})$]. Clearly, instead of trying to guess the wave function, as in Hartree-Fock theory, and minimize this function of 4-$M$ variables, it is advantageous to try to find how the energy depends on $n(\mathbf{r})$ and minimize this as a functional of $n(\mathbf{r})$. In fact, subsequent to the work of Hohenberg and Kohn, it has been shown that this result also holds if the ground state is degenerate.

Hohenberg and Kohn went on to show that all the terms in the expression for the total energy may be evaluated as functionals of the charge density:

$$E[n] = T[n] + E_{e-i}[n] + E_H[n] + E_{xc}[n] + E_{i-i}. \tag{14}$$

Clearly, $E_{e-i}[n]$ and $E_H[n]$ are manifestly functionals of the charge density:

$$E_{e-i}[n] = -\int n(\mathbf{r}) \sum_a \frac{Z_a}{|\mathbf{r} - \mathbf{R}_a|} d\mathbf{r}, \quad E_H[n] = \frac{1}{2} \int \frac{n(\mathbf{r}_1)n(\mathbf{r}_2)}{|\mathbf{r}_1 - \mathbf{r}_2|} d\mathbf{r}_1 d\mathbf{r}_2,$$

and, as before, the ion–ion term is given by

$$E_{i-i} = \frac{1}{2} \sum_{a \neq b} \frac{Z_a Z_b}{|\mathbf{R}_a - \mathbf{R}_b|}.$$

The main difficulty is to write down expressions for the exchange-correlation and kinetic energies as functionals of the charge density. The exchange correlation is apparently the more challenging of the two, as this is describing the complicated many-body interactions that take place. An exact expression for this is not available, and in practice a number of approximate forms are used, the most common being the local density approximation (LDA) and the local spin density approximation (LSDA). LDA is for systems that are not spin-polarized, and the exchange correlation energy is written as

$$E_{xc}[n] = \int n(\mathbf{r}) \varepsilon_{xc}(n) d\mathbf{r},$$

where the exchange-correlation energy density, $\varepsilon_{xc}(n)$ is the function obtained for the homogeneous electron gas and detailed in the previous section. It is readily seen that this is a *local* approximation—it is assumed that for any small region in the system, the contribution made to the exchange-correlation energy is just the same as in a uniform electron gas with the same density.

In the LSDA, we write

$$E_{xc}[n] = \int n(\mathbf{r})\varepsilon_{xc}(n_\uparrow, n_\downarrow) d\mathbf{r},$$

which is similar in principle to the LDA but uses the expression for the energy density of a polarized homogeneous electron gas. The use of this approach in a real system (i.e., with a varying charge density) is an approximation, but it is one that has been shown to be successful and to have acceptable accuracy for modeling materials containing atoms from many parts of the periodic table.

There has been interest in going beyond local expressions for exchange and correlation and to include some terms dependent on the gradient of the density (see Kutzler and Painter, 1992, for applications to first-row diatomic molecules). The general experience is that the energy is improved but at the cost of an inferior structure.

It is possible to write down a functional for the kinetic energy $T[n]$ using the same approach as used for the exchange correlation (i.e., using the result for the homogeneous electron gas):

$$T[n] = \int \frac{3n(\mathbf{r})}{10} (3\pi^2 n)^{2/3} d\mathbf{r}.$$

We then arrive at a similar result to the Thomas-Fermi theory. Unfortunately, this is not accurate enough to describe the small changes in total energies that take place on chemical bonding. This problem was solved by Kohn and Sham (1965). They introduced a set of orthonormal orbitals as a basis for the charge density. In the spin-polarized theory, the spin densities would be written in terms of these as

$$n_s(\mathbf{r}) = \sum_\lambda \delta(s_\lambda, s) |\psi_\lambda(r)|^2.$$

This is effectively claiming that the charge density can always be written as that derived from a wave function consisting of a single Slater determinant. It should be emphasized that this is wholly different to Hartree-Fock theory, in which the many-particle wave function was written as a single determinant and used as a variational function. In the Kohn-Sham procedure, this wave function is used only as a means of expanding the charge density, and at no stage is the energy considered to be obtainable from an expression of the form $\langle \Psi | H | \Psi \rangle$. Strictly, therefore, we cannot interpret the Kohn-Sham orbitals as single-particle states.

In terms of these, the kinetic energy can be written down as

$$T = -\frac{1}{2} \sum_{\lambda,s} \int \psi_\lambda^* \nabla^2 \psi_\lambda d\mathbf{r},$$

which completes the terms that make up the total energy.

The Kohn-Sham orbitals will be determined by the requirement that $E$ is minimized with respect to $n_s$ subject to the total number of electrons $M$ and spin $S$ being fixed, where

$$M = \sum_s \int n_s(\mathbf{r})d\mathbf{r}, \qquad S = \int (n_\uparrow - n_\downarrow)d\mathbf{r}.$$

The Kohn-Sham equations are now derived by a variational principle, remembering that the orbitals $\psi_\lambda$ are orthonormal. Thus the quantity

$$E - \sum_\lambda E_\lambda \left\{ \sum_s \int |\psi_\lambda(r)|^2 d\mathbf{r} - 1 \right\},$$

is minimized with respect to $\psi_\lambda$, $\psi_\lambda^*$, and $E_\lambda$. This yields:

$$\left\{ -\frac{1}{2}\nabla^2 - \sum_a \frac{Z_a}{|\mathbf{r} - \mathbf{R}_a|} + V^H(\mathbf{r}) + V_{s_\lambda}^{xc}(n_\uparrow, n_\downarrow) - E_\lambda \right\} \psi_\mu(r) = 0$$

$$\sum_s \int |\psi_\lambda(r)|^2 d\mathbf{r} = 1.$$

Here,

$$V_s^{xc} = \frac{d(n\varepsilon_{xc})}{dn_s}.$$

The main differences with the Hartree-Fock equations are the exchange-correlation term and the interpretation of the energy levels $E_\lambda$. In the case of jellium, the density of states $N(E)$ is no longer zero at $E_f$, and hence identifying the Kohn-Sham energies $E_\lambda$ with quasi-particle energies is natural. This, however, is not strictly correct. Janak's Theorem (Janak, 1978) asserts that if we change the occupancy of level $\lambda$ by $\delta f_\lambda$, then the change of energy is $E_\lambda \delta f_\lambda$. This is not the same as the energy change that results from the addition or subtraction of a single electron. Such quasi-particle energies should be derived from extensions of the theory, the $GW$ approximation of Hedin (1969) being the most successful. Implementation of this theory has given fundamental energy gaps to within a few tenths of an eV (Delsole *et al.*, 1994).

The exchange-correlation potential $V_s^{xc}$ depends on the densities of both up and down spins. Consequently the spin up solutions are not the same as the spin down ones if $S$ is nonzero. This means that there are two sets of equations, each corresponding to a particular spin state.

An alternative formula for $E$ is found by multiplying by $\psi_\lambda^*(r)$ and integrating, followed by a sum over the occupied orbitals $\lambda$.

$$E = \sum_\lambda E_\lambda + E_{i-i} - E_H + \sum_s \int n_s(\mathbf{r}) \{\varepsilon_{xc} - V_s^{xc}\} d\mathbf{r}. \tag{15}$$

This formula is the starting point of approximate methods such as tight binding schemes (Sutton *et al.*, 1988). It is argued that the terms $E_{i-i}$ and $E_H$ largely cancel, as they must for jellium. For simple systems, the sum of energy eigenvalues acts as an attractive potential—it gets more positive as the separation between atoms increases—but the other three terms combine to act as a repulsive potential. It is often assumed that the repulsive one is short-ranged and falls off quickly to zero. The matrix elements of a tight-binding Hamiltonian are either constructed by fitting to a band structure derived by a combination of *ab initio* theory and experiment, or evaluated from a localized basis set (Porezag *et al.*, 1995; Seifert and Eschrig, 1985; Seifert *et al.*, 1986). The short-ranged repulsive potential is then fitted so that the structures of representative systems are reproduced. We have not followed this approach and will confine ourselves to this brief comment here.

The derivation of density functional theory presented here shows it to be essentially a ground-state theory. Properties of the ground state can be obtained with quantitative accuracy. To consider the ground state, the lowest lying Kohn-Sham orbitals are filled. However, a problem arises when we have several degenerate orbitals, and not all should be occupied. Examples of this occur in atoms (e.g., in carbon, two electrons need to be placed into the three degenerate $2p$ states) and defects (e.g., the vacancy in diamond, where two electrons need to be placed into three degenerate states of $t_2$ symmetry). This is essentially a multiplet problem and should lie beyond density functional theory. However, a method for obtaining approximate multiplet energies has been given by von Barth (1979). In this approach, the density functional energy is taken to be the energy of a single Slater determinant, the contents of which are governed by the choice of Kohn-Sham states to be filled. The energies of different multiplets, which can be built out of determinants corresponding to different configurations, may be found by writing each determinant, $|D\rangle$, as a linear combination of multiplets, $|M_a\rangle$, as

$$|D\rangle = \sum_a c_a |M_a\rangle.$$

If we operate on this with the Hamiltonian and identify the expectation value of the left-hand side as the energy $E$ found using density functional theory, then

$$E = \sum_a |c_a|^2 \langle M_a|H|M_a\rangle.$$

We can now choose different configurations—each corresponding to single Slater determinants—and then deduce several equations relating the multiplet energies, which can, in favorable cases, be solved.

### III. Pseudopotential Theory

Pseudopotentials are a very important component of first-principles calculations, as they remove the need to consider core electrons—only the valence electrons need be considered. This is extremely important if one wants to be able to treat all the elements of the periodic table. Using the full Coulomb potential can cause considerable problems. The total energy then becomes extremely large, and since one is interested in differences in energies between similar-sized clusters, there can be a significant loss of accuracy. Second, the fitting of core wave functions with Gaussian orbitals, and even more so with plane waves, is extremely difficult, and small errors can make large differences in the core eigenvalues. Figure 2 shows the $4s$ wave function and pseudo-wave functions in Ni. It is clear that the pseudo-wave function is a much simpler and smoother function to approximate than the all-electron wave function. Third, for the heavier atoms, relativistic effects are important and the Dirac equation is required. However, the valence electrons can continue to be treated nonrelativistically and a spin–orbit potential can be introduced that describes polarized valence electrons.

The justification for the use of a pseudopotential lies in the fact that the highly localized core wave function cannot take part in the bonding of

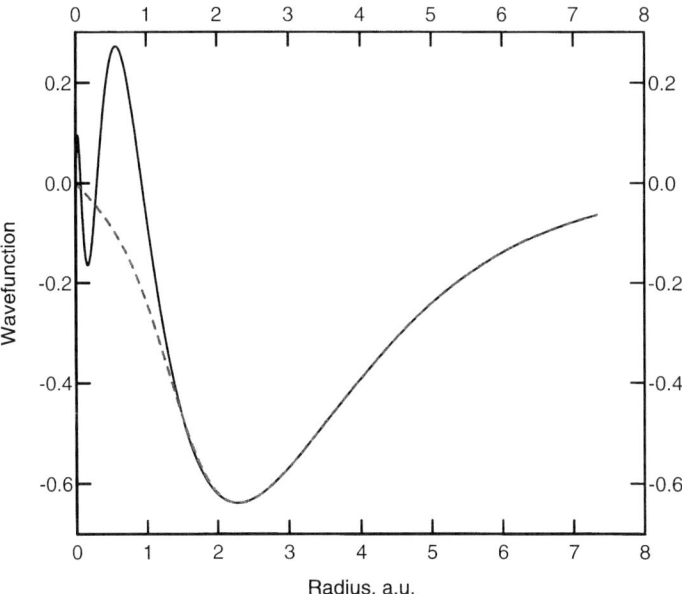

FIG. 2. The 4s (full) and pseudo- (dashed) radial wave function (a.u.) for the Ni atom.

atoms. Nevertheless, the valence electrons undergo exchange interactions with core ones, and this makes the problem of constructing pseudopotentials nontrivial.

The precise description of generating a pseudopotential is a complicated procedure, and there are many different prescriptions given in the literature. These prescriptions differ because the resulting potentials have different uses. Methods using plane waves require pseudopotentials whose momentum matrix elements decay as quickly as possible. This is not an important requirement for the real-space basis used here, although it is of great advantage to deal with smooth wave functions, which have as few nodes as possible, for such functions are easier to represent in terms of Gaussian basis sets. Here, we follow closely the prescription given by Bachelet et al. (1982), who have produced a comprehensive table of pseudopotentials for all the elements between H and Pu. These potentials are called norm-conserving because they yield the exact atomic charge density outside the core. This is an important property for self-consistent calculations.

The first step is the solution of the Kohn-Sham equations for all the electrons in an atom. This is done by choosing a configuration leading to a spherically symmetric charge density, and hence the atomic Kohn-Sham levels are labeled by angular momentum numbers for light elements and by $j = l \pm \frac{1}{2}$ symbols for heavy elements when the Dirac equation must be used. The final pseudopotentials will possess the same valence eigenvalues and give pseudo-wave functions that agree exactly with the true ones outside a core radius. Now, if one filled up the Kohn-Sham energy levels in ascending order, choosing for carbon, for example, the $1s^2 2s^2 2p^2$ configuration, then the $3d$ level would be empty. It might seem then that these unoccupied $d$ levels do not have to be considered in describing the binding of C atoms with other elements. This, however, is not strictly correct. The wave functions for states in solids will be made up of combinations of all atomic states, including $d$ states, and although the atomic $d$ states may not play a major role, it is not clear that their influence can be neglected. The pseudopotential then should possess the same $s$, $p$, and $d$ valence energy levels as the all-electron atom for the valence states. To accommodate this, the atom is solved in the ground-state configuration $1s^2 2s^2 2p^2$ for $l = 0$ and 1, but for $l = 2$, one chooses an ionized excited-state configuration such as $1s^2 2s^{0.75} 2p^1 d^{0.25}$, in which these $d$ levels are occupied with a small amount of charge. The fractional occupancy of the $s$ shell is chosen to eliminate "bumps" in the potential. Different configurations are used for other elements. In Si, for example, the $d$ pseudopotential is derived from the configuration $3s^1 3p^{0.75} 3d^{0.25}$.

The Kohn-Sham equations for the atom using these configurations, $v$, yield the all-electron wave functions and energy levels. The spin densities can then be found from the wave functions as well as the all-electron potential $V^v(r)$. This is the sum of the nuclear, Hartree, and exchange-correlation potentials and possesses a Coulomb singularity at $r = 0$. A first-step pseudopotential for each configuration and angular momentum index $l$ (or $j = l \pm \frac{1}{2}$ for the Dirac equation) is then constructed, eliminating the Coulomb singularity by defining

$$V_l(r) = V^v(r)(1 - f(r/r_{c,l})) + c_l^v f(r/r_{c,l}).$$

Here, $f(x)$ is a function that is unity at the origin and vanishes rapidly for $x$ much bigger than 1 (e.g., $e^{-x^{3.5}}$). Hence, for $r$ close to the origin, $V_l$ is a constant, $c_l^v$, which is chosen so that the lowest energy level for each $l$ is exactly the same as the solution of the all-electron atom for the same $l$.

The corresponding normalized pseudo-wave function, $w_{1l}^v$, is clearly equal (up to a normalization factor) to the all-electron wave function for large $r$, as the potential $V_l(r)$ is exactly the same as $V^v(r)$ there. The value of $r_{c,l}$ is called the core radius, and it determines when the all-electron wave function aproaches the pseudo-wave function. Clearly it must not be too big, but if chosen too small, it lies in a region where the wave function is rapidly varying and difficult to represent by any basis. It is usually chosen to be about half-way between to the outermost node and the outermost extrema of the all-electron wave function.

The next step is to modify $w^v_{1l}$ by introducing a second wave function, $w^v_{2l}$, by

$$u^v_{2l} = \gamma^v_l \{u^v_{1l}(r) + \delta^v_l r^{l+1} f(r/r_{cl})\}.$$

The constants $\gamma^v_l$ and $\delta^v_l$ are selected so that the normalized function $w^v_{2l}$ agrees exactly with the all-electron wave function outside the core and is not just proportional to it. The potential giving rise to the $w^v_{2l}$ wave functions is then found by inverting the Schrödinger equation using the energy levels that agree with the all-electron values. This potential then has the correct eigenvalues and a wave function that agrees exactly with the all-electron one outside the core. Finally, the contribution of the Hartree and exchange-correlation potentials arising from the pseudo-wave functions $w^v_{2l}$ are subtracted, leaving a bare ion potential $V_i(r)$. This last step is exact for the Hartree potential, as this is linear in the core and valence charge densities. However, it is an approximation for the nonlinear exchange-correlation potential. The approximation can be improved, along with the transferability of the pseudopotential, by subtracting instead the exchange-correlation potential corresponding to the all-electron charge density and spin polarization (Louie *et al.*, 1982).

For relativistic atoms, where the states are labeled by $j \pm \frac{1}{2}$, an average pseudopotential is defined:

$$V_l(r) = \frac{1}{2l+1}\{lV_{l-1/2}(r) + (l+1)V_{l+1/2}(r)\}.$$

This is called the scalar relativistic potential. The spin–orbit potential is

$$V^{so}_l(r) = \frac{2}{2l+1}\{V_{l+1/2}(r) - V_{l-1/2}(r)\},$$

and the full pseudopotential is then

$$V^{ps}(\mathbf{r}) = \sum_l |l\rangle\{V_l(r) + V^{so}_l(r)\mathbf{L}\cdot\mathbf{S}\}\langle l|. \tag{16}$$

The potentials have been parametrized by fitting them to simple functions in the following way:

$$V_l(r) = -\frac{Z_v}{r}\left\{\sum_{i=1}^{2} c_i^c \operatorname{erf}(\sqrt{\alpha_i^c}r)\right\} + \sum_{i=1}^{3}\{A_{i,l} + r^2 A_{i+3,l}\}e^{-\alpha_{i,l}r^2},$$

$$V_l^{so}(r) = \sum_{i=1}^{3}\{B_{i,l} + r^2 B_{i+3,l}\}e^{-\alpha_{i,l}r^2}.$$

Here, $Z_v$ is the valence charge, $\alpha_i^c$ is the inverse of the extent of the core charge density, and erf is the error function. The coefficients $c_i^c$ are independent of $l$, and hence this first term is a simple function called the local pseudopotential. The sum of the coefficients $c_i^c$ is unity, so the local potential gives rise to a potential $-Z_v/r$ for $\alpha_i^c r^2 \gg 1$. The second term does depend on $l$ and is called the nonlocal pseudopotential. Tables of values of $Z_v$, $c_i^c$, $\alpha_i^c$, $A_{i,l}$, $B_{i,l}$, and $\alpha_{i,l}$ are given in Bachelet et al. (1982).

A crucial property of the pseudo-wave function is that it can accurately describe different bonding configurations. One test is to compare the energy differences between configurations corresponding to the promotion of valence electrons. For example, the energy difference between C in the $s^2p^2$ and $sp^3$ configurations is 8.23 eV when the all-electron theory is used and 8.25 eV using the pseudopotential. The agreement is not quite as good for Ni, as the corresponding $d^8s^2 \to d^9s^1$ energies are $-1.66$ and $-1.36$ eV, respectively.

## IV. The Real-Space Cluster Method

We now discuss applying LSDF theory to a cluster of atoms. The wave functions of the cluster are expanded in a basis of localized orbitals $\phi_i(\mathbf{r} - \mathbf{R}_i)$ as

$$\psi_\lambda(\mathbf{r}, s) = \lambda_\alpha(s) \sum_i c_i^\lambda \phi_i(\mathbf{r} - \mathbf{R}_i). \tag{17}$$

In this way, the Kohn-Sham differential equations are converted to matrix equations for $c_i^\lambda$. The localized orbitals are often taken to be Gaussian ones of the form

$$(x - R_{ix})^{n_1}(y - R_{iy})^{n_2}(z - R_{ix})^{n_3} e^{-a_i(\mathbf{r}-\mathbf{R}_i)^2},$$

where $n_1$, $n_2$, and $n_3$ are integers. If these are all zero, they correspond to s orbitals of spherical symmetry. Orbitals of *p* symmetry correspond to one of these integers being unity and the others zero, whereas five *d*-like and one *s*-like orbital can be generated if $\Sigma_i n_i = 2$.

The advantage with Gaussian orbitals is that the many integrals required can be evaluated analytically, but their disadvantages are that they quickly become "over-complete," and unlike Slater orbitals, they do not individually approximate solutions to the Kohn-Sham equations. The over-completeness is exemplified by the singular nature of the overlap matrix when two orbitals with similar exponents are sited too close together. The basis functions are real, and hence all matrix elements, as well as the coefficients $c_i^\lambda$, are real. We can therefore drop complex conjugates from the equations.

The density for each spin is then given in terms of the density matrix, $b_{ij,s}$,

$$n_s(\mathbf{r}) = \sum_{ij} b_{ij,s} \phi_i(\mathbf{r} - \mathbf{R}_i) \phi_j(\mathbf{r} - \mathbf{R}_j), \qquad b_{ij,s} = \sum_\lambda \delta(s, s_\lambda) c_i^\lambda c_j^\lambda. \qquad (18)$$

The sum is over-occupied orbitals $\lambda$ with spin *s*. The charge density $n(\mathbf{r})$ can be written

$$n(\mathbf{r}) = \sum_s n_s(\mathbf{r}) = \sum_{ij} b_{ij} \phi_i(\mathbf{r} - \mathbf{R}_i) \phi_j(\mathbf{r} - \mathbf{R}_j), \qquad b_{ij} = \sum_s b_{ij,s}.$$

Let us now consider the various terms in the LSDF expression for the energy *E* in Eq. (14) when this basis of localized orbitals is used. The kinetic energy and pseudo-potential terms involve integrals of the form

$$T_{ij} = -\frac{1}{2} \int \phi_i(\mathbf{r} - \mathbf{R}_i) \nabla^2 \phi_j(\mathbf{r} - \mathbf{R}_j) d\mathbf{r},$$

$$V_{ij}^{ps} = \int \phi_i(\mathbf{r} - \mathbf{R}_i) \sum_a V_a^{ps}(\mathbf{r} - \mathbf{R}_a) \phi_j(\mathbf{r} - \mathbf{R}_j) d\mathbf{r},$$

which are easily found. The evaluation of the Hartree energy requires $O(N^4)$ integrals, where *N* is the number of basis functions, which is prohibitively large for a cluster where *N* might be 1000 or more. Many of these integrals are negligible, particularly those associated with basis functions with fast decay rates. However, many remain. For example, if the smallest exponent $a_i$ is about 0.1 a.u., then the "overlap" of two such orbitals will be nonnegligible for separations of centers less than about 15 a.u. Thus, for Si and diamond, there would be between 100 and 500 atoms in a sphere of this size. This means for most of the clusters considered, *all* of these integrals need to be evaluated. For this reason, it is essential to approximate the Hartree energy in some way. One way is to carry out a numerical integration over a finely meshed grid (Pederson *et al.*, 1991; Chen *et al.*, 1995). However, unless the mesh is chosen very carefully, the resulting matrix elements of the Hartree and exchange-correlation potentials will not transform correctly under the point group operations of the cluster. This will mean that the eigenfunctions of the Hamiltonian, and the normal coordinates of vibrational modes, will not have the correct symmetries, nor the eigenvalues the correct degeneracy. There is then a great advantage in being able to compute these integrals using an analytic formula that preserves any point group symmetry.

Usually this is done by introducing an approximate but analytic expression for the Hartree and exchange-correlation energies, from which the corresponding potentials can be easily found. In addition, it is essential to be able to differentiate them with respect to the positions of the nuclei so that the forces acting on each atom can also be found analytically. These approximate expressions are written in terms of an approximate density for each spin $\tilde{n}_s$, which is expanded in a set of basis functions (Dunlap *et al.*, 1979; Jones and Sayyash, 1986):

$$\tilde{n}_s(\mathbf{r}) = \sum_k d_{k,s} g_k(\mathbf{r}), \qquad \tilde{n}(\mathbf{r}) = \sum_s \tilde{n}_s(\mathbf{r}).$$

We now consider these approximate expressions for the Hartree and exchange-correlation energies.

1. THE HARTREE ENERGY

The exact Hartree energy is, from Eq. (30),

$$E_H = \frac{1}{2} \int \frac{n(\mathbf{r}_1) n(\mathbf{r}_2)}{|\mathbf{r}_1 - \mathbf{r}_2|} d\mathbf{r}_1 d\mathbf{r}_2,$$

and is replaced by an approximate value, $\tilde{E}_H$, involving an approximate charge density $\tilde{n}$:

$$\tilde{E}_H = \int \frac{n(\mathbf{r}_1) \tilde{n}(\mathbf{r}_2)}{|\mathbf{r}_1 - \mathbf{r}_2|} d\mathbf{r}_1 d\mathbf{r}_2 - \frac{1}{2} \int \frac{\tilde{n}(\mathbf{r}_1) \tilde{n}(\mathbf{r}_2)}{|\mathbf{r}_1 - \mathbf{r}_2|} d\mathbf{r}_1 d\mathbf{r}_2.$$

The replacement is exact when $\tilde{n} = n$. Now, we expand the density in terms of a basis set $g_k(\mathbf{r})$ so that

$$\tilde{n}(\mathbf{r}) = \sum_k c_k g_k(\mathbf{r}),$$

and $c_k$ is chosen to minimize the error in estimating the Hartree energy:

$$E_H - \tilde{E}_H = \frac{1}{2} \int \frac{\{n(\mathbf{r}_1) - \tilde{n}(\mathbf{r}_1)\}\{n(\mathbf{r}_2) - \tilde{n}(\mathbf{r}_2)\}}{|\mathbf{r}_1 - \mathbf{r}_2|} d\mathbf{r}_1 d\mathbf{r}_2.$$

Differentiating this with respect to $c_k$ to determine the minimum gives

$$\sum_l G_{kl} c_l = \sum_{lj} t_{ijk} b_{ij}. \tag{19}$$

Here,

$$t_{ijk} = \int \phi_i(\mathbf{r} - \mathbf{R}_i)\phi_j(\mathbf{r} - \mathbf{R}_j)g_k(\mathbf{r}_2) \frac{1}{|\mathbf{r}_1 - \mathbf{r}_2|} d\mathbf{r}_1 d\mathbf{r}_2,$$

$$G_{kl} = \int g_k(\mathbf{r}_1)g_l(\mathbf{r}_2) \frac{1}{|\mathbf{r}_1 - \mathbf{r}_2|} d\mathbf{r}_1 d\mathbf{r}_2.$$

We notice that $\tilde{E}_H$ is always bounded above by $E_H$, and the quality of the fit can be assessed by the increase in $\tilde{E}_H$ when the number of basis functions is increased.

We now consider the choice of $g_k$. The simplest consists of Gaussian functions $e^{-b_k(\mathbf{r} - \mathbf{R}_k)^2}$ defined by a site $\mathbf{R}_k$ and an exponent $b_k$. The sites need not correspond to the location of atoms but can include, for example, bond centers. All the integrals can be computed analytically, which leads to considerable time saving. However, for clusters of less than about 100 atoms, it is the evaluation of the $t_{ijk}$ that is often the most time-consuming procedure. The number of basis functions $g_k$ is usually proportional to the number of basis functions $\phi_i$ (i.e., $N$), and hence there are $O(N^3)$ integrals of the type $t_{ijk}$. These cannot be stored in main memory and must either be evaluated once and stored on disk or, for very fast processors, be repeatedly evaluated during each of the self-consistent cycles. There is then an advantage in choosing a set of $g_k$ that leads to a simple analytical form for $t_{ijk}$, which can be evaluated very quickly.

This can be done when $g_k$ is divided into two sets. The first set has $g_k$ defined by

$$g_k = \left\{1 - \frac{2b_k}{3}(\mathbf{r} - \mathbf{R}_k)^2\right\} e^{-b_k(\mathbf{r} - \mathbf{R}_k)^2}. \tag{20}$$

These functions give a potential of Gaussian form:

$$\int \frac{g_k(\mathbf{r}_1)}{|\mathbf{r}-\mathbf{r}_1|} d\mathbf{r}_1 = \frac{3b_k}{2\pi} e^{-b_k(\mathbf{r}-\mathbf{R}_k)^2},$$

and thus integrals in $t_{ijk}$ involve a product of three Gaussian functions and can be evaluated very quickly. Also, many of these are now zero, as all three Gaussians must overlap to give a nonzero value. However, to get the short-ranged Gaussian potential and to avoid the long-ranged Coulomb potential, the integral of $g_k$ must vanish. It is readily verified that this indeed happens for the functions in Eq. (20). It is therefore necessary to select additional functions $g_k$ that are purely Gaussian and whose integrals do contribute to the total number of electrons. It is, however, sometimes possible to choose the coefficients of these as fixed quantities related to the anticipated total charge on the atom or ion. Thus, their contribution to the Hamiltonian does not change during the self-consistent cycle, and they behave in the same way as the external potential of the nuclei. This leads to a considerable speed-up in the code.

## 2. The Exchange-Correlation Energy

In the same way, the exact expression for this energy,

$$E_{xc} = \int \varepsilon_{xc}(n_\uparrow, n_\downarrow) n \, d\mathbf{r},$$

is replaced by an approximate one $\tilde{E}_{xc}$ involving an approximate density, $\tilde{n}_s$.

$$\tilde{E}_{xc} = \int \varepsilon_{xc}(\tilde{n}_\uparrow, \tilde{n}_\downarrow) \tilde{n} \, d\mathbf{r}. \tag{21}$$

Clearly, the error we make is negligible if $\tilde{n}_s$ is close to $n_s$. The first step then is to fit $n_s$ to a set of functions. It is possible to choose the same $g_k(\mathbf{r})$ as was used in the construction of $\tilde{E}_H$. However, the least-squares procedure used there minimizes the electrostatic energy of the error in the charge density (i.e., $n - \tilde{n}$), and it does not mean that at each value of $\mathbf{r}$, $n(\mathbf{r})$ and $\tilde{n}(\mathbf{r})$ are almost equal. Moreover, the choice of $g_k$ was selected to reflect the difficulty of working out the integrals $t_{ijk}$. Hence, in dealing with $\tilde{E}_{xc}$, it is better to use a sum of simple Gaussian functions $h_k$ so that

$$\tilde{n}_s(\mathbf{r}) = \sum_k d_{k,s} h_k(\mathbf{r}), \quad (22)$$

where $d_{k,s}$ is found from minimizing

$$\int \{n_s(\mathbf{r}) - \tilde{n}_s(\mathbf{r})\}^2 d\mathbf{r}.$$

Differentiating this with respect to the coefficients $d_{k,s}$ leads to the equations

$$\sum_l H_{kl} d_{l,s} = \sum_{ij} u_{ijk} b_{ij,s}, \quad (23)$$

where

$$H_{kl} = \int h_k(\mathbf{r}) h_l(\mathbf{r}) d\mathbf{r}, \quad (24)$$

$$u_{ijk} = \int \phi_i(\mathbf{r} - \mathbf{R}_i) \phi_j(\mathbf{r} - \mathbf{R}_j) h_k(\mathbf{r}) d\mathbf{r}. \quad (25)$$

We note that the integrals are the same for each spin index $s$ and that $u_{ijk}$ are simply proportional to $t_{ijk}$ if $g_k$ is chosen, as in Eq. (20). This saves a considerable amount of computer time.

We consider first the nonpolarized or spin-averaged case in which we can dispense with the spin label and write

$$\tilde{E}_{xc} = \sum_k d_k \int h_k(\mathbf{r}) \varepsilon_{xc}(\tilde{n}) d\mathbf{r}.$$

If $h_k$ is chosen to be a positive definite localized function, such as a Gaussian, then each integral is proportional to the average value of the exchange-correlation density under $h_k$,

$$\langle \varepsilon_{xc}(\tilde{n}) \rangle_k.$$

We next note from Fig. 1 that $\varepsilon_{xc}(n)$ varies slowly with $n$, and hence we expect

$$\langle \varepsilon_{xc}(\tilde{n}) \rangle_k \approx \varepsilon_{xc}(\langle \tilde{n} \rangle_k),$$

$$\langle \tilde{n} \rangle_k = \frac{\Sigma_l d_l \int h_k h_l d\mathbf{r}}{I_k}, \quad (26)$$

where $I_k$ is simply the integral of $h_k$. This approximation is tantamount to replacing the exact exchange-correlation density at **r** by its homogeneous electron gas value for the average density $\langle \tilde{n} \rangle_k$. We can improve on this approximation as follows. Now, as discussed in Section II.4, the exchange-correlation density behaves with high accuracy as a power series in $n$,

$$\varepsilon_{xc}(n) = An^s$$

with $s = 0.30917$. Let us now consider the function $f(s)$, where

$$f(s) = \ln\left(\frac{\langle \tilde{n}^s \rangle_k}{\langle \tilde{n} \rangle_k^s}\right).$$

Clearly, for $s = 0$ or 1, $f(s)$ is 0, while for $s = 2$, $f(s)$ must be a positive quantity. Since we are interested in values of $s$ around 0.3, we can approximate $f(s)$ by

$$f(s) = s(s-1)f(s)/2.$$

The right-hand side of this equation is found from the second moment of $\tilde{n}$:

$$\langle \tilde{n}^2 \rangle_k = \frac{\sum_{lm} d_l d_m \int h_k h_l h_m d\mathbf{r}}{I_k}. \tag{27}$$

These integrals can all be evaluated analytically. We have, finally,

$$\tilde{E}_{xc} = \sum_k d_k \varepsilon_k, \tag{28}$$

where

$$\varepsilon_k = I_k \varepsilon_{xc}(\langle \tilde{n} \rangle_k)e^{f_k},$$

$$f_k = \frac{1}{2}s(s-1)\ln\left(\frac{\langle \tilde{n}^2 \rangle_k}{\langle \tilde{n} \rangle_k^2}\right).$$

This theory has been extended to the spin-polarized case (Lister and Jones, 1988). The spin-polarized exchange-correlation energy is written as in Eq. (13),

$$E^{xc}(n_\uparrow, n_\downarrow) = \sum_{i,s} A_i \int n_s^{p_i+1} n_{1-s}^{q_i} d\mathbf{r},$$

and we replace $n_s$ on the right-hand side by $\tilde{n}_s$, obtaining

$$\tilde{E}_{xc} = \sum_{ks} d_{k,s}\varepsilon_{k,s}, \tag{29}$$

where

$$\varepsilon_{k,s} = \sum_i A_i I_k \langle \tilde{n}_s^{p_i} \tilde{n}_{1-s}^{q_i} \rangle_k.$$

Now, we define the quantity $f$ by

$$\langle \tilde{n}_s^p \tilde{n}_{1-s}^q \rangle_k = \langle \tilde{n}_s \rangle_k^p \langle \tilde{n}_{1-s} \rangle_k^q e^{f(p,q)}$$

$$f(p, q) = \ln\left(\frac{\langle \tilde{n}_s^p \tilde{n}_{1-s}^q \rangle_k}{\langle \tilde{n}_s \rangle_k^p \langle \tilde{n}_{1-s} \rangle_k^q}\right).$$

We now approximate $f$ by the formula

$$f(p, q) = \frac{1}{2} p(p-1) f(2, 0) + \frac{1}{2} q(q-1) f(0, 2) + pq f(1, 1),$$

which interpolates $f$ between the known integer values. In this way, the spin-polarized exchange-correlation energy is evaluated.

## 3. Matrix Formulation

In terms of the approximate Hartree and exchange-correlation energies, the total energy can now be written

$$E = \sum_{ij} \{T_{ij} + V_{ij}^{ps}\} b_{ij} + \tilde{E}_H + \tilde{E}_{xc} + E_{i-1}, \tag{30}$$

where

$$b_{ij} = \sum_\lambda c_i^\lambda c_j^\lambda$$

$$\tilde{E}_H = \frac{1}{2} \sum_{kl} c_k c_l G_{kl}$$

$$\tilde{E}_{xc} = \sum_{ks} d_{k,s} \varepsilon_{k,s},$$

and $E_{i-i}$ is given by Eq. (8). The fitting coefficients $c_k$ and $d_{k,s}$ are defined in terms of $b_{ij,s}$ by Eqs. (19) and (23).

$E$ is minimized subject to an orthonormal set of wave functions; that is,

$$\sum c_i^\lambda c_j^\mu S_{ij} = \delta_{\lambda\mu},$$

where the overlap matrix $S$, is defined by

$$S_{ij} = \int \phi_i(\mathbf{r} - \mathbf{R}_i) \phi_j(\mathbf{r} - \mathbf{R}_j) d\mathbf{r}. \tag{31}$$

This can be done by introducing Lagrange undetermined multipliers, $E_\lambda$, so that we minimize without constraint,

$$\sum_{ij\lambda} c_i^\lambda \{T_{ij} + V_{ij}^{ps} - E_\lambda S_{ij}\} c_j^\lambda + \tilde{E}_H + \tilde{E}_{xc} + E_{i-i}, \tag{32}$$

with respect to $c_i^\lambda$. Now, this introduces the matrix elements of the Hartree and exchange-correlation potentials through

$$\frac{\partial \tilde{E}_H}{\partial c_i^\lambda} = \sum_j V_{ij}^H c_j^\lambda, \quad V_{ij}^H = \sum_{kl} G_{kl} c_l \frac{\partial c_k}{\partial b_{ij}}$$

$$\frac{\partial \tilde{E}_{xc}}{\partial c_i^\lambda} = \sum_j V_{ij,s\lambda}^{xc} c_j^\lambda, \quad V_{ij,s}^{xc} = \sum_k \left\{ \varepsilon_{k,s} + \sum_l d_{l,s} \frac{\partial \varepsilon_{l,s}}{\partial d_{k,s}} \right\} \frac{\partial d_{k,s}}{\partial b_{ij,s}}$$

From Eqs. (19) and (23), we find

$$\sum_l G_{kl} \frac{\partial c_l}{\partial b_{ij}} = t_{ijk}$$

$$\sum_l H_{kl} \frac{\partial d_{l,s}}{\partial b_{ij,s}} = u_{ijk}.$$

Differentiating Eq. (32) with respect to $c_i^\lambda$, we get the Kohn-Sham equations:

$$\sum_j \{T_{ij} + V_{ij}^{ps} + V_{ij}^H + V_{ij,s\lambda}^{xc} - E_\lambda S_{ij}\} c_j^\lambda = 0. \tag{33}$$

We note that the total number of electrons $M$ and the spin $S$ are given by

$$M = \sum_{ijs} S_{ij} b_{ij,s}, \tag{34}$$

$$S = \sum_{ij} S_{ij}(b_{ij,\uparrow} - b_{ij,\downarrow}).$$

We can now write Eq. (33) more compactly in matrix form. Equation (33) is written in terms of two generalized eigenvalue problems, one for each spin, as

$$\sum_j (H_{ij} - ES_{ij})c_j = 0,$$

or in matrix notation as

$$(H - ES)c = 0.$$

For the cluster sizes and values of the exponents of the basis sets typically used, the matrices $H$ and $S$ are not sufficiently sparse to warrant special numerical techniques, and consequently, the overlap matrix $S$ is written in terms of an upper triangular matrix using a Choleski decomposition:

$$S = U^t U.$$

$U$ and its inverse can be evaluated in $O(N^3)$ operations. We then define a vector $d$ by $Uc = d$, and the generalized eigenvalue problem is converted into the usual one:

$$\{(U^{-1})^t H U^{-1} - E\}d = 0$$

The eigenvalues of this can be found by a standard Householder scheme, which first reduces the matrix to tridiagonal form, from which the eigenvalues can be found. As the number of occupied states is much smaller than the dimension of the Hamiltonian, the eigenvectors are best found by inverse iteration.

All these matrix operations can be carried out on a parallel computer (Briddon, 1996) using the PBLAS and SCALAPACK libraries, which provide routines to carry out all the required operations, provided the matrices are correctly distributed between nodes. In fact, the matrices are divided up into square blocks, and these are allocated to different nodes in a specified manner. In this way, efficient parallel code is easily written. The time-dominant step for large clusters is the computation of the eigenvalues and eigenvectors that scale as $N^3$ for the dense matrices found in practice.

## V. Self-Consistency and Atomic Forces

### 1. SELF-CONSISTENCY

Self-consistency is the situation obtained after a successful solution of the Kohn-Sham equations when the charge density that would be produced by the Kohn-Sham orbitals gives rise to the same potential as was used in the equation that determined them. In short, it is the process by which charge is distributed around the cluster, minimizing its energy for a fixed structure. The self-consistency cycle is initiated by choosing sets of charge-density coefficients $c_k$ and $d_{k,s}$ taken from either neutral atoms or a previous run. The Kohn-Sham equations given in Eq. (33) are then solved, and the density matrix $b_{ij,s}$ is found. Equations (19) and (23) are then used to determine the output charge-density coefficients, $c_k^0$ and $d_{k,s}^0$. The next step consists of selecting a new input charge density $c'_k$, defined in terms of $c_k^0$ and $c_k$. This is done by using a weighted combination, as in

$$c'_k = c_k + w(c_k^0 - c_k).$$

The same weighting is used to define the new spin-density coefficients $d'_{k,s}$. If the process of generating the output charge density is denoted by the (nonlinear) operation

$$c_k^0 = L_k(c),$$

where we have written $c$ to stand for the vector $c_k$, then

$$c_k^{0'} = L_k(c') = L_k(c + w(c^0 - c)).$$

Now, provided $w$ is small enough, we can linearize this equation to get

$$c_k^{0'} = L_k(c) + w \sum_l D_{kl}(c_l^0 - c_l)$$

$$= c_k^0 + w \sum_l D_{kl}(c_l^0 - c_l).$$

The condition for self-consistency is that the input and output charge densities are equal; that is,

$$c_k^{0'} = c'_k,$$

or

$$c_k + w(c_k^0 - c_k) = c_k^0 + w \sum_l D_{kl}(c_l^0 - c_l).$$

Hence

$$(w^{-1} - 1)(c_k - c_k^0) = \sum_l D_{kl}(c_l^0 - c_l). \tag{35}$$

The right-hand side can be determined by choosing a small value of $w$, say $w_1$, and the output charge density $c_{1k}^0$ then found. This gives

$$c_{1k}^0 = c_k + w_1 \sum_l D_{kl}(c_l^0 - c_l).$$

Hence,

$$\sum_l D_{kl}(c_l^0 - c_l) = (c_{1k}^0 - c_k)/w_1.$$

Inserting this into Eq. (35) gives us an equation for $w$ that is solved by a least-squares procedure. We denote the difference between the sides of this equation as

$$e_k = (1 - w)(c_k - c_k^0)/w - (c_{1k}^0 - c_k)/w_1,$$

and choose $w$ by minimizing the electrostatic energy of the "charge density" $e_k$ defined by $\Sigma_k e_k g_k(\mathbf{r})$. Thus, the energy defined by

$$\frac{1}{2} \sum_{kl} e_k G_{kl} e_l$$

is made as small as possible. It is possible to generalize this procedure so that the predicted charge density is built up from several previous iterates $c_{lk}$; that is,

$$c_k' = \sum_l \{c_{lk} + w_l(c_{lk}^0 - c_{lk})\}.$$

In practice, the self-consistency cycle converges exponentially and quickly, taking between four to ten iterations, with the difference in the input and output Hartree energies typically becoming less than $10^{-5}$ a.u. Convergence is particularly rapid when there is a gap between the highest filled and lowest empty level, but problems can arise when this gap is very small or vanishes. These are often related to an attempted crossing of an occupied and unoccupied energy level, whereupon the charge density changes discontinuously. This can be avoided by "smearing out" the occupation of levels by using Fermi statistics. Thus, we suppose that the level $E_\lambda$ is occupied by $f_\lambda$ electrons. This means that the energy to be minimized now includes an entropy term as well as a term constraining the total number of electrons to $M$:

$$F = E + k_B T \sum_\lambda \{f_\lambda \ln f_\lambda + (1 - f_\lambda)\ln(1 - f_\lambda)\} - \mu \left\{\sum_\lambda f_\lambda - M\right\}. \quad (36)$$

Here the sum is now over all orbitals $\lambda$. Minimizing the free energy $F$ with respect to $f_\lambda$ and $\mu$ gives

$$f_\lambda = \frac{1}{e^{(E_\lambda - \mu)/k_B T} + 1}$$

and

$$\sum_\lambda f_\lambda = M.$$

Eq. (18) must also be generalized to

$$b_{ij,s} = \sum_\lambda \delta(s, s_\lambda) f_\lambda c_i^\lambda c_j^\lambda.$$

In practice, $k_B T$ is taken to be about 0.04 eV. Often, where we have two energy levels separated by 0.1 eV that "cross" in the approach to self-consistency, this will remove the discontinuous change, but when self-consistency is achieved, provided the final splitting is more than 0.04 eV, one state is found to be fully occupied and the other empty. In this sense, we are using variable filling purely as a computational tool and are not attempting to simulate materials at finite temperatures.

It is worth pointing out here that incorrect use of Fermi statistics can lead to incorrect structures. This occurs when a Jahn-Teller effect operates, as, for example, for a substitutional $Ni^-$ impurity in Si (Jones et al., 1995b). Here, the gap contains $t_2^4$ levels, with the upper one containing two electrons. The system distorts, leading to a lowering of symmetry and a splitting of the $t_2$ levels into $a_1$, $b_1$, and $b_2$. The occupied levels will be displaced downwards and the unoccupied level upwards, leading to a lowering in the energy, provided the strain energy arising from the distortion is less than the lowering in the occupied level. Fermi statistics, however, result in an equal occupation of the levels, and the driving force for the distortion vanishes. This can be overcome by, for example, occupying the $b_1$ and $b_2$ levels throughout the self-consistency cycle, even though the unoccupied $a_1$ level might lie below one or both of the occupied levels during part of the cycle. Such an approach is found to successfully model the defect.

There are other troublesome cases in which even the use of Fermi statistics is unable to give a self-consistent solution. Often, this means that the starting structure is physically unreasonable but the problem disappears once partial structural relaxation has occurred.

## 2. EVALUATION OF FORCES

Once the self-consistent charge density has been found, the force acting on each atom can be evaluated. It is essential to determine the forces accurately in order to relax the cluster and calculate its vibrational modes. The force on an atom $a$, in direction $l$, is given by

$$f_{la} = -\frac{\partial F}{\partial R_{la}} = -\frac{\partial E}{\partial R_{la}}.$$

This can be evaluated by considering the change to each term in the energy in Eq. (32) when $R_{la}$ is displaced by $\Delta R_{la}$. Thus,

$$\Delta E = \sum_{ij} b_{ij}\Delta\{T_{ij} + V_{ij}^{ps}\} + \sum_{ij} \{T_{ij} + V_{ij}^{ps}\}\Delta b_{ij} + \Delta\tilde{E}_H + \Delta\tilde{E}_{xc} + \Delta E_{i-i}$$

$$\Delta\tilde{E}_H = \sum_{kl} c_k G_{kl}\Delta c_l + \frac{1}{2}\sum_{kl} c_k c_l \Delta G_{kl}$$

$$\Delta\tilde{E}_{xc} = \sum_{k,s} \varepsilon_{k,s}\Delta d_{k,s} + \sum_{k,s} d_{k,s}\Delta\varepsilon_{k,s}.$$

$\Delta c_k$ can be evaluated from Eq. (19):

$$\sum_l G_{kl}\Delta c_l = \sum_{ij}\{t_{ijk}\Delta b_{ij} + b_{ij}\Delta t_{ijk}\} - \sum_l c_l \Delta G_{kl}.$$

In the same way, $\Delta d_{k,s}$ can be evaluated from Eq. (23):

$$\sum_l H_{kl}\Delta d_{l,s} = \sum_{ij}\{u_{ijk}\Delta_{ij,s} + b_{ij,s}\Delta u_{ijk}\} - \sum_l d_{l,s}\Delta H_{kl}.$$

Now, if we gather together the terms in $\Delta b_{ij}$ and $\Delta b_{ij,s}$ we get,

$$\sum_{ij}\{T_{ij} + V_{ij}^{ps} + V_{ij}^H\}\Delta b_{ij} + \sum_{ijs} V_{ij,s}^{xc}\Delta b_{ij,s}.$$

From the Kohn-Sham Eq. (33), this equals $\sum_{ij\lambda} E_\lambda S_{ij}\Delta c_i^\lambda c_j^\lambda$, which can be written as

$$\sum_\lambda E_\lambda \Delta \left\{\sum_{ij} c_i^\lambda c_j^\lambda S_{ij}\right\} - \sum_{ij\lambda} E_\lambda c_i^\lambda c_j^\lambda \Delta S_{ij}.$$

The first term on the right-hand side vanishes as Eqs. (18) and (34) show that the expression in braces is the total number of electrons $M$, which is constant. Thus, the force does not contain any derivatives in the wave function coefficients—as required by the Hellmann-Feynman theorem (Slater, 1960).

The term $\Delta \varepsilon_{k,s}$ contains $\Delta \langle \tilde{n}_s \rangle_k$ and $\Delta \langle \tilde{n}_s^2 \rangle_k$. These can be found from Eqs. (26) and (27):

$$\Delta \langle \tilde{n}_s \rangle_k = \frac{1}{I_k} \sum_l \{H_{kl} \Delta d_{l,s} + d_{l,s} \Delta H_{kl}\}$$

$$\Delta \langle \tilde{n}_s^2 \rangle_k = \frac{1}{I_k} \sum_{lm} \{2 u_{klm} d_{l,s} \Delta d_{m,s} + d_{l,s} d_{m,s} \Delta u_{klm}\}.$$

Terms involving the matrix elements $T_{ij}$ and $S_{ij}$ depend on $R_{la}$ only through the basis functions $\phi_i(\mathbf{r} - \mathbf{R}_a)$, but the pseudo-potential term has an additional dependence arising from $V_a^{ps}(\mathbf{r} - \mathbf{R}_a)$. This can be evaluated by integrating by parts:

$$\int \phi_i(\mathbf{r} - \mathbf{R}_i) \Delta V_a^{ps}(\mathbf{r} - \mathbf{R}_a) \phi_j(\mathbf{r} - \mathbf{R}_j) d\mathbf{r}$$

$$= -\int \{\phi_j(\mathbf{r} - \mathbf{R}_j) \Delta \phi_i(\mathbf{r} - \mathbf{R}_i) + \phi_i(\mathbf{r} - \mathbf{R}_i) \Delta \phi_j(\mathbf{r} - \mathbf{R}_j)\} V_a^{ps}(\mathbf{r} - \mathbf{R}_a) d\mathbf{r}.$$

Despite the complexity of the equations, the time taken to evaluate the forces is small in comparison with that taken to determine the self-consistent energy.

## VI. Structural Optimization

### 1. Unconstrained Relaxation

Usually, the positions of atoms in the starting cluster do not correspond to the positions in equilibrium. The first step then is to determine the energy and forces acting on them, but once this has been achieved, it is necessary to consider efficient algorithms that allow the equilibrium sites to be found. The optimization strategy that is often used is a conjugate gradient one (Press et al., 1987). This requires only the forces to be known at any stage. The atoms are moved to a new set of positions whose energy is lower than that of the previous set. Suppose in some configuration the forces acting on atom $a$ are $f'_{la}$ in direction $l$. Then the atoms are moved along a conjugate direction $d'_{la}$ so that the new atomic position is

$$R'_{la} = R_{la} + w d'_{la}.$$

Here, $w$ is chosen so that the free energy $F$ in Eq. (36) is least. This is usually accomplished by quadratic or cubic interpolation. The directions $d'_{la}$ are related to the forces $f_{la}$ through

$$d'_{la} = f'_{la} - x d_{la},$$

where $d_{la}$ is the previous search direction, where the force was $f_{la}$. The value of $x$ is just

$$x = \frac{\sum_{la} f'_{la}(f'_{la} - f_{la})}{\sum_{la} f_{la}^2},$$

and is set to zero initially.

The efficiency of the optimization strategy depends on the number of constraints. For an atom strongly bonded to at least three others in a nonplanar configuration, relaxation is very fast as the atom is over-constrained. Thus, we find that about 10 iterations are required to relax the inner atoms of a tetrahedrally bonded cluster, reducing the forces on each atom to less than 0.001 a.u. On the other hand, if the atoms have low coordination, the structure is more floppy and the number of relaxations required is much greater. This can happen if the surface $H$ atoms are allowed to move. In such cases, the movement of the $H$ atoms sets up an elastic wave in the bulk, the reflections of which repeatedly affect the surface. This can be overcome by attaching springs to the $H$ atoms, simulating the outer crystal. The choice of spring constants is, however, somewhat arbitrary.

Other problems occur with clusters of water molecules where weak hydrogen bonds co-exist with strong covalent intramolecular O-H bonds. In this case, the choice of a single quantity, $w$, may not be the best strategy, and it would be desirable to include information on the derivatives of the forces.

The most serious problem with the optimization strategy is that it finds a local minimum in the energy. There may be—and often are—other lower minima separated by barriers from the one found. The only way to reach these with the static relaxation method described here is to start the calculation from different structures, but even then there can be no guarantee that the global minimum has been found. The global minimum, of course, may not really be of any physical interest. For example, the global minimum for a vacancy or dislocation in silicon corresponds to the defects lying on the surface of the cluster.

## 2. Constrained Relaxation

For some purposes, it is important to relax the cluster with constraints, as, for example, in determining the saddle point for defect migration or reorientation. At the saddle point, there is at least one direction along which the energy falls when the atoms are displaced along it, while along other directions, the energy rises. A commonly used procedure for finding the saddle point would be to average the atomic coordinates corresponding to the beginning and end points of the migration path and then to calculate the Hessian or energy second derivatives at this point. If the structure is close to the saddle point, the Hessian matrix has at least one negative eigenvalue. The eigenvector corresponding to this eigenvalue gives a direction, $d_{la}$, in which the energy decreases. The cluster is then relaxed so that the displaced coordinates, $\Delta R_{la}$, lie orthogonal to this direction; that is,

$$\sum_{la} d_{la} \Delta R_{la} = 0.$$

The saddle point is then located by moving along the direction $d_{la}$ so that the energy *increases* to a maximum. However, this procedure is not a practical one because of the time taken to evaluate the Hessian. Clearly, some constraints must be imposed on the coordinates; otherwise the conjugate gradient algorithm would push the coordinates away from the saddle point. To deal with this problem, it is important to select a few variables for which the energy varies rapidly. These include the bond lengths nearest to the defect core. For example, in oxygen migration, they would include the Si-O and Si-Si lengths nearest the defect. The relevant variables are held fixed, while the remainder are allowed to vary, minimizing the energy. If variables that are not relevant are selected — for example, the position of an atom outside the defect core or some angle — then the structure could slide from one configuration into the other, rapidly passing the barrier, which often manifests as a cusp. Again, it is important to be able to calculate the forces on all the atoms, allowing for the constraint. The procedure that we have found successful is as follows.

Suppose that atom $a$ is hopping from one site to another, during which one bond $a$–$b$ is broken and the bond $a$–$c$ is created. Then the constraint used involves the relative bond lengths, $|\mathbf{R}_a - \mathbf{R}_b|$ and $|\mathbf{R}_a - \mathbf{R}_c|$, and the cluster is relaxed, maintaining this constraint. For technical reasons, the actual constraint used is:

$$x = (\mathbf{R}_a - \mathbf{R}_b)^2 - (\mathbf{R}_a - \mathbf{R}_c)^2 = \text{constant}. \tag{37}$$

This provides a linear equation for one of the Cartesian components, say $l$, of the "central" atom $a$, which can be solved for any value of $x$.

$$R_{la} = \frac{2\Sigma_{k \neq l}(R_{kb} - R_{kc})R_{ka} + \Sigma_k(R_{kb}^2 - R_{kc}^2) - x}{2(R_{lb} - R_{lc})}. \tag{38}$$

Now, if the atoms are moved in any way, the change to the energy is

$$\Delta E = - \sum_{m,d} f_{md} \Delta R_{md}.$$

Now,

$$(\mathbf{R}_c - \mathbf{R}_b) \cdot \Delta \mathbf{R}_a = (\mathbf{R}_a - \mathbf{R}_b) \cdot \Delta \mathbf{R}_b - (\mathbf{R}_a - \mathbf{R}_c) \cdot \Delta \mathbf{R}_c,$$

and hence the term in $\Delta R_{la}$ in Eq. (39) can be written in terms of $\Delta R_{ka}$, $k \neq l$, $\Delta \mathbf{R}_b$, and $\Delta \mathbf{R}_c$ and hence be eliminated from the expression for $\Delta E$. This modifies the forces on the atoms $a$, for $k \neq l$ as well as the atoms $b$ and $c$, but these changes can now be easily found. The value of $l$ is selected to give the greatest value of the denominator in Eq. (38). The analysis can be generalized to deal with several constraints simultaneously, and the method has proved successful in dealing with a number of problems in which bonds are switched between different configurations.

## VII. Determination of Vibrational Modes

### 1. Energy Second Derivatives and Musgrave-Pople Potentials

The vibrational modes of clusters can be found from the dynamical matrix $E_{la,mb}$, as described in Section II.1. This is related to the second derivative of the energy with respect to displacements of atom $a$ at $\mathbf{R}_a$, along the Cartesian direction $l$, and atom $b$ at $\mathbf{R}_b$, along direction $m$. The cluster is first relaxed so that the forces on all the atoms, or those around the defect, are essentially zero. Then atom $a$ is displaced by $\varepsilon$ ($\approx 0.025$ a.u.) along the $l$ axis. The electrons will attempt to "follow" this displacement so that the self-consistent charge density will be different from that in the equilibrium configuration. This new self-consistent charge density must then be found. When this has been done, the forces on all the atoms of the cluster are evaluated, and these will no longer be zero because of the change in charge density and structure. Suppose that force $f_{mb}^+(l, a)$ acts on atom $b$ in direction $m$. This adiabatic force includes the effect of the screening charge density seeking to oppose the change caused by moving atom $a$. The whole process is now repeated by moving atom $a$ by $-\varepsilon$ along the same direction $l$, producing forces $f_{mb}^{-}(l, a)$. The energy second derivative is then, up to second order in $\varepsilon$,

$$D_{la,mb} = (f_{mb}^{+}(l, a) - f_{mb}^{-}(l, a))/2\varepsilon.$$

It is important to realize that these are not infinitesimal derivatives. They include contributions from all even powers of $\varepsilon$, and the frequencies that they give rise to contain anharmonic contributions. For this reason, the latter are sometimes called *quasi-harmonic* frequencies (Jones et al., 1994b).

Only some of the entries of the dynamical matrix of a large cluster can be found in this way. This is because it is a very time-consuming procedure to evaluate the second derivatives, and those for atoms near the surface are irrelevant for the frequencies of vibration for bulk solids. The next step is to fit the calculated derivatives to those arising from a valence force potential. The potential can then be used to generate the dynamical matrix for any type of cluster or unit cell composed of the same elements and bonding configuration.

One could choose many types of potential, but one that is particularly useful is due to Musgrave and Pople (1962). This includes all possible bond-length and bond-angle distortions up to second order. The potential for atom $a$ is

$$V_a = \frac{1}{4} \sum_b k_r^{(a)}(\Delta r_{ab})^2 + r_0^2/2 \sum_{b>c} k_\theta^{(a)}(\Delta \theta_{bac})^2 + r_0 \sum_{c>b} k_{r\theta}^{(a)}(\Delta r_{ab} + \Delta r_{ac})\Delta \theta_{bac}$$
$$+ \sum_{c>b} k_{rr}^{(a)} \Delta r_{ab} \Delta r_{ac} + r_0^2 \sum_{d>c>b} k_{\theta\theta}^{(a)} \Delta \theta_{bac} \Delta \theta_{cad}.$$

Here, $\Delta r_{ab}$ and $\Delta \theta_{bac}$ are the changes in the length of the $a$–$b$ bond and angle between the $a$–$b$ and $a$–$c$ bonds, respectively. The sums are over the nearest-neighbor atoms of atom $a$. The Musgrave-Pople potential is superior to a Keating potential, for example, since it includes correlations between bond stretch and bend.

This potential can be used to derive phonon dispersion curves for the bulk solid. This has been done in a number of cases, such as diamond (Jones, 1988), Si (Jones, 1987), Ge (Berg Rasmussen et al., 1994), GaAs (Jones and Öberg, 1991a), AlAs (Jones and Öberg, 1994a), InP (Ewels et al., 1996a), and quartz (Purton et al., 1992). The potential usually gives the highest frequencies accurate to a few wave numbers, although it cannot account for the splitting of the longitudinal and transverse optic modes due to a long-range Coulomb field. The worst errors occur at low frequencies, where the assumption that forces beyond second-shell atoms are zero is inadequate. Moreover, the elastic constants derived from the potential are in poor agreement with the experiment. Nevertheless, their main use is in describing the local modes of defects, and the disagreement arising at low frequencies is not important.

Let us now consider how the local and resonant modes of a defect are calculated. The second derivatives of the cluster containing the defect are found in the same way as described previously. These derivatives are usually evaluated between the defect atoms and their nearest neighbors. Other entries in the dynamical matrix are then found from the Musgrave-Pople potential. The normal modes and their frequencies are then found by direct diagonalization of the dynamical matrix with the masses of the terminating $H$ atoms set to infinity.

This procedure works well for frequencies well away from the one-phonon spectrum. Usually, infrared absorption spectroscopy is able to detect only local modes, and for such problems, this method is satisfactory. However, there are cases in which it is the resonant modes that are of interest, as, for example, nitrogen-related defects in diamond. We shall discuss their evaluation in the next section. Usually, the number of modes calculated for a defect exceeds the number observed. The remaining ones are not detected for a number of reasons. They might fall into a part of the spectrum where there is strong absorption arising from an overtone or combination band from the bulk or substrate; their lifetimes might be very short due to anharmonic interactions, and this implies a very broad spectrum, as, for example, the highest modes of the interstitial carbon dimer, $C_i$-$C_s$, in Si (Jones et al., 1995a; Leary et al., 1996); or, finally, the defect may possess a very small transition dipole moment. It is this last issue that can be addressed with the method.

## 2. Effective Charges

Leigh and Szigeti (1967) give the integrated intensity of absorption due to a defect as

$$\frac{2\pi^2 \rho}{ncM'}\eta^2.$$

Here, $\eta$ is the effective charge, $c$ is the velocity of light, $n$ is the refractive index of the material, and $M'$ and $\rho$ are the mass and concentration of the impurity, respectively. $\eta^2$ is given by the sum over any degenerate modes of

$$M'\left(\frac{\partial M_x}{\partial Q_i}\right)^2, \qquad (40)$$

where $M_x$ is the dipole moment in the direction of the polarization of the electromagnetic field. $Q_i$ is the normal coordinate of the mode $i$. That is, the displacement of each atom is $Q_i u^i_{ta}/\sqrt{M_a}$. The induced dipole can be evaluated from the changes in the dipole moment of the cluster,

$$\mathbf{M} = \sum_a Z_a \mathbf{R}_a - \int \mathbf{r} n(\mathbf{r}) d\mathbf{r},$$

when the atoms are subjected to this displacement.

There are several points to note here. The effective charge depends on the mode and its displacement pattern. It can be very different for different modes of the same defect. This is illustrated by the H wag mode of the passivated C acceptor in GaAs (Jones and Öberg, 1991a). Although the effective charge for the stretch mode is almost unity (Kozuch et al., 1990), that of the wag mode is almost zero, as it was originally undetected by infrared absorption measurements but had been observed by Raman scattering (Wagner et al., 1995).

The effective charge, in general, depends on the polarization of the electrical field and the orientation of the defect. If thermal equilibrium prevails, leading to defects, assuming all possible orientations that have degenerate energies, then $\eta^2$ must be averaged over these orientations. The important case of a defect with $C_{3v}$ symmetry in cubic crystals has been considered by Clerjaud and Côte (1992), who showed that the average effective charges are given by

$$\eta^2_{A_1} = \frac{M'}{3} \sum_l \left(\frac{\partial M_l}{\partial Q_{A_1}}\right)^2,$$

for the $A_1$ nondegenerate mode and

$$\eta^2_E = \frac{2M'}{3} \sum_l \left(\frac{\partial M_l}{\partial Q_E}\right)^2,$$

for the twofold degenerate $E$ mode. These effective charges are independent of the direction of the polarization of light. The assumption of equal numbers of point defects in different orientations related by symmetry is not always valid, even in the absence of stress, as, for example, the case of single passivated substitutional C dimer in GaAs. Here, the defects form during growth on the surface and are frozen in during cooling. The $C_{As}$-Ga-$C_{As}$ unit is oriented along only one of the two possible $\langle 110 \rangle$ orientations for a (001) growth plane (Cheng et al., 1994; Davidson et al., 1994).

The effective charge is independent of the mass $M'$ in only simple cases, such as a H stretch and wag mode. In these cases, essentially only the H atom is undergoing a displacement and the derivative of $M_x(R_{la} + Q_i u_{la}^i / \sqrt{M_a})$ scales as $1/\sqrt{M'}$. In this case, $\eta$ is the same for H as for D, and the integrated intensity is then a factor 2 smaller for D than for H. Of course, this argument neglects anharmonic effects, which are more important for H.

### 3. Resonant Modes

A Green function method has been developed to analyze these modes. The Green function for the bulk crystal can be evaluated directly from the dynamical matrix constructed from the Musgrave-Pople potential. If the atoms in the unit cell are located at $\mathbf{R}_{\tau_a}$, then the Green function for a vector $\mathbf{k}$ in the Brillouin zone is given by

$$G^0_{l\tau_a,m\tau_b}(\mathbf{k}) = \sum_i \frac{u^i_{l\tau_a}(\mathbf{k}) u^i_{m\tau_b}(\mathbf{k})}{w^2 - w_i^2(\mathbf{k})}.$$

We can find the crystal Green function in real space between atoms $a$ and $b$ related to basis atoms by

$$\mathbf{R}_a = \mathbf{R}_{\tau_a} + \mathbf{R}_L, \quad \mathbf{R}_b = \mathbf{R}_{\tau_b} + \mathbf{R}_M$$

$$G^0_{la,mb} = \frac{1}{\Omega} \sum_{\mathbf{k}} e^{i\mathbf{k}\cdot(\mathbf{R}_L - \mathbf{R}_M)} G^0_{l\tau_a,m\tau_b}(\mathbf{k}).$$

The Green function in the presence of a defect, whose contribution to the dynamical matrix differs by $V_{la,mb}$, is

$$G_{la,mb} = G^0_{la,mb} + \sum_{nc,pd} G^0_{la,nc} V_{nc,pd} G_{pd,mb}.$$

This equation can be readily solved for a point defect since the elements of $V_{la,mb}$ are taken to be zero for either $a$ or $b$ outside the second shell of neighbors surrounding the defect. The density of phonon states projected onto atom $a$ is then found from the trace of the imaginary part of this Green function; that is,

$$-\frac{2\omega}{\pi} \text{Imag.} \sum_{l} G_{la,la}.$$

In this way, the contribution of the resonance to the local density of states is found. However, the infrared absorption is controlled by the induced dipole moment, and it is better to evaluate this assuming reasonable values of the charge distributed over atoms of the defect. If the charge $q_a$ is located on atom $a$, then the induced dipole is

$$M_l = -\frac{2\omega}{\pi} \text{Imag.} \sum_{a} q_a G_{la,la}.$$

The procedure has been used to discuss the absorption of resonant modes due to complex N defects in diamond (Jones et al., 1992b). In this case, 27,000 points in the Brillouin zone were used to construct $G^0$.

## VIII. Practical Considerations

### 1. Choice of Basis Sets

There are two different basis sets used in the method. The first is a basis used to describe the wave functions. This is invariably a set of Gaussian functions defined by an exponent, $a_i$, and sited on an atom or at the center of a bond or some other location, $R_i$. As described in Section IV.1, this Gaussian is multiplied by a polynomial in $x - R_{xi}$, $y - R_{yi}$, $z - R_{zi}$. For spherically symmetric s functions, the polynomial is trivially unity. For p orbitals, the three possible polynomials are $x - R_{xi}$, $y - R_{yi}$, or $z - R_{zi}$. For d orbitals, all six polynomials of degree 2 are used, which generates a linear combination of five d orbitals and one s orbital. The code also includes f orbitals generated by polynomials of degree 3. The complete basis is then a linear superposition of these orbitals for different exponents $a_i$ and centers $R_i$.

The use of bond-centered orbitals seems unique to the AIMPRO code. They serve a useful function of representing the pseudo-wave function in the region where it is often large. They make it unnecessary to use atom-centered d orbitals for Si, Ge, and GaAs. They are particularly important in dealing with some impurities, such as oxygen in silicon, where the strong strain effects on atoms distant from the impurity make it necessary to use a basis that gives as accurate as possible elastic constants for the host. For other materials (e.g., diamond), relaxation effects are often very small and bond-centered orbitals have very little effect. Although the location of the bond centers is often kept fixed at a specified point (e.g., the midpoint) of a bond, it is possible to allow them to relax or float with the atoms of the cluster until the minimum energy is found. This has not been commonly used because they often move close to atoms, leading to instabilities.

The optimum exponents $a_i$, $i = 1, 2 \ldots m$, for a particular atom can be found by minimizing the energy of the pseudo-atom as a function of $a_i$. This procedure also generates the coefficients of the wave function: $c_i^\lambda$, $i = 1, 2, \ldots$, $m$, for each valence state $\lambda$. For example, it generates a set of coefficients for an $s$ orbital and a set for the three $p$ orbitals. When an application is made to a large cluster, the same fixed linear combination of the Gaussian orbitals with different exponents can then be used. This gives a basis of 4 orbitals for each Si atom, for example, and 10 for a transition element such as Ni. Such a basis is called a minimum one. In many applications, the minimum atomic basis is used for atoms far away from the core. For other atoms, the coefficients that multiply the Gaussian orbitals are treated as variational parameters, as described in Section IV.

A second basis is used to expand the charge density. This, again, is a set of Gaussian functions, or modified Gaussian functions, as described in Section IV, defined by an exponent $b_k$ and center $\mathbf{R}_{fk}$. Again, the centers $\mathbf{R}_{fk}$ can be chosen to lie at nuclei, bond centers, or other locations. The optimum basis consists of exponents and sites that *maximize* the estimated Hartree energy $\tilde{E}_H$, as described in Section IV.1.

It is always desirable to locate the Gaussian orbitals at a symmetrical site or the set of sites generated by symmetry, since otherwise the energy levels, vibrational modes, and so forth, will not possess the required degeneracy. It is expedient to define the basis in terms of $N - MX$, which means that a basis of $N$ Gaussian $s$, $p$, or $d$ orbitals are placed at the location of each atom of type $X$ to describe the wave functions, while a basis of $M$ Gaussian $s$ functions are used to describe the charge density. In addition, the sites treated in terms of a minimal basis set need to be defined, as well as any orbitals and fitting functions placed at bond centers. A minimal basis is often placed on the surface H atoms.

One basis set that has been of occasional use is an icosahedral set of "bond centers" sited close to an atom. This has an advantage that the $d$ and $f$ degeneracy of an atom is not compromised, but, in general, the point group symmetry of a defect will be lost.

The basis size has a significant effect on calculated properties—with structures being least sensitive and energies and wave functions being most sensitive. It is not possible to converge total energies with the same degree of rigor as is occasionally obtained in plane-wave treatments. This is because simply increasing the number of exponents used to describe the basis eventually results in a numerical instability for the Choleski decomposition of the overlap matrix. However, in practice, it is energy *differences* that are important—for example, between a H atom at a bond-centered and tetrahedral interstitial site. In this case, the dependence of the total energy difference can be easily checked.

## 2. The Construction of a Suitable Cluster

In dealing with defects within semiconductors, H-terminated clusters have invariably been used. These saturate the dangling bonds at the surface of the cluster, leading to widely separated filled and empty surface states for "bulk" clusters (i.e., clusters composed of the same stoichiometry and atomic arrangement as the bulk semiconductor). If the surface H bond lengths are close to their equilibrium values, the bandgaps are much greater than those of the bulk solids, with the exception of diamond. Values for representative clusters are given in Table II. These were calculated for tetrahedral clusters with an 8-8 basis on the inner five atoms and a minimal basis on all the others. Two bond-centered Gaussian basis functions with different exponents were sited on all the bond centers between host atoms.

TABLE II

Lowest Kohn-Sham Level, $E_1$, Highest Occupied Level $E_v$, Ccalculated Bandgap $E_g$, and Experimental Gap for Various Clusters in eV

|  | Cluster Size |  | $E_1$ | $E_v$ | $E_g$ | Exptal. Gap |
|---|---|---|---|---|---|---|
| Diamond | 71 | $C_{35}H_{36}$ | $-23.42$ | $-6.09$ | 5.01 | 5.5 |
|  | 131 | $C_{71}H_{60}$ | $-23.69$ | $-5.36$ | 5.92 |  |
|  | 297 | $C_{181}H_{116}$ | $-24.17$ | $-4.70$ | 5.37 |  |
| Silicon | 71 | $Si_{35}H_{36}$ | $-16.31$ | $-6.48$ | 3.82 | 1.17 |
|  | 131 | $Si_{71}H_{60}$ | $-16.77$ | $-6.41$ | 3.13 |  |
|  | 297 | $Si_{181}H_{116}$ | $-16.91$ | $-5.96$ | 2.51 |  |
| Germanium | 71 | $Ge_{35}H_{36}$ | $-16.70$ | $-6.12$ | 3.53 | 0.75 |
|  | 131 | $Ge_{71}H_{60}$ | $-16.92$ | $-5.93$ | 2.70 |  |
|  | 297 | $Ge_{181}H_{116}$ | $-17.25$ | $-5.58$ | 2.14 |  |
| Gallium arsenide | 71 | $(Ga_{19}As_{16}H_{36})^{3-}$ | $-10.91$ | 0.31 | 2.95 | 1.42 |
|  | 131 | $(Ga_{31}As_{40}H_{60})^{9+}$ | $-34.92$ | $-22.59$ | 2.27 |  |
|  | 297 | $(Ga_{89}As_{92}H_{116})^{3+}$ | $-22.76$ | $-10.10$ | 1.92 |  |
|  | 71 | $(As_{19}Ga_{16}H_{36})^{3+}$ | $-24.92$ | $-13.13$ | 2.83 |  |
|  | 131 | $(As_{31}Ga_{40}H_{60})^{9-}$ | $-1.47$ | 10.16 | 2.64 |  |
|  | 297 | $(As_{89}Ga_{92}H_{116})^{3-}$ | $-14.24$ | $-1.752$ | 2.09 |  |

These gaps are much larger than those found using density functional theory in supercells, which are in turn smaller than experimental gaps. The cluster band gaps vary only slightly with the basis size but become smaller if longer H bonds are allowed. It is not advisable to use long H bonds, as this imposes a strain on inner bonds around defects and can certainly modify their structure, seriously perturbing the local vibrational modes. The bandgap also decreases slowly with cluster size.

Despite the large bandgaps, some information on the position of energy levels can be obtained. There are two common ways of "correcting" the bandgap to make allowance for the difference with experiment. The first is to simply scale defect levels by the bandgap. Clearly, this is simply pushing both valence and conduction band states closer together. The second is to use "scissors" operator. This is added to the Hamiltonian and displaces the unoccupied states of the "perfect" clusters upwards by $V$. The scissor operator is

$$\Delta(r, r') = V \sum_{\lambda'} \psi_{\lambda'}(r)\psi_{\lambda'}(r'),$$

where the sum is over unoccupied levels. It can also be expressed in terms of the occupied states and, for the spin-averaged case,

$$\Delta(\mathbf{r}, \mathbf{r}') = V \sum_{ij} (\delta_{i,j} - b_{ij})\phi_i(\mathbf{r} - \mathbf{R}_i)\phi_j(\mathbf{r}' - \mathbf{R}_j).$$

$V$ is chosen to give the correct bandgap. This is then applied to a cluster containing a defect. Few calculations have been carried out using this method.

Energy differences between different defect states can be found using the Slater transition method (Slater, 1960). Here, the difference in total energies of the configurations where an electron is promoted from the $\lambda$ to the $\mu$ orbital is found from the eigenvalues of the configuration corresponding to half of the promoted electron being placed in each orbital. Then the total energy difference is accurately given by $E_\mu - E_\lambda$. This differs from the zero phonon line of the optical transition by the relaxation energy of the defect. This method has been used to treat optical transitions for vacancy impurity complexes in diamond (Goss et al., 1996a).

The size of the cluster used varies with the application, and, for point defects, at least one and usually two shells of host atoms surround the defect. In the compound semiconductors, such as GaAs, one can choose between stoichiometric clusters that contain as many Ga as As atoms and are chemically neutral, or others possessing a greater number of As than Ga atoms, for example. In both cases, for clusters representing bulk material, it is important to occupy them with $M$ electrons where $M$ is twice the number of covalent bonds. This results in all the bonding states being filled and the antibonding ones being empty. Again, there is a large gap between the two. Such a procedure results in bond lengths close to experimental values. If the cluster is not stoichiometric, this procedure necessarily leads to charged clusters. This arises as the number of protons is determined by the numbers of As, Ga, and terminating H atoms, but it is the topology that determines the number of electrons, and clearly for fourfold coordinated atoms, the number of electrons is exactly the same as if the cluster were made from Si atoms and terminated by H. The charged clusters do not appear to cause serious problems. It is especially desirable to use charged nonstoichiometric clusters when the defect has a high symmetry and consequently has degenerate energy levels or vibrational modes (e.g., a C atom substituting for As in AlAs). A disadvantage with charged clusters is that the energy levels are shifted up for negative and down for positively charged clusters, and the position of a defect level with respect to a band edge becomes more difficult to assess. For this reason, a stoichiometric cluster is preferred. However, this is not free from difficulty, as then the cluster has more As atoms in its upper half than Al ones, for example, and leads to a dipole moment. This has an effect on the bond lengths parallel and perpendicular to this direction. For example, an 86-atom stoichiometric cluster, $Al_{22}As_{22}H_{42}$, has $C_{3v}$ symmetry and is centered on the middle of an Al-bond. It has a dipole moment along the $C_3$ axis, leading to Al-As bond length of 2.476 Å compared with 2.428 Å for the other six bonds (Jones and Öberg, 1994a). These are all within 2% of the experimental value of 2.43 Å.

2. MULLIKEN POPULATIONS

In many cases, it is important to understand the nature of gap states and hybridization state they refer to. For example, in the $C_i$ defect in Si, the gap states are localized on $p$ orbitals on the C and Si atom sharing a lattice site (Leary et al., 1996). The simplest way of determining the character of a state is to plot its wave function. However, the use of pseudo-potentials implies that the amplitude is invariably small near a nucleus. It is not then easy to deduce from which atoms gap states arise. One way that gives some information is to evaluate the Mulliken bond populations $m_i^\lambda$. These are defined from the integral of the square of the wave function. Equations (17) and (31) show

$$\sum_s \int \psi_\lambda^2(\mathbf{r}, s) d\mathbf{r} = \sum_{ij} c_i^\lambda c_j^\lambda S_{ij} = \sum_i m_i^\lambda$$

where

$$m_i^\lambda = c_i^\lambda \sum_j S_{ij} c_j^\lambda.$$

If the state is localized on atom $a$, then $c_i^\lambda$, and hence $m_i^\lambda$, will be large for basis functions $i$ localized there. One problem is that $m_i^\lambda$ can be large and negative because the phases of Gaussian orbitals with different exponents, but centered on atom $a$, are rarely the same.

### 4. Radiative Lifetimes

For complicated defects (e.g., Ni in diamond), there are many gap levels, whereas only one or two optical transitions related to the defect have been observed. There is then a problem in assigning the transition. The calculated radiative lifetimes of the various transitions can be very different, and the most intense transition will be associated with the smallest lifetime. It is then necessary to calculate this quantity.

The rate of electrical dipole transitions between two states, $\lambda$ and $\mu$, can be found using the expression (Svelto, 1976)

$$\frac{1}{\tau_{\lambda\mu}} = \frac{4n\omega^3}{3c^3\hbar} \frac{e^2}{4\pi\varepsilon_0} |\mathbf{r}_{\lambda\mu}|^2, \tag{41}$$

where

$$\mathbf{r}_{\lambda\mu} = \sum_s \int \psi_\lambda(\mathbf{r}, s) \mathbf{r} \psi_\mu(\mathbf{r}, s) d\mathbf{r}$$

is the dipole matrix element, $n$ the refractive index, $e$ the electron charge, $c$ the speed of light and $\omega$ the transition frequency. Estimates of the radiative lifetime are sensitive to the transition energy and spatial extent of the wave function. We shall describe an application to the Si-V center in diamond in the next section.

### IX. Applications

#### 1. General

Many applications of the formalism have been made to molecules as well as point, line, and surface defects in large clusters simulating bulk material. Investigations into molecular entities such as fullerenes (Estreicher *et al.*,

1992; Eggen et al., 1996), water dimers, and octamers (Heggie et al., 1996b) and surface problems associated with the growth of CVD diamond (Latham et al., 1994) will not be reviewed here. We shall instead emphasize some of the applications to point and line defects in bulk solids that have brought about a deeper insight into the properties of the defect.

2. POINT DEFECTS IN BULK SOLIDS

a. *Diamond*

Nitrogen is one of the most important impurities in diamond, occurring in concentrations as large as $10^{20}$ cm$^{-3}$ (Evans et al., 1981). It readily complexes with itself and with other impurities and intrinsic defects, and the resulting complexes are often important optical centers. The high solubility of N is attributed to the low misfit energy of inserting an N atom with N-C bond lengths of 1.47 Å into the diamond lattice where the C-C bonds are 1.54 Å. The neutral substitutional defect exhibits trigonal symmetry, as convincingly demonstrated by electron paramagnetic resonance (EPR) (Smith et al., 1959; Cook et al., 1966). In type Ib or synthetic diamonds, $N_s$ is present as an isolated defect, but in annealed synthetic diamonds, the nitrogen aggregates to give complexes with more than one N atom. These complexes are also found in the great majority of natural (type Ia) diamonds. It is a long-standing problem to elucidate the final fate of N aggregation in diamond when it is annealed for long periods. The *ab initio* calculations have helped to clarify the properties of many of these nitrogen complexes.

The substitutional defect was investigated by Briddon et al. (1991) and led to an explanation of the "anomalous" vibrational mode associated with the defect. This local mode at 1344 cm$^{-1}$ was observed (Collins et al., 1982) in an infrared absorption study on type Ib diamonds and its intensity correlated with the EPR signal due to $N_s$, suggesting that it is associated with the vibrations of $N_s$. Surprisingly, however, the mode does not shift with $^{15}$N doping. The cluster calculation revealed that for the neutral substitutional defect, not only N was displaced from a lattice site along [111] by 0.2 Å, but also the neighboring C atom was displaced along [$\bar{1}\bar{1}\bar{1}$] by the amount, thus leading to back C-C bonds about 5% shorter than the normal C-C bonds. This was independently found by plane-wave pseudo-potential calculations (Kajihara et al., 1991). There are two gap levels of $a_1$ symmetry: a bonding state between N and the unique C atom and an antibonding state containing one electron. The presence of two $a_1$ levels is consistent with stress studies on the 4.059-eV optical center associated with $N_s$ (Koppitz et al., 1986; Vaz et al., 1987). The absence of degenerate levels for the metastable $T_d$ defect suggests that the mechanism for the off-site distortion is a chemical rebonding one rather than a Jahn-Teller distortion (Bachelet et al., 1981). The vibrational modes of the defect were found from the Green function method discussed in Section VII.3. This gave three bands centered at 1320, 1122, and 1032 cm$^{-1}$ in good agreement with observed ones at 1344, 1130, and 1080 cm$^{-1}$ (Collins et al., 1982). The highest mode was localized on the unique C atom and its C neighbors and does not shift with a change in the N isotope. This explains the anomolous mode. The N-related modes fell below the Raman frequency at 1332 cm$^{-1}$. It is not surprising, with hindsight, to understand the character of the C-related mode arising from the $sp^2$ bonding of the unique C atom. The reorientation energy of the defect has also been calculated (Breuer and Briddon, 1996) and found to be 0.7 eV — the same as that observed experimentally by Ammerlaan and Burgemeister (1981).

The $N_s$ defect is not stable during prolonged annealing at high temperatures and aggregates first into A centers, which are $N_s$ dimers (Davies, 1976) and second into B centers, which are believed to be vacancies surrounded by four N atoms (Loubser and Van Wyk, 1981). The calculated (Jones et al., 1992b) vibrational modes of the defects are in reasonable agreement with observation giving further support to the assignments. Furthermore, they give a clue as to why N atoms should aggregate. The highest filled level in the A center is around mid-gap, which is considerably lower than that of the $N_s$ donor. Thus, the driving force for aggregation is the lowering of the one-electron energy. It is an insight such as this that makes the theory so useful. The B defect also has deep mid-gap states, which should make it optically active, and this has prompted investigations into vacancies and vacancy-impurity defects, which are also known to be very important optical centers in diamond.

Vacancies and interstitials were investigated by Breuer and Briddon (1995), who confirmed the importance of many-body effects and the need to determine the energies of different multiplets. The theory found that V$^-$ was a spin $\frac{3}{2}$ defect, in agreement with experiment, and the calculated optical transition energy agreed well with the observed value. However, the Von Barth procedure, discussed in Section V, yields too few equations to find the multiplets for the neutral vacancy, and hence the theory is unable to describe this important case. In many cases, vacancies will complex with impurities, and study of N-V and Si-V centers (Goss et al., 1996a) concluded that they possess very different structures. Whereas the N-V defect has $C_{3v}$ symmetry, Si-V possesses $D_{3d}$ symmetry where the Si atom sits midway between two adjacent vacancies. This finding explains the surprising optical properties of the defect. The dangling bonds on each of the two sets of three

C atoms nearest Si form $a_1$ and $e$ states. The two sets of $e$ states combine to form bonding and antibonding $e$ levels around mid-gap, the highest of which is occupied by two electrons. Now, in synthetic or type Ib diamonds, the higher $e$ level traps an additional electron from N donors, leading to a $^2E$ ground state. An internal optical transition can then take place with an electron promoted from the lower $e$ level. This $^2E \to {}^2E$ transition leads to four close-by luminescence lines if the $e$ levels are split by a Jahn-Teller effect, or possibly a spin–orbit interaction. The presence of three isotopes of Si causes a slight shift in the zero-point energy and leads to the appearance of a remarkable 12-line spectrum (Clark et al., 1995). The computed radiative lifetime of 3 ns is in very good agreement with the experimental values around 1 to 4 ns (Sternschulte et al., 1994). This example shows that the method is able to explain very simply a complicated optical spectrum that would otherwise be difficult to understand.

b. *Silicon*

A great deal of effort has been devoted to understanding the properties of the light impurities H, B, C, O, and N in silicon, and especially their vibrational frequencies, as local mode spectroscopy has been such a valuable experimental tool. Calculations have been made for the various substitutional defects: $VH_n$ (Bech Nielsen et al., 1995), $B_s$, $C_s$ (Jones and Öberg, 1992d), $VO_n$ (Ewels et al., 1995), $N_s$ (Jones et al., 1994c), with substantial agreement obtained with experimental results for the local vibrational modes in each case. In addition, interstitial defects such as H (Jones, 1991b), $C_i$ (Jones et al., 1995a; Leary et al., 1996), $O_i$ (Jones et al., 1992c), and $N_i$ (Jones et al., 1994c), and complexes of these, have also been investigated.

One example is the $C_i$-$O_i$ defect, which turned out to have very surprising structure. The C interstitial on its own takes the form of a [100] oriented split interstitial, as shown in Fig. 3 (Watkins, 1964; Zheng et al., 1994). Two of the C-Si bonds along [011] are 1.8 Å long and pull in the Si neighbors lying there. This leaves the Si-Si bonds along this direction extended, and hence they are favorable sites for attack by oxygen (Trombetta and Watkins, 1987). The calculation (Jones and Öberg, 1992a) shows that O does not lie at a bond-center site within these dilated bonds but rather moves toward the Si radical shown in Fig. 3. The reason for this is that the electronegativity of C exceeds Si, rendering the Si radical positively charged. This in turn attracts the O atom so that its becomes overcoordinated, leading to rather long Si-O bonds. The three Si-O bonds are by no means equal in strength. A consequence is that the O-related vibrational mode lies well below that of interstitial and even substitutional oxygen. The same process occurs for

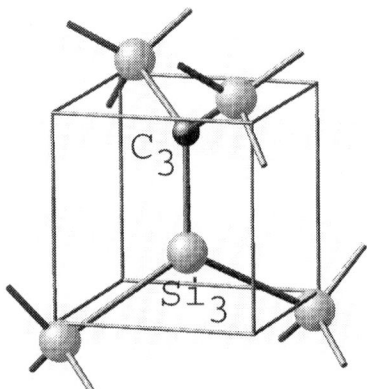

FIG. 3. The $C_i$ defect.

interstitial N. But now the state arising from the dangling bond on the $Si_3$ atom is occupied. This has led to a remarkable finding (Ewels et al., 1996b): the O atom, being negatively charged, squeezes itself into dilated Si-Si bonds adjacent to N and pushes up the donor level due to $Si_3$. For $N_i$-$O_2$, the level is displaced almost to the conduction band. This defect might explain the occurrence of shallow thermal donors, which arise when Czochralski Si, containing N, is annealed to 650°C (Suezawa et al., 1986).

This "wonderbra" mechanism of deep to shallow level conversion is not unique to N. A shallow donor level also arises when a C-H unit replaces N. Of course, this begs the question as to whether an interstitial C-H defect with this structure is stable at these temperatures. But thus unit is known to be a constituent of the T center, which is stable to 600°C (Minaev and Mudryi, 1981). This photoluminescent center has an emission band at 0.9351 eV and a rich spectrum of local-mode satellites. The isotope shifts of these local modes with $^{13}C$ and D have been calculated and agree very well with the observed ones (Safonov et al., 1996). The defect has a gap level occupied by a single electron, and it would be interesting to know if this level is displaced upward, becoming a shallow donor, if O atoms cluster around this defect.

The N interstitial referenced here is not the dominant N defect in Si. This consists of a close-by pair of [100] split interstitials. The evidence comes partly from channeling experiments showing that each N atom is displaced about 1 Å from lattice sites; partly from infrared spectroscopy showing that the N atoms are equivalent and the high frequencies are due to an interstitial complex, and partly from the theoretical modeling (Jones et al., 1994c). The model refutes earlier suggestions that nitrogen forms molecules within silicon. A complex of the N pair with oxygen, yielding an electrically inactive NNO defect, has also been investigated both experimentally and theoretically (Jones et al., 1994d; Berg Rasmussen et al., 1995).

### c. Compound Semiconductors

The cluster calculations were the first to describe and detail the structure and modes of H-passivated Si donors and Be acceptors in GaAs (Briddon and Jones, 1990). Subsequent studies of trigonal C-H complexes in GaAs have been particularly fruitful. The local modes of the defect (Jones and Öberg, 1991a) exhibit several unusual properties. The $E^-$ mode, which involves a movement of H perpendicular to the $C_3$ axis and out of phase with C, was placed around $715\,cm^{-1}$. The C-related $A_1$ and $E^+$ modes, which involve motion of H in phase with C in respective directions parallel and perpendicular to the C-H bond, were calculated to lie at 413 and $380\,cm^{-1}$, respectively. Infrared spectroscopy on GaAs, containing high concentrations of C and H grown by molecular beam epitaxy and chemical vapor deposition methods, located modes at 453 (X) and $563\,cm^{-1}$ (Y) (Woodhouse et al., 1991). Both were subsequently shown to be due to the C-H defects, as they exhibited shifts with C and H isotopes. A Raman scattering experiment (Wagner et al., 1991) assigned the 453-$cm^{-1}$ mode to C-$A_1$. Y is now believed to be the $E^+$ mode (Davidson et al., 1993). The $E^-$ mode was not observed in these early experiments. However, in deuterated samples, the $E^-$ mode was detected at $637\,cm^{-1}$. This must imply that the unobserved $E^-$ mode in the H samples lies above $637\,cm^{-1}$, and a simple force constant model (Davidson et al., 1993) predicted it to lie at $745.2\,cm^{-1}$. The failure of the early infrared experiments to locate the H-$E^-$ mode was explained by the *an initio* cluster theory as the consequence of a small transition dipole moment. This mode has been detected at $739\,cm^{-1}$ by Raman scattering experiments (Wagner et al., 1995). Similar calculations have been carried out for C in AlAs (Jones and Öberg, 1994a). The effect of anharmonicity on the stretch mode has also been investigated (Jones et al., 1994b).

It is possible to grow heavily C-doped GaAs by chemical beam epitaxy, using $CBr_4$ as a doping source so that the films are free of hydrogen. C is a very electronegative element and naturally favors an As site. The calculations (Jones and Öberg, 1994a) show a large build-up of charge around C, and there is no evidence that C can occupy Ga or Al sites and behave as a donor. However, when C-rich samples are annealed at 850°C, there is a loss of C from As sites, together with a reduction in the hole density. It was first suggested by Jones and Öberg (1994e), and independently by Cheong and

Chang (1994), that rather than $C_{Ga}$ defects being formed, a [100] oriented C-C dimer located at an As site, and which acts as a single donor, could be created. Thus, for every pair of C atoms lost from As sites, there would be a loss of three holes. The computed stretch frequency of the C-C dimer in GaAs is $1799\,\text{cm}^{-1}$ and is Raman but not infrared active. Two dimers have been detected by Raman scattering (Wagner et al., 1996), with modes at 1742 and $1858\,\text{cm}^{-1}$ in the annealed material. The hole density is about 10% of its preannealed value ($2.5 \times 10^{20}\,\text{cm}^{-3}$), but the concentration of $C_{As}$ dropped to 30% of its preannealed value, also about $2.5 \times 10^{20}\,\text{cm}^{-3}$. Hence, some donors or hole traps must have been introduced by the annealing. If only $(C-C)_{As}$ dimers were introduced by the annealing, we would require their concentration to be $5 \times 10^{19}\,\text{cm}^{-3}$ to account for the carrier density, but that would result in about $8 \times 10^{19}\,\text{cm}^{-3}$ of carbon unaccounted for. Presumably this forms the second type of dimer, which possibly lies at an interstitial $T_d$ site and acts as a single acceptor.

Other types of dimers are present in epitaxially deposited material containing H. A pair of substitutional C atoms at neighboring As sites traps H or $H_2$ with modes slightly shifted from those of $C_{As}$-H. An unexpected finding is that the dimers are preferentially oriented along one of the two [011] directions perpendicular to the (100) growth surface, probably because of kinetic reasons (Cheng et al., 1994; Davidson et al., 1994). Calculations (Goss et al., 1996b) confirm that H lies at a site accounting for the observed polarization in the $C_2$-H defect, but in the case of $C_2$-$H_2$, the theory predicts one mode polarized along [011] and another — the higher mode — along [0$\bar{1}$1]. This has not yet been observed.

The C-H defect is unusual in possessing a resonant electron trap, which has profound consequences for the dissociation of the defect. The calculated activation energy for dissociation is above 1 eV lower in the presence of minority carriers, which can be trapped in the resonant level (Breuer et al., 1996). This calculation anticipated experimental results (Fushimi and Wada, 1996), confirming this reduction in the activation energy.

### 3. Line Defects

In addition to the work carried out on point defects, there has been a considerable attempt to understand the structure and kinetics of dislocations in groups IV and III–V semiconductors. In these materials, dislocations are dissociated into partials separated by a stacking fault. Commonly occurring partials are 90° and 30° ones. The cluster theory was the first *ab initio* one to reveal that 90° degree partial dislocations in Si (Heggie et al., 1991) and GaAs (Öberg et al., 1995) are reconstructed, as shown in Fig. 4.

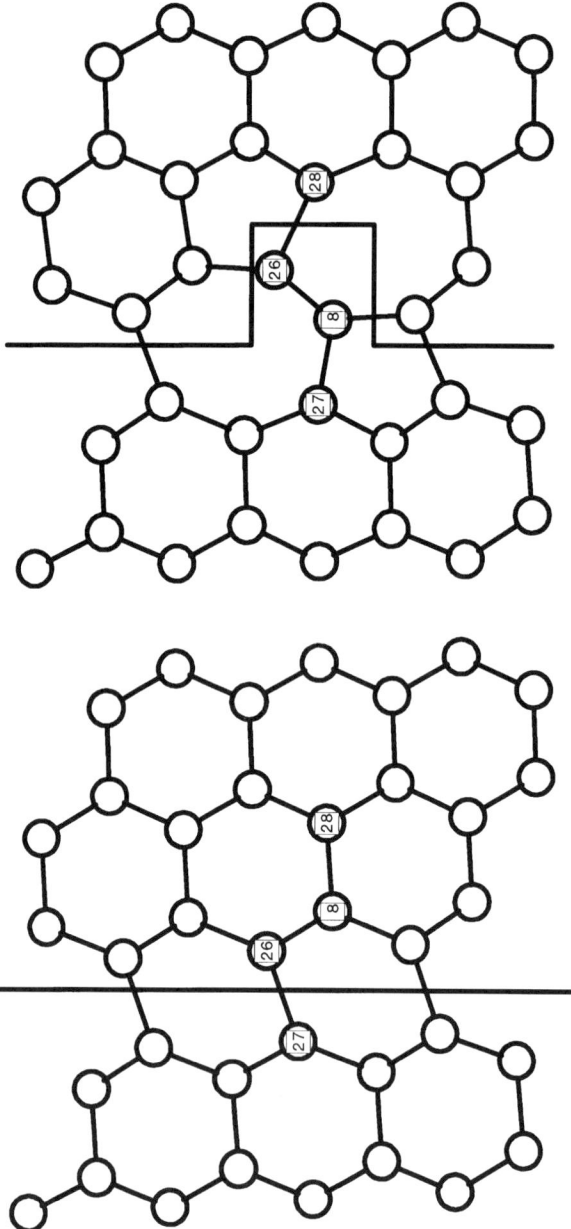

FIG. 4. The reconstructed 90° dislocation and double kink in silicon. Vertical axis is $[01\bar{1}]$; horizontal axis is $[\bar{2}11]$.

The reconstruction leads to electrical inactivity of the line and is to be contrasted with earlier models of deep states arising from a line of dangling bonds. Intriguingly, impurities such as P and N have a pronounced effect on the reconstruction in Si and actually break it (Heggie et al., 1993; Sitch et al., 1994). This effect might explain the very strong locking effect of these impurities — especially N — which has important technological implications.

An important question concerns the mobility of dislocations, as this controls their rate of growth and ultimately their density in the crystal. This is especially important, as dislocations bind point defects such as vacancies and interstitials as well as impurities, all of which possess deep gap levels, which can greatly affect the electronic and optical properties of the material. Now, it is believed that dislocations propagate by creating double kinks, as shown in Fig. 4, which then expand under the influence of stress, leading to motion of the dislocation. The energetics of this process can be followed by embedding the dislocation in a cluster. The kink formation energy was found (Öberg et al., 1995) to be a very small value in these materials (about 0.1 eV), whereas the activation energy necessary to break the reconstructed bonds was considerable. The total activation energy for dislocation motion was found to be 1.9 eV in Si and 1.4 and 0.8 eV for $\beta$ and $\alpha$ partials, respectively, in GaAs. These energies are in fair agreement with observations: 2.1 eV in Si (Imai and Sumino, 1983) and 1.24 to 1.57 eV for $\beta$, and 0.89 to 1.3 eV for $\alpha$ partials in GaAs (Matsui and Yokoyama, 1986; Yonenaga and Sumino 1989). Moreover, these activation energies are sensitive to the Fermi level. This arises because during the transition to the saddle point structure, a level moves from close to the band edge to become deep in the gap. Clearly, then, the activation energy will depend on whether this level is occupied. In this way, the pronounced reduction in the activation energy for $\beta$ partials in $p$-type material, and $\alpha$ partials in $n$-type material, can be explained. A similar effect occurs for SiC and has led to predictions of the effect of doping on dislocations in that material (Sitch et al., 1995).

## X. Summary

The cluster theory that we have described has led to significant advances in understanding defects in bulk solids, atomic processes in molecules, and interaction effects of hydrogen on diamond surfaces. The method is a straightforward application of local-density functional theory, with a localized basis, to large clusters. It is remarkably stable with bonding patterns quite insensitive to cluster size and has been remarkable for the accuracy of the predicted local and resonant vibrational modes. It is perhaps this aspect that has caught the greatest attention of experimental groups, several of which have sought help in the understanding of defects of interest to themselves. In many cases, this collaboration has been very successful, and the theory has built upon experimental findings to elucidate the detailed geometry of a defect or a key ingredient in an atomic process.

The future advances in computing power—especially the development of cheaper parallel processor machines—will enable clusters as large as 1000 atoms to be *routinely* relaxed and investigated. This will pave the way for an exploration of the structure of larger clusters and extended defects, such as interstitial aggregates. However, this will take the theory into areas where few experiments can probe the microstructure, and the results described in outline from here must provide the underlying confidence in any predictions that emerge.

ACKNOWLEDGMENTS

The many developments and applications of the formalism presented here have been carried out by a great number of collaborators. Among these, Sven Öberg, Malcolm Heggie, Paul Sitch, Chris Latham, Steve Breuer, Chris Ewels, Jon Goss, Paul Leary, Bernd Eggen, Deepak Jain, Grenville Lister, Andrey Umerski and Vitor Torres are especially to be thanked.

REFERENCES

Ammerlaan, C. A. J. and Burgemeister, E. A. (1981). *Phys. Rev. Lett.* **47**, 954.
Bachelet, G. B., Baraff, G. A., and Schlüter, M. (1981). *Phys. Rev. B* **24**, 4736.
Bachelet, G. B., Hamann, D. R., and Schlüter, M. (1982). *Phys. Rev. B* **26**, 4199.
Bech Nielsen, B., Hoffmann, L., Budde, M., Jones, R., Goss, J., and Öberg, S. (1995). *Mat. Sci. Forum* **196–201**, 933.
Berg Rasmussen, F., Jones, R., and Öberg, S. (1994). *Phys. Rev. B* **50**, 4378.
Berg Rasmussen, F., Jones, R., Öberg, S., Ewels, C., Goss, J., Miro, J., and Deák, P. (1995). *Mat. Sci. Forum* **196–201**, 791.
Born, M. and Huang, K. (1954). *Dynamical Theory of Crystal Lattices.* Oxford University Press, London.
Breuer, S. J. and Briddon, P. R. (1995). *Phys. Rev. B* **51**, 6984.
Breuer, S. J. and Briddon, P. R. (1996). *Phys. Rev. B* **53**, 7819.
Breuer, S. J., Jones, R., Briddon, P. R., and Öberg, S. (1996). *Phys. Rev. B* **53**, 16289.
Briddon, P. R. and Jones, R. (1990). *Phys. Rev. Lett.* **64**, 2535.
Briddon, P. R., Heggie, M. I., and Jones, R. (1991). *Mat. Sci. Forum* **83–87**, 457.
Briddon, P. R. (1996). Unpublished.

Ceperley, D. M. and Alder, B. J. (1980). *Phys. Rev. Lett.* **45**, 566.
Chen, X. J., Langlois, J. M., and Goddard, W. A. (1995). *Phys. Rev. B* **52**, 2348.
Cheng, Y., Stavola, M., Abernathy, C. R., Pearton, S. J., and Hobson, W. S. (1994). *Phys. Rev. B* **49**, 2469.
Cheong, B. H. and Chang, K. J. (1994). *Phys. Rev. B* **49**, 17436.
Clark, C. D., Kanda, H., Kiflawi, I., and Sittas, G. (1995). *Phys. Rev. B* **51**, 16681.
Clerjaud, B., and Côte, D. (1992). *J. Phys. Condens. Matter* **4**, 9919.
Collins, A. T., Stanley, M., and Woods, G. S. (1982). *Phil. Mag. A* **46**, 77.
Cook, R. J. and Whiffen, D. J. (1966). *Proc. Roy. Soc. A* **295**, 99.
Davidson, B. R., Newman, R. C., Bullough, T. J., and Joyce, T. B. (1993). *Phys. Rev. B* **48**, 17106.
Davidson, B. R., Newman, R. C., Kaneto, T., and Naji, O. (1994). *Phys. Rev. B* **50**, 12250.
Davies, G. (1976). *J. Phys. C: Solid State Phys.* **9**, L537.
Delsole, R., Reining, L., and Godby, R. W. (1994). *Phys. Rev. B* **49**, 8024.
Doukhan, J. C. and Trepied, L. (1985). *Bull. Mineral.* **108**, 97.
Dunlap, B. I., Connolly, W. J., and Sabin, J. R. (1979). *J. Chem. Phys.* **71**, 4993.
Eggen, B. R., Heggie, M. I., Jungnickel, G., Latham, C. D., Jones, R., and Briddon, P. R. (1996). *Science* **272**, 87.
Estreicher, S. K., Latham, C. D., Heggie, M. I., Jones, R., and Öberg, S. (1992). *Chem. Phys. Lett.* **196**, 311.
Evans, T., Qi, Z., and Maguire, J. (1981). *J. Phys. C: Solid State Phys.* **14**, L379.
Ewels, C., Jones, R., and Öberg, S. (1995). *Mat. Sci. Forum* **196–201**, 1297.
Ewels, C. P., Öberg, S., Jones, R., Pajot, B., and Briddon, P. R. (1996a). *Semicond. Sci. Technol.* **11**, 502.
Ewels, C. P., Jones, R., Öberg, S., Miro, J., and Deák, P. (1996b). *Phys. Rev. Lett.* **77**, 865.
Fushimi, H. and Wada, T. (1996). Private communication.
Goss, J., Resende, A., Jones, R., Öberg, S., and Briddon, P. R. (1995). *Mat. Sci. Forum* **196–201**, 67.
Goss, J. P., Jones, R., Breuer, S. J., Briddon, P. R., and Öberg, S. (1996a). *Phys. Rev. Lett.*, in press.
Goss, J. P., Jones, R., and Öberg, S. (1996b). Unpublished.
Griggs, D. T. and Blacic, J. D. (1965). *Science* **147**, 292.
Hedin, L. (1969). *Solid State Phys.* **23**, 1–180.
Heggie, M. and Jones, R. (1987). *Phil. Mag. Lett.* **55**, 47.
Heggie, M. I., Jones, R., and Umerski, A. (1991). *Inst. Phys. Conf. Ser.* **117**, 125.
Heggie, M. I., Jones, R., and Umerski, A. (1993). *Phys. Stat. Solidi (a)* **138**, 383.
Heggie, M. I., Briddon, P. R., and Jones, R., (1996a). Unpublished.
Heggie, M. I., Latham, C. D., Maynard, S. C. P., and Jones, R. (1996b). *Chem. Phys. Lett.* **249**, 485.
Hohenberg, P. and Kohn, W. (1964). *Phys. Rev. B* **136**, 864.
Janak, J. F. (1978). *Phys. Rev. B* **18**, 7165.
Jones, R. and Sayyash, A. (1986). *J. Phys. C: Solid State Phys.* **19**, L653.
Jones, R. (1987). *J. Phys. C: Solid State Phys.* **20**, 271.
Jones, R. (1988). *J. Phys. C: Solid State Phys.* **21**, 5735.
Jones, R. and Öberg, S. (1991a). *Phys. Rev. B* **44**, 3673.
Jones, R. (1991b). *Physica B* **170**, 181.
Jones, R. and Öberg, S., (1992a). *Phys. Rev. Lett.* **68**, 86.
Jones, R., Briddon, P. R., and Öberg, S. (1992b). *Phil. Mag. Lett.* **66**, 67.
Jones, R., Umerski, A., and Öberg, S. (1992c). *Phys. Rev. B* **45**, 11321.
Jones, R. and Öberg, S. (1992d). *Semicond. Sci. Technol.* **7**, 27.
Jones, R. and Öberg, S. (1994a). *Phys. Rev. B* **49**, 5306.

Jones, R., Goss, J., Ewels, C., and Öberg, S. (1994b). *Phys. Rev. B* **50**, 8378.
Jones, R., Öberg, S., Berg Rasmussen, F., and Bech Nielsen, B. (1994c). *Phys. Rev. Lett.* **72**, 1882.
Jones, R., Ewels, C., Goss, J., Miro, J., Deák, P., Öberg, S., and Berg Rasmussen, F. (1994d). *Semicond. Sci. Technol.* **9**, 2145.
Jones, R. and Öberg, S. (1994e). *Mat. Sci. Forum* **143-147**, 253.
Jones, R., Leary, P., Öberg, S., and Torres, V. J. T. (1995a). *Mat. Sci. Forum* **196-201**, 785.
Jones, R., Öberg, S., Goss, J., Briddon, P. R., and Resende, A. (1995b). *Phys. Rev. Lett.* **75**, 2734.
Kajihara, S. A., Antonelli, A., Bernholc, J., and Car, R. (1991). *Phys. Rev. Lett.* **66**, 2010.
Kohn, W. and Sham, L. J. (1965). *Phys. Rev. A* **140**, 1133.
Koppitz, J., Schirmer, O. F., and Seal, M. (1986). *J. Phys. C: Solid State Phys.* **19**, 1123.
Kozuch, G. M., Stavola, M., Pearton, S. J., Abernathy, C. R., and Lopata, J. (1990). *Appl. Phys. Lett.* **57**, 2561.
Kutzler, F. W. and Painter, G. S. (1992). *Phys. Rev. B* **45**, 3236.
Latham, C. D., Heggie, M. I., Jones, R., and Briddon, P. R. (1994). *Diamond Related Mat.* **3**, 1370.
Leary, P., Jones, R., Öberg, S., and Torres, V. J. B. (1996). *Phys. Rev. B*, in press.
Leigh, R. S. and Szigeti, B. (1967). *Proc. Roy. Soc. A* **301**, 211.
Lister, G. M. S. and Jones, R. (1988). Unpublished.
Loubser, J. H. N. and van Wyk, J. (1981). *Proceedings of the Diamond Conference*, Reading, U.K. Unpublished.
Louie, S. G., Froyen, S., and Cohen, M. L. (1982). *Phys. Rev. B* **26**, 1738.
Matsui, M. and Yokoyama, T. (1986). *Inst. Phys. Conf. Ser.* **79**, 13.
Minaev, N. S. and Mudryi, A. V. (1981). *Phys. Stat. Solidi A* **68**, 561.
Musgrave, M. J. P. and Pople, J. A. (1962). *Proc. Roy. Soc.* **A268**, 474.
Öberg, S., Sitch, P. K., Jones, R., and Heggie, M. I. (1995). *Phys. Rev. B* **51**, 13138.
Pederson, M., R., Jackson, K. A., and Pickett, W. E. (1991). *Phys. Rev. B* **44**, 3891.
Perdew, J. P. and Zunger, A. (1981). *Phys. Rev. B* **23**, 5048.
Porezag, D., Frauenheim, Th., Köhler, Th., Seifert, G., and Kaschner, R. (1995). *Phys. Rev. B* **51**, 12947.
Press, W. H., Flannery, B. P., Teukolsky, S. A., Vetterling, W. T. (1987). *Numerical Recipes.* Cambridge University Press, Cambridge.
Purton, J., Jones, R., Heggie, M., Öberg, S., and Catlow, C. R. A. (1992). *Phys. Chem. Minerals* **18**, 389.
Safonov, A. N., Lightowlers, E. C., Davies, G., Leary, P., Jones, R., and Öberg. S. (1996). Unpublished.
Seifert, G. and Eschrig, H. (1985). *Phys. Stat. Solidi B* **127**, 573.
Seifert, G., Eschrig, H., and Bieger, W. (1986). *Z. Phys. Chem. (Leipzig)* **267**, 529.
Sitch, P., Jones, R., Öberg, S., and Heggie, M. I. (1994). *Phys. Rev. B* **50**, 17717.
Sitch, P., Jones, R., Öberg, S., and Heggie, M. I. (1995). *Phys. Rev. B* **52**, 4951.
Slater, J. C. (1960). *Quantum Theory of Atomic Structure*, Vol. 2. McGraw Hill, New York.
Smith, W. V., Sorokin, P. P., Gelles, L. L., and Lasher, G. J. (1959). *Phys. Rev.* **115**, 1546.
Sternschulte, H., Thonke, K., Sauer, R., Münzinger, P. C., and Michler, P. (1994). *Phys. Rev. B* **50**, 14554.
Stoneham, M. (1975). *Defects in Solids.* Oxford University Press, London.
Suezawa, M., Sumino, K., Harada, H., and Abe, T. (1986). *Jpn. J. Appl. Phys.* **25**, L859.
Sutton, A. P., Finnis, M. W., Pettifor, D. G., and Ohta, Y. (1988). *J. Phys. C: Solidi State Phys.* **21**, 35.
Svane, A. and Gunnarsson, O. (1990). *Phys. Rev. Lett.* **65**, 1148.
Svelto, O. (1976). *Principles of Lasers.* Plenum Press, New York.

Szotek, Z., Temmerman, W. M., and Winter, H. (1993). *Phys. Rev. B* **47**, 11533.
Trombetta, J. M. and Watkins, G. D. (1987). *Appl. Phys. Lett.* **51**, 1103.
Vaz de Carvalho, M. H. and das Neves, A. T. (1987). *J. Phys. C.: Solid State Phys.* **20**, 2713.
Von Barth, U. and Hedin, L. (1972). *J. Phys. C.: Solid State Phys.* **5**, 1629.
Von Barth, U. (1979). *Phys. Rev. A* **20**, 1693.
Wagner, J., Maier, M., Lauterback, Th., Bachem, K. H., Ashwin, M. J., Newman, R. C., Woodhouse, K., Nicklin, R., and Bradley, R. R. (1992). *Appl. Phys. Lett.* **60**, 2546.
Wagner, J., Bachem, K. H., Davidson, B. R., Newman, R. C., Bullough, T. J., and Joyce, T. B (1995). *Phys. Rev. B* **51**, 4150.
Wagner, J., Newman, R. C., Davidson, B. R., Westwater, S. P., Bullough, T. J., Joyce, T. B., Latham, C. D., Jones, R., and Öberg, S. (1996). Unpublished.
Watkins, G. D. (1964). *Radiation Damage in Semiconductors*, edited by P. Barach. Dunod, Paris, 97.
Woodhouse, K., Newman, R. C., deLyon, T. J., Woodall, J. M., Scilla, G. J., and Cardone, F. (1991). *Semicond. Sci. Technol.* **6**, 330.
Yin, M. T. and Cohen, M. L. (1980). *Phys. Rev. Lett.* **45**, 1004.
Yonenaga, I. and Sumino, K. (1989). *J. Appl. Phys.* **65**, 85.
Zheng, J. F., Stavola, M., and Watkins, G. D. (1994). *The Physics of Semiconductors*, edited by D. J. Lockwood. World Scientific, Singapore, 2363.

# Index

## A

A centers, 73, 75
*Ab initio* cluster methods. *see also* Hartree-Fock theory
    advantages, 345–346
    AIMPRO, 289, 332
    applications, 337–338
    line defects, 343, 344$f$, 345
        point defects in bulk solids, 338–343
    choice of basis sets, 332–334
    cluster construction, 334$t$, 334–336
    density functional theory, 289, 300–306
    future directions, 346
    Mulliken populations and, 336–337
    radiative lifetimes and, 337
    real-space cluster, 310–319
ACAR (angular correlation of annihilation radiation), 225, 226$f$, 227
AIMPRO, 289, 332
AlAs/GaAs. *see* Aluminum arsenide/gallium arsenide
ALC (avoided level-crossing). *see* Muon level-crossing resonance
Aluminum, interstitial, annealing in Si, 16–17, 17$f$
Aluminum arsenide/gallium arsenide (AlAs/GaAs)
    mismatch, 102
    Si-doped, 116–120
Aluminum gallium arsenide (AlGaAs)
    Be acceptors, 130–131
    $Ga_i$ in, 122–125, 123$f$, 124$f$
$Al_xGa_{1-x}As$ layers, As-grown n-type, metastable defects in, 275–279, 276$f$, 277$f$, 279$f$
$Al_xGa_{1-x}$/GaAs heterostructures, group VI donors, 120–121
Angular correlation of annihilation radiation (ACAR), 225, 226$f$, 227
Annealing
    of interstitial Al, in Si, 16–17, 17$f$
    temperature effects, 37
Arsenic
    growth of GaAs, native defects in, 244$f$, 244–246
    vacancy
        in n-type GaAs, 258–261
        in SI GaAs, 262–268, 263$f$, 265$f$–267$f$
Average positron lifetime, 221
Avoided level-crossing (ALC). *see* Muon level-crossing resonance
Azimuthal quantum number ($m_j$), 8

## B

Berylium acceptors (Be), in epilayers, 130–131, 133
Binary semiconductors. *see also* specific binary semiconductors
    effective-mass donors in, 115–116
Boltzmann distribution function, 30
Born-Oppenheimer approximation, for many-body problem, 291–292

## C

CE (charge exchange), LF-SR and, 183–187
CESR (conduction-electron spin resonance), 114
Charge exchange (CE), LF-SR and, 183–187

Chemical shift, 116
Clusters, H-terminated, 334–336
Compound semiconductors. *see also* specific compound semiconductors
  point defects, 342–343
Conduction-electron spin resonance (CESR), 114
Copper-doped gallium phosphide (GaP:Cu), 128–129
Core-electron momentum spectroscopy, 231–232
Cross-relaxation (CR), 170. *see also* Muon level-crossing resonance

## D

DAPs. *see* Donor-acceptor pairs
Deep donor level (DX)
  in As-grown n-type $Al_xGa_{1-x}As$ layers, 275–279, 276$f$, 277$f$, 279$f$
  atomic structure, 279–281, 281$f$
  for low-temperature PL experiments, 115
Deep-level transient spectroscopy (DLTS), 34
Defects. *see also* specific types of defects
  alignment, of vacancy-vacancy axis, 32$f$, 32–33
  charge-state changes, 35–36
  compensating, in highly Si-doped GaAs layers, 250–254, 251$f$, 253$f$, 254$f$
  correlation, with optical dichroism, 33
  detection. *see* specific detection methods
  effective-mass, 99
  electrical-level determination, 33–34
  electrically active, 2
  emission bands, 47
  excited states, 35
  line, 343, 344$f$, 345
  migration, activation energy for, 33
  mismatch-induced, 102–104, 103$f$
  nonmagnetic, 2
  open volume, 282
  optical absorption, 47
  optical alignment, 36
  oxygen-vacancy pair, 73, 75
  paramagnetic, 2, 4–5
  perturbation, 2
  point. *see* Point defects
  production, 39–40, 40$f$
  spatial distribution, measurement, 66–68, 67$f$
  vacancy, charge state of, 282
Density functional theory, 289
  advantages, 300
  atom cluster applications. *see* Real-space cluster method
  derivation, 300–306
  disadvantages, 300–301
Diamond
  bond-centered orbitals, 333
  $^{13}C$ enriched, resolution of nuclear hyperfine interaction in, 163$f$, 163–164
  clusters, 334$t$
  lattice vacancy, 26
  neutral defect, excited state of, 35
  nitrogen defects, 337
  point defects, 338–340
DLTS (deep-level transient spectroscopy), 34
Donor-acceptor pairs (DAPs)
  deep, 99–100
  recombination, 94–96, 95$f$
Donor-acceptor recombination model, 78–82, 80$f$
Doppler broadening spectroscopy, 282
  core annihilation spectrum, sensitivity of, 239–240, 240$f$
  core-electron momentum, 231$f$, 231–232
  experimental aspects, 227–228
  momentum parameters, 228–231, 229$f$, 230$f$, 234
DX. *see* Deep donor level

## E

$E_{12}$ (valley-orbit splitting), 116
EDENDOR (electrical detection of electron-nuclear double-resonance), 84–86, 85$f$, 86$f$, 88
EDEPR. *see* Electrical detection of electron paramagnetic resonance
Effective mass, hydrogenic, 115–116
Effective-field approximation, SR, 152–153
Effective-mass acceptors, 129–130, 130$t$
Effective-mass defects, g-tensors, 99
Effective-mass donors
  α, hydrostatic strain and, 113$t$, 113–114

X, axial strain and, 114f, 114–122, 115f, 119f–121f
EL2 defect, atomic structure, 279–281, 281f
EL2 midgap donor, 268–272, 269f
Electrical detection of electron paramagnetic resonance (EDEPR)
  concentration dependence, 82–84, 83f
  donor-acceptor recombination model, 78–82, 80f
  experimental observation, 73–78, 74f, 75f
  signal intensities, 88f, 88–89
  signal-to-noise ratio, 88f, 88–89
  spectrometer, low-frequency, 89–90, 90f
  spectrum, angular dependence of, 89, 89f
  temperature dependence, 82–84, 83f
  via photoconductivity, 86–87, 87f
  vs. EPR, 76
Electrical detection of electron-nuclear double-resonance (EDENDOR), 84–86, 85f, 86f, 88
Electrical-level determination, of defect, 33–34
Electron paramagnetic resonance (EPR)
  applications, 1–2, 45–46
  auxiliary techniques, 28–29
    applied uniaxial stress, 29–3530f–32f
    defect production, 39–40
    optical illumination *in situ*, 35–36
    temperature and, 37f, 37–39
  of defects
    intrinsic, in wide-bandgap semiconductors, 24–28
    in irradiated silicon, 16–24, 17f–23f
    as microwave-induced changes, 47–48
  electrical detection of. *see* Electrical detection of electron paramagnetic resonance
  excitation spectroscopy, 47
  experiment, 3–4
  MCDA tagged, 61–66, 62f, 65f, 66f
  MCDA-detected
    example of, 53–61, 54f, 56f–61f
    principles in, 48–53, 50f
  microwave power dependence, 76–78, 77f, 78f
  nitrogen impurities, in diamond, 338
  optical detection. *see* Optical detection of electron paramagnetic resonance
  with photoluminescence, 93
  sensitivity, 45–46, 88

spectrometer, 3–4
theory, 15–16
  fine-structure terms for S· 1/2, 13–15, 14f
  g-tensor and, 4–7, 5f, 7f
  hyperfine interactions and, 7–12, 9f
  transitions, selection rule for, 15
vs. EDEPR, 76
Electron-nuclear double-resonance (ENDOR), 1–2
  advantages, 107–108
  of defects
    intrinsic, in wide-bandgap semiconductors, 24–28
    in irradiated silicon, 16–24, 17f–23f
  electrical detection of, 84–86, 85f, 86f, 88
  experiment, 3–4
  MCDA method, 70–73, 71f–73f
  optical detection, 72–73, 73f, 107–108
  spectrometer, 3–4
  theory, 15–16
    g-tensor and, 4–7, 5f, 7f
    hyperfine interactions and, 12–13
ENDOR. *see* Electron-nuclear double-resonance
Energy second derivatives, 327–329
Epitaxial layers (epilayers)
  acceptors
    Be, 130–131, 133
    effective-mass, 129–130, 130t
    in GaN, 131f, 131–133, 132f
  As-grown n-type $Al_xGa_{-x}As$, metastable defects in, 275–279, 276f, 277f, 279f
  deep centers, 128–129
  donors in N-doped ZnSe/GaAs, 125–128, 126f, 127f
  $Ga_i$ in AlGaAs, 122–125, 123f, 124f
  effective-mass donors
    donors, hydrostatic strain and, 113t, 113–114
    X, axial strain and, 114f, 114–122, 115f, 119f–121f
Epitaxy
  mismatch-induced defects and, 102–104
  perfect, 101–102, 102f
EPR. *see* Electron paramagnetic resonance
Exchange-correlation energy, real-space cluster method and, 314–317
Excited states, photo-induced, of defects, 35

## F

Fermi contact term, 10
Forces, evaluation of, 323–324

## G

GaAs. *see* Gallium arsenide
Gallium (Ga)
  interstitial, in AlGaAs, 122–125, 123*f*, 124*f*
  vacancy
    in GaAs, 235*f*, 235–238, 237*f*
    in GaP, 24*f*, 24–25, 34
Gallium arsenide (GaAs)
  As-grown, native defects in, 244*f*, 244–246
  Be acceptors, 130–131
  clusters, 334*t*
  defects, MCDA-detected EPR of, 53–55, 54*f*
  donor-acceptor pair recombination, 95
  electron-irradiated, negative vacancies and ions in, 246–250, 247*f*, 249*f*
  heavily-doped n-type, Mu-center in, 164–167, 165*f*, 166*t*, 167*f*
  high-resistivity
    diamagnetic center in, 160*f*, 160–163, 161*t*, 162*t*
    Mu$^{RT}$ quantum diffusion in, 181–182, 183*f*
  ion-implanted, depth profiles of vacancies in, 254–257, 255*f*, 257*f*
  MCDA tagging by EPR, 61–66, 62*f*, 65*f*, 66*f*
  muonium as deep recombination center in, 189–191, 190*f*
  neutral muonium states in, 142
  n-type
    arsenic vacancy in, 258–261
    Mu- electronic structure in, 172–173, 173*f*
  point defects, 342–343
  positron annihilation spectroscopy, 282
  Si-doped layers, compensation of defects in, 250–254, 251*f*, 253*f*, 254*f*
  ZnSe on, nitrogen doped, 125–128, 126*f*, 127*f*
Gallium nitride (GaN)
  acceptors in, 131*f*, 131–133, 132*f*
  Mg-doped, 132*f*, 132–133
Gallium nitride silicon carbonate (GaN/SiC), 103
Gallium phosphide (GaP)
  copper-doped, 128–129
  gallium vacancy in, 24*f*, 24–25, 34
  high-resistivity, Mu$^0$BC center in, 174–176, 175*f*
  P$_{Ga}$ antisite in, 25–26, 26*f*
GaN/Al$_2$O$_3$ system, mismatch, 103–104
GaN/SiC (gallium nitride silicon carbonate), 103
GaP. *see* Gallium phosphide
Germanium clusters, 334*t*
Green function method
  point defects, vibrational modes of, 339
  for resonant modes determination, 331–332
G-tensors
  effective-mass acceptors, 130, 130*t*
  effective-mass defects, 99
  effective-mass donors, 113, 113*t*
  EPR and, 4–7, 5*f*, 7*f*

## H

Hartree energy, 312–314
Hartree-Fock theory, 289
  many-body problem and, 292–295
  *vs.* Kohn-Sham equations, 304
  *vs.* Thomas-Fermi theory, 303
  wave functions, 11
HEM (hydrogenic effective-mass), 115–116
Homoepitaxy, 101
Homogeneous electron gas, many-body problem and, 295–297
Hund's rule, 21
Hydrogen
  charge states, 138
  experimental conditions, 141–142
  passivation complexes, 203–204
  physical properties, 140–141, 141*t*
  *vs.* muonium, 203
Hydrogenic effective-mass (HEM), 115–116
Hydrostatic strain, effective-mass donors and, 113*t*, 113–114
Hyperfine interactions, 99
  calculations, with known electronic wave function, 10–11

dipole-dipole, 10
ENDOR, 12–13, 107–108
EPR, 7–12, 9f
muon, 139
nuclear or superhyperfine
  definition of, 154–155
  resolution in $^{13}$C enriched diamond, 163f, 163–164
  with Si, satellites from, 9–10, 17–18
  with single nucleus, 9

## I

Impurities. *see also specific impurities*
  transition element, 27–28
Indium phosphide (InP)
  antisites
    properties of, 109–110, 110t
    structure of, 109, 109f
  donor-acceptor pair recombination, 95
  Zn-doped, 104f, 104–105
    effective-mass acceptor ODMR, 110–112, 111f
    effective-mass donor ODMR, 105, 106, 106f
    ODMR, 112
    $P_{In}$ ODMR, 106–110, 107f–109f, 110t
  vacancy-impurity complexes, 242–243, 243f
InP:Zn. *see* Indium phosphide, Zn-doped
Ion implantation, for defect production, 40

## J

Jahn-Teller distortion
  effective-mass acceptors and, 130
  magnitude, 31
  orthorhombic, 27
  quantitative measure, 30, 30f
  relaxation time, 31, 31f
  reorientation, thermally activated, 37–39
  sign of, 29–30
  silicon divacancy and, 23
  silicon lattice vacancy and, 20–21
  uniaxial stress applications and, 29–32, 30f, 31f
Jellium, many-body problem and, 295–297

## K

Kink formation energy, 345
Kohn-Sham energy levels, 303–304, 307–308
Kohn-Sham equations
  derivation, 304
  self-consistency and, 319–322
Kubo-Toyabe functions, 176–178, 177f

## L

LCAO-MO (linear combination of atomic orbitals), 11, 19–20
LDA (local density approximation), 302–303
Lifetime spectroscopy, positron, 222–225, 224f
Line defects, 343, 344f, 345
Linear combination of atomic orbitals (LCAO-MO), 11
Linewidth, temperature effects, 37f, 37–39
Local density approximation (LDA), 302–303
Local pseudo-potential, 310
Local spin density approximation (LSDA), 302–303
Local spin-density functional theory (LSDF)
  advantages, 300
  atom cluster applications. *see* Real-space cluster method
  derivation, 300–306
  disadvantages, 300–301
LSDA (local spin density approximation), 302–303
LSDF. *see* Local spin-density functional theory

## M

Magnesium-doped gallium nitride (GaN:Mg), 132f, 132–133
Magnetic circular dichroism of absorption (MCDA)
  ENDOR-detected, 70–73, 71f–73f
  in EPR detection. *see* Electron paramagnetic resonance, MCDA-detected

Magnetic circular dichroism of absorption (*continued*)
  spatially resolved, 66–68, 67*f*
  spin state determination, 68–69, 70*f*
Magnetic resonance
  of epitaxial layers, future directions, 133–134
  optically detected. *see* Optical detection of magnetic resonance
Magneto-optical detection, 47, 48–73
Many-body problem
  Born-Oppenheimer approximation, 291–292
  density functional theory, 300–306
  Hartree-Fock theory, 292–295
  homogeneous electron gas and, 295–297
  Schrödinger equation for, 290–291
  spin-polarized electron gas and, 297–300, 298*f*, 299*f*
Mapping, spatial distribution of defects, 66–68, 67*f*
MCDA. *see* Magnetic circular dichroism of absorption
Metastable defects, 282
Midgap donor EL2, 268–272, 269*f*
Minimum atomic basis, 333
$m_j$ (azimuthal quantum number), 8
Motional narrowing phenomenon, 37
Mott insulators, 289
Mulliken populations, 336–337
Muon level-crossing resonance (LCR), 157
  applications, 168
  counters, 167
  electronic structure of Mu- in n-type GaAs, 172–173, 173*f*
  $Mu_{BC}^0$ center, in high-resistivity GaP, 174–176, 175*f*
  principle, 168–172, 170*f*, 171*f*
  spectroscopy features, 171–172
Muon spin relaxation, rotation and resonance. *see* SR
Muonium
  center formation, 138
  charge state interconversions, in silicon, 196–200, 197*f*, 198*f*, 200*t*
  charge states, 138
  charge-spin exchange cycling with free carriers, 140
  diffusion, 140
  dynamics, 143, 203

electronic structure, 143
evidence as deep recombination center in GaAs, 189–191, 190*f*
experimental conditions, 141–142
experiments, timescale for, 204
interstitial sites, 138–139, 139*f*
physical properties, 140–141, 141*t*
repolarization, 192–193
usefulness, 142
*vs.* hydrogen, 203
Muons
  charge states, TF-SR detections, 160*f*, 160–163, 161*t*, 162*t*
  implantation, 142
  level-crossing resonances, 154
  Mu-
    center, in heavily-doped n-type GaAs, 164–167, 165*f*, 166*t*, 167*f*
    diffusion, in n-type GaAs, 178–180, 179*f*
    electronic structure, in n-type GaAs, 172–173, 173*f*
  $Mu_{BC}^0$ center, in high-resistivity GaP, 174–176, 175*f*
  $Mu_{RT}^0$, quantum diffusion, in high-resistivity GaAs, 181–182, 183*f*
  physical properties, 140–141, 141*t*
  positive, 145
  production, 142
  spin direction, 144, 145
  spin polarization, 142–143
  spin polarization function calculation, 145–149
    diamagnetic centers in absence of nuclear spins, 149–151, 150*f*
    paramagnetic centers in absence of nuclear spins, 151–152, 152*t*
  spin-polarized
    decay of, 144*f*, 145
    production of, 144–145
  surface, 144–145
Musgrave-Pople potentials, 328–329

N

Neutral paramagnetic centers, 138–140
Nitrogen defects
  in diamond, 337, 338–339
  in silicon, 341–342

Nitrogen doped ZnSe/GaAs, 125–128, 126f, 127f
Nuclear hyperfine interactions
  defined, 154–155
  resolution in $^{13}$C-enriched diamond, 163f, 163–164
Nuclear spin, influence on SR, 153–155

## O

ODENDOR. see Optical detection of electron-nuclear double-resonance
ODEPR. see Optical detection of electron paramagnetic resonance
ODMR. see Optical detection of magnetic resonance
Optical detection of electron paramagnetic resonance (ODEPR), 46
  absorption-detected. see Magneto-optical detection
  MCDA excitation spectra of, 61–66, 62f, 65f, 66f
  selectivity, 46–47
  sensitivity, 46
  spatially resolved, 66–68, 67f
Optical detection of electron-nuclear double-resonance (ODENDOR)
  6H-SiC crystals, 72–73, 73f
  on magnetic circular dichroic absorption, 109
  $^{31}$P, 107f, 107–108, 108f
Optical detection of magnetic resonance (ODMR), 93–94
  acceptors in GaN, 131f, 131–133, 132f
  model, 96–99, 97f, 98f
  on photoluminescence, 94–101
  resonance parameters, 99–101
  spectrometers, 100f, 100–101
Optical dichroism, correlation with defect, 33
Optimization strategy, 324–327
Orbital angular momentum, 6, 15, 34–35

## P

Paramagnetic defects, 2, 4–5
Photoluminescence, with EPR, 93
Pion decay, 144, 144f
Point defects

anisotropic, optical transitions, 33
  in bulk solids, 338–343
  production, 39–40
Positron. see also Positron annihilation spectroscopy
  lifetime, 210, 282
    in defect identification, 232–234, 233t
    open volume and, 234
    spectroscopy of, 222–225, 224f
    temperature and, 235f, 235–236
    vacancy charge state and, 232–233, 233t
  in solids, 211–212
    annihilation characteristics, 216–217
    bulk studies, 212
    diffusion of, 213–215, 214f
    layer studies, 212–213
    mobility of, 215
    states, 215–216
    wave functions, 215–216, 216f
  trapping
    kinetic model, 220–222
    rate and coefficient for, 218–220, 219f, 236–238, 237f
    temperature dependence, 237f, 238
Positron annihilation spectroscopy, 282
  advantages, 210–211
  characteristics, in solids, 216–217
  GaAs, 282
  identification
    open volume, 233–234
    positron lifetime, 232–234, 233t
    vacancy charge state, 235f, 235–238, 237f
    vacancy sublattice, 240–242, 241f
    of vacancy-impurity complexes, 242–243, 243f
  impact, 211
  low-energy beam
    compensating defects in highly Si-doped GaAs layers, 250–254, 251f, 253f, 254f
    depth profiles of vacancies in ion-implanted GaAs, 254–257, 255f, 257f
  metastable defects, 268
    As-grown n-type $Al_xGa_{1-x}As$ layers, 275–279, 276f, 277f, 279f
    As-grown SIGaAs, 268–272, 269f
    DX, atomic structure of, 279–281, 281f

Positron annihilation spectroscopy, metastable defects, (*continued*)
  EL2, atomic structure of, 279–281, 281*f*
  in electron-irradiated SiGaAs, 273*f*, 273–275, 274*f*
  principles, 210
  sensitivity, 210
  shallow positron traps, negative ions as, 243
    electron-irradiated GaAs, 246–250, 247*f*, 249*f*
    native defects in As-grown GaAs, 244*f*, 244–246
  techniques
    angular correlation, 225, 226*f*, 227
    core-electron momentum, 231*f*, 231–232
    Doppler broadening, 227–232
    lifetime, 222–225, 224*f*
  vacancy ionization levels, 258
    optical transitions, 262–268, 263*f*, 265*f*–267*f*
    thermal ionization and, 258–261, 259*f*, 261*f*
Proton, physical properties, 140–141, 141*t*
Pseudo-potential theory, 306–310, 307*f*

## Q

Quartz, hydrolytic weakening, 288
Quasi-harmonic frequencies, 328

## R

Real-space cluster method, 310–312
  exchange-correlation energy and, 314–317
  Hartree energy and, 312–314
  matrix formulation, 317–319
Relaxation
  constrained, 326–327
  unconstrained, 324–325
Resonant modes, 329, 331–332

## S

Saddle point, 326

Satellites, from Si hyperfine interactions, 9–10, 17–18
Scalar relativistic potential, 309
Self-consistency, 319–322
Silicon (Si)
  AA9 center, 138
  charge-state dynamics, at high temperatures, 187–189, 188*f*
  clusters, 334*t*
  defect-vacancy complexes, 21–22, 22*f*
  divacancy, 22–24, 23*f*, 29, 39, 40
  doped GaAs layers, compensation of defects in, 250–254, 251*f*, 253*f*, 254*f*
  doping of AlAs/GaAs, 116–120
  irradiated, defects in, 16–24, 17*f*–23*f*
  lattice vacancy, 17–21, 18*f*–20*f*
  muonium states interconversions in, 196–200, 197*f*, 198*f*, 200*t*
  nitrogen defect, 341
  n-type, spin-exchange scattering in, 191–192, 192*f*
  oxygen impurities in, 332, 340–341
  point defects, 340–342, 341*f*
  vanadium-containing 6H-SiC crystals, MCDA-detected EPR spectra of, 55–59, 56*f*–58*f*
Semi-insulating gallium arsenide (SI GaAs)
  arsenic vacancy in, 262–268, 263*f*, 265*f*–267*f*
  As-grown, metastable defects in, 268–272, 269*f*
  electron-irradiated, metastable defects in, 273*f*, 273–275, 274*f*
Slater transition method, 335
Spectrometers
  EDEPR, low-frequency, 89–90, 90*f*
  ENDOR, 3–4
  EPR, 3–4, 4*f*
  ODMR, 100*f*, 100–101
Spectroscopy
  angular correlation, 225, 226*f*, 227
  core-electron momentum, 231–232
  deep-level transient, 34
  Doppler broadening. *see* Doppler broadening spectroscopy
  excitation, 47
  muon level-crossing resonance, 171–172
  positron annihilation. *see* Positron annihilation spectroscopy
  positron lifetime, 222–225, 224*f*

Spin-density functional theory, 289
Spin-exchange scattering
  LF-SR and, 185–187
  in n-type silicon, 191–192, 192$f$
Spin-polarized electron gas, many-body problem and, 297–300, 298$t$, 299$f$
Spin-polarized theory, 303
SR, 138, 143
  detector configurations, 146$f$, 147
  effective-field approximation, 152–153
  experimental techniques, 155–157
  facilities, comparison of, 201–202
  level-crossing resonance technique. *see* Muon level-crossing resonance
  longitudinal-field, 150, 155–156, 180–181
    charge-state dynamics in Si at high temperatures, 187–189, 188$f$
    detector configuration, 146$f$, 147
    evidence of muonium as deep recombination center in GaAs, 189–191, 190$f$
    influence of spin and charge exchange, 183–187, 185$f$, 186$f$
    quantum diffusion of $Mu_{RT}^0$ in high-resistivity GaAs, 181–182, 183$f$
    repolarization of muonium, 192–193
    spin-exchange scattering in n-type silicon, 191–192, 192$f$
  nuclear spin influence, 153–155
  radio-frequency-field, 156
    counter arrangement, 193–194
    interconversion between muonium states in silicon, 196–200, 197$f$, 198$f$, 200$t$
    principles of, 194–196, 196$f$
    resonance condition, 194
  techniques, comparison of, 201–202
  time-integrated polarization, 147, 148
  transverse-field, 148, 150, 155–156
    arrangement for, 157–159
    counters, 158–159, 159$f$
    detector configuration, 146$f$, 147
    diamagnetic center in high-resistivity GaAs, 160–163
    Mu- center in heavily-doped n-type GaAs, 164–167, 165$f$, 166$t$, 167$f$
    muon charge state detections, 160–163
    resolution of nuclear hyperfine interaction, 163$f$, 163–164
  zero-field, 156

counter arrangement, 176, 177
diffusion of Mu- in n-type GaAs, 178–180, 179$f$
  Kubo-Toyabe functions, 176–178, 177$f$
Stationary trapping equations, 221
Strain
  axial, X effective-mass donors and, 114$f$, 114–122, 115$f$, 119$f$–121$f$
  hydrostatic, effective-mass donors and, 113$t$, 113–114
Stress, uniaxial
  effective-mass acceptors and, 129–130, 130$t$
  EPR auxiliary techniques and, 29–35, 30$f$–32$f$
Superhyperfine interactions
  defined, 154–155
  resolution in $^{13}$C-enriched diamond, 163$f$, 163–164

## T

Tagging
  EPR excitation spectroscopy, 47
  MCDA, by EPR, 61–66, 62$f$, 65$f$, 66$f$
Thermal double donors (TDDs), 74–75
Thermal ionization, 258–261
Thomas-Fermi theory, 303
Tight binding schemes, 305
Transition element impurities, 27–28

## V

Valley-orbit splitting ($E_{12}$), 116
VCA (virtual crystal approximation), 119
Vibrational modes
  anomalous, nitrogen impurities in diamond and, 338
  of clusters, 327–332
  effective charges, 329–331
  local, 329
  resonant, 329, 331–332
Virtual crystal approximation (VCA), 119

## W

Wide-bandgap semiconductors, intrinsic defects, 24–28

## Z

Zeeman energy differences, of defects, 3
Zeeman Hamiltonian, 129
Zeeman interaction, of paramagnetic defect, 4–5
Zero phonon lines (ZPLs), 55, 62–63
Zinc selenium gallium arsenide (ZnSe/GaAs)
  mismatch, 103
  nitrogen doped, 125–128, 126$f$, 127$f$
Zinc selenium (ZnSe)
  defect production, 40
  nitrogen doped, on GaAs, 125–128, 126$f$, 127$f$
Zinc-blend, interstitial sites, 138–139, 139$f$
ZnSe. *see* Zinc selenium
ZnSe/GaAs. *see* Zinc selenium gallium arsenide
ZPLs (zero phonon lines), 55, 62–63

# Contents of Volumes in This Series

### Volume 1    Physics of III–V Compounds

C. Hilsum, Some Key Features of III–V Compounds
Franco Bassani, Methods of Band Calculations Applicable to III–V Compounds
E. O. Kane, The k-p Method
V. L. Bonch-Bruevich, Effect of Heavy Doping on the Semiconductor Band Structure
Donald Long, Energy Band Structures of Mixed Crystals of III–V Compounds
Laura M. Roth and Petros N. Argyres, Magnetic Quantum Effects
S. M. Puri and T. H. Geballe, Thermomagnetic Effects in the Quantum Region
W. M. Becker, Band Characteristics near Principal Minima from Magnetoresistance
E. H. Putley, Freeze-Out Effects, Hot Electron Effects, and Submillimeter Photoconductivity in InSb
H. Weiss, Magnetoresistance
Betsy Ancker-Johnson, Plasma in Semiconductors and Semimetals

### Volume 2    Physics of III–V Compounds

M. G. Holland, Thermal Conductivity
S. I. Novkova, Thermal Expansion
U. Piesbergen, Heat Capacity and Debye Temperatures
G. Giesecke, Lattice Constants
J. R. Drabble, Elastic Properties
A. U. Mac Rae and G. W. Gobeli, Low Energy Electron Diffraction Studies
Robert Lee Mieher, Nuclear Magnetic Resonance
Bernard Goldstein, Electron Paramagnetic Resonance
T. S. Moss, Photoconduction in III–V Compounds
E. Antoncik ad J. Tauc, Quantum Efficiency of the Internal Photoelectric Effect in InSb
G. W. Gobeli and I. G. Allen, Photoelectric Threshold and Work Function
P. S. Pershan, Nonlinear Optics in III–V Compounds
M. Gershenzon, Radiative Recombination in the III–V Compounds
Frank Stern, Stimulated Emission in Semiconductors

## Volume 3    Optical of Properties III–V Compounds

*Marvin Hass*, Lattice Reflection
*William G. Spitzer*, Multiphonon Lattice Absorption
*D. L. Stierwalt and R. F. Potter*, Emittance Studies
*H. R. Philipp and H. Ehrenveich*, Ultraviolet Optical Properties
*Manuel Cardona*, Optical Absorption above the Fundamental Edge
*Earnest J. Johnson*, Absorption near the Fundamental Edge
*John O. Dimmock*, Introduction to the Theory of Exciton States in Semiconductors
*B. Lax and J. G. Mavroides*, Interband Magnetooptical Effects
*H. Y. Fan*, Effects of Free Carries on Optical Properties
*Edward D. Palik and George B. Wright*, Free-Carrier Magnetooptical Effects
*Richard H. Bube*, Photoelectronic Analysis
*B. O. Seraphin and H. E. Bennett*, Optical Constants

## Volume 4    Physics of III–V Compounds

*N. A. Goryunova, A. S. Borschevskii, and D. N. Tretiakov*, Hardness
*N. N. Sirota*, Heats of Formation and Temperatures and Heats of Fusion of Compounds $A^{III}B^V$
*Don L. Kendall*, Diffusion
*A. G. Chynoweth*, Charge Multiplication Phenomena
*Robert W. Keyes*, The Effects of Hydrostatic Pressure on the Properties of III–V Semiconductors
*L. W. Aukerman*, Radiation Effects
*N. A. Goryunova, F. P. Kesamanly, and D. N. Nasledov*, Phenomena in Solid Solutions
*R. T. Bate*, Electrical Properties of Nonuniform Crystals

## Volume 5    Infrared Detectors

*Henry Levinstein*, Characterization of Infrared Detectors
*Paul W. Kruse*, Indium Antimonide Photoconductive and Photoelectromagnetic Detectors
*M. B. Prince*, Narrowband Self-Filtering Detectors
*Ivars Melngalis and T. C. Harman*, Single-Crystal Lead-Tin Chalcogenides
*Donald Long and Joseph L. Schmidt*, Mercury-Cadmium Telluride and Closely Related Alloys
*E. H. Putley*, The Pyroelectric Detector
*Norman B. Stevens*, Radiation Thermopiles
*R. J. Keyes and T. M. Quist*, Low Level Coherent and Incoherent Detection in the Infrared
*M. C. Teich*, Coherent Detection in the Infrared
*F. R. Arams, E. W. Sard, B. J. Peyton, and F. P. Pace*, Infrared Heterodyne Detection with Gigahertz IF Response
*H. S. Sommers, Jr.*, Macrowave-Based Photoconductive Detector
*Robert Sehr and Rainer Zuleeg*, Imaging and Display

## Volume 6    Injection Phenomena

*Murray A. Lampert and Ronald B. Schilling*, Current Injection in Solids: The Regional Approximation Method
*Richard Williams*, Injection by Internal Photoemission
*Allen M. Barnett*, Current Filament Formation

*R. Baron and J. W. Mayer*, Double Injection in Semiconductors
*W. Ruppel*, The Photoconductor-Metal Contact

## Volume 7  Application and Devices
### Part A

*John A. Copeland and Stephen Knight*, Applications Utilizing Bulk Negative Resistance
*F. A. Padovani*, The Voltage-Current Characteristics of Metal-Semiconductor Contacts
*P. L. Hower, W. W. Hooper, B. R. Cairns, R. D. Fairman, and D. A. Tremere*, The GaAs Field-Effect Transistor
*Marvin H. White*, MOS Transistors
*G. R. Antell*, Gallium Arsenide Transistors
*T. L. Tansley*, Heterojunction Properties

### Part B

*T. Misawa*, IMPATT Diodes
*H. C. Okean*, Tunnel Diodes
*Robert B. Campbell and Hung-Chi Chang*, Silicon Junction Carbide Devices
*R. E. Enstrom, H. Kressel, and L. Krassner*, High-Temperature Power Rectifiers of $GaAs_{1-x}P_x$

## Volume 8  Transport and Optical Phenomena

*Richard J. Stirn*, Band Structure and Galvanomagnetic Effects in III–V Compounds with Indirect Band Gaps
*Roland W. Ure, Jr.*, Thermoelectric Effects in III–V Compounds
*Herbert Piller*, Faraday Rotation
*H. Barry Bebb and E. W. Williams*, Photoluminescence I: Theory
*E. W. Williams and H. Barry Bebb*, Photoluminescence II: Gallium Arsenide

## Volume 9  Modulation Techniques

*B. O. Seraphin*, Electroreflectance
*R. L. Aggarwal*, Modulated Interband Magnetooptics
*Daniel F. Blossey and Paul Handler*, Electroabsorption
*Bruno Batz*, Thermal and Wavelength Modulation Spectroscopy
*Ivar Balslev*, Piezopptical Effects
*D. E. Aspnes and N. Bottka*, Electric-Field Effects on the Dielectric Function of Semiconductors and Insulators

## Volume 10  Transport Phenomena

*R. L. Rhode*, Low-Field Electron Transport
*J. D. Wiley*, Mobility of Holes in III–V Compounds
*C. M. Wolfe and G. E. Stillman*, Apparent Mobility Enhancement in Inhomogeneous Crystals
*Robert L. Petersen*, The Magnetophonon Effect

## Volume 11  Solar Cells

*Harold J. Hovel*, Introduction; Carrier Collection, Spectral Response, and Photocurrent; Solar Cell Electrical Characteristics; Efficiency; Thickness; Other Solar Cell Devices; Radiation Effects; Temperature and Intensity; Solar Cell Technology

## Volume 12  Infrared Detectors (II)

*W. L. Eiseman, J. D. Merriam, and R. F. Potter*, Operational Characteristics of Infrared Photodetectors
*Peter R. Bratt*, Impurity Germanium and Silicon Infrared Detectors
*E. H. Putley*, InSb Submillimeter Photoconductive Detectors
*G. E. Stillman, C. M. Wolfe, and J. O. Dimmock*, Far-Infrared Photoconductivity in High Purity GaAs
*G. E. Stillman and C. M. Wolfe*, Avalanche Photodiodes
*P. L. Richards*, The Josephson Junction as a Detector of Microwave and Far-Infrared Radiation
*E. H. Putley*, The Pyroelectric Detector–An Update

## Volume 13  Cadmium Telluride

*Kenneth Zanio*, Materials Preparations; Physics; Defects; Applications

## Volume 14  Lasers, Junctions, Transport

*N. Holonyak, Jr. and M. H. Lee*, Photopumped III–V Semiconductor Lasers
*Henry Kressel and Jerome K. Butler*, Heterojunction Laser Diodes
*A Van der Ziel*, Space-Charge-Limited Solid-State Diodes
*Peter J. Price*, Monte Carlo Calculation of Electron Transport in Solids

## Volume 15  Contacts, Junctions, Emitters

*B. L. Sharma*, Ohmic Contacts to III–V Compounds Semiconductors
*Allen Nussbaum*, The Theory of Semiconducting Junctions
*John S. Escher*, NEA Semiconductor Photoemitters

## Volume 16  Defects, (HgCd)Se, (HgCd)Te

*Henry Kressel*, The Effect of Crystal Defects on Optoelectronic Devices
*C. R. Whitsett, J. G. Broerman, and C. J. Summers*, Crystal Growth and Properties of $Hg_{1-x}Cd_xSe$ alloys
*M. H. Weiler*, Magnetooptical Properties of $Hg_{1-x}Cd_xTe$ Alloys
*Paul W. Kruse and John G. Ready*, Nonlinear Optical Effects in $Hg_{1-x}Cd_xTe$

## Volume 17  CW Processing of Silicon and Other Semiconductors

*James F. Gibbons*, Beam Processing of Silicon
*Arto Lietoila, Richard B. Gold, James F. Gibbons, and Lee A. Christel*, Temperature Distribu-

tions and Solid Phase Reaction Rates Produced by Scanning CW Beams
*Arto Leitoila and James F. Gibbons*, Applications of CW Beam Processing to Ion Implanted Crystalline Silicon
*N. M. Johnson*, Electronic Defects in CW Transient Thermal Processed Silicon
*K. F. Lee, T. J. Stultz, and James F. Gibbons*, Beam Recrystallized Polycrystalline Silicon: Properties, Applications, and Techniques
*T. Shibata, A. Wakita, T. W. Sigmon, and James F. Gibbons*, Metal-Silicon Reactions and Silicide
*Yves I. Nissim and James F. Gibbons*, CW Beam Processing of Gallium Arsenide

## Volume 18  Mercury Cadmium Telluride

*Paul W. Kruse*, The Emergence of $(Hg_{1-x}Cd_x)Te$ as a Modern Infrared Sensitive Material
*H. E. Hirsch, S. C. Liang, and A. G. White*, Preparation of High-Purity Cadmium, Mercury, and Tellurium
*W. F. H. Micklethwaite*, The Crystal Growth of Cadmium Mercury Telluride
*Paul E. Petersen*, Auger Recombination in Mercury Cadmium Telluride
*R. M. Broudy and V. J. Mazurczyck*, (HgCd)Te Photoconductive Detectors
*M. B. Reine, A. K. Soad, and T. J. Tredwell*, Photovoltaic Infrared Detectors
*M. A. Kinch*, Metal-Insulator-Semiconductor Infrared Detectors

## Volume 19  Deep Levels, GaAs, Alloys, Photochemistry

*G. F. Neumark and K. Kosai*, Deep Levels in Wide Band-Gap III–V Semiconductors
*David C. Look*, The Electrical and Photoelectronic Properties of Semi-Insulating GaAs
*R. F. Brebrick, Ching-Hua Su, and Pok-Kai Liao*, Associated Solution Model for Ga-In-Sb and Hg-Cd-Te
*Yu. Ya. Gurevich and Yu. V. Pleskon*, Photoelectrochemistry of Semiconductors

## Volume 20  Semi-Insulating GaAs

*R. N. Thomas, H. M. Hobgood, G. W. Eldridge, D. L. Barrett, T. T. Braggins, L. B. Ta, and S. K. Wang*, High-Purity LEC Growth and Direct Implantation of GaAs for Monolithic Microwave Circuits
*C. A. Stolte*, Ion Implantation and Materials for GaAs Integrated Circuits
*C. G. Kirkpatrick, R. T. Chen, D. E. Holmes, P. M. Asbeck, K. R. Elliott, R. D. Fairman, and J. R. Oliver*, LEC GaAs for Integrated Circuit Applications
*J. S. Blakemore and S. Rahimi*, Models for Mid-Gap Centers in Gallium Arsenide

## Volume 21  Hydrogenated Amorphous Silicon
## Part A

*Jacques I. Pankove*, Introduction
*Masataka Hirose*, Glow Discharge; Chemical Vapor Deposition
*Yoshiyuki Uchida*, di Glow Discharge
*T. D. Moustakas*, Sputtering
*Isao Yamada*, Ionized-Cluster Beam Deposition
*Bruce A. Scott*, Homogeneous Chemical Vapor Deposition

Frank J. Kampas, Chemical Reactions in Plasma Deposition
Paul A. Longeway, Plasma Kinetics
Herbert A. Weakliem, Diagnostics of Silane Glow Discharges Using Probes and Mass Spectroscopy
Lester Gluttman, Relation between the Atomic and the Electronic Structures
A. Chenevas-Paule, Experiment Determination of Structure
S. Minomura, Pressure Effects on the Local Atomic Structure
David Adler, Defects and Density of Localized States

## Part B

Jacques I. Pankove, Introduction
G. D. Cody, The Optical Absorption Edge of a-Si:H
Nabil M. Amer and Warren B. Jackson, Optical Properties of Defect States in a-Si:H
P. J. Zanzucchi, The Vibrational Spectra of a-Si:H
Yoshihiro Hamakawa, Electroreflectance and Electroabsorption
Jeffrey S. Lannin, Raman Scattering of Amorphous Si, Ge, and Their Alloys
R. A. Street, Luminescence in a-Si:H
Richard S. Crandall, Photoconductivity
J. Tauc, Time-Resolved Spectroscopy of Electronic Relaxation Processes
P. E. Vanier, IR-Induced Quenching and Enhancement of Photoconductivity and Photoluminescence
H. Schade, Irradiation-Induced Metastable Effects
L. Ley, Photoelectron Emission Studies

## Part C

Jacques I. Pankove, Introduction
J. David Cohen, Density of States from Junction Measurements in Hydrogenated Amorphous Silicon
P. C. Taylor, Magnetic Resonance Measurements in a-Si:H
K. Morigaki, Optically Detected Magnetic Resonance
J. Dresner, Carrier Mobility in a-Si:H
T. Tiedje, Information about band-Tail States from Time-of-Flight Experiments
Arnold R. Moore, Diffusion Length in Undoped a-Si:H
W. Beyer and J. Overhof, Doping Effects in a-Si:H
H. Fritzche, Electronic Properties of Surfaces in a-Si:H
C. R. Wronski, The Staebler-Wronski Effect
R. J. Nemanich, Schottky Barriers on a-Si:H
B. Abeles and T. Tiedje, Amorphous Semiconductor Superlattices

## Part D

Jacques I. Pankove, Introduction
D. E. Carlson, Solar Cells
G. A. Swartz, Closed-Form Solution of I–V Characteristic for a a-Si:H Solar Cells
Isamu Shimizu, Electrophotography
Sachio Ishioka, Image Pickup Tubes

*P. G. LeComber and W. E. Spear*, The Development of the a-Si:H Field-Effect Transistor and Its Possible Applications
*D. G. Ast*, a-Si:H FET-Addressed LCD Panel
*S. Kaneko*, Solid-State Image Sensor
*Masakiyo Matsumura*, Charge-Coupled Devices
*M. A. Bosch*, Optical Recording
*A. D'Amico and G. Fortunato*, Ambient Sensors
*Hiroshi Kukimoto*, Amorphous Light-Emitting Devices
*Robert J. Phelan, Jr.*, Fast Detectors and Modulators
*Jacques I. Pankove*, Hybrid Structures
*P. G. LeComber, A. E. Owen, W. E. Spear, J. Hajto, and W. K. Choi*, Electronic Switching in Amorphous Silicon Junction Devices

## Volume 22  Lightwave Communications Technology
## Part A

*Kazuo Nakajima*, The Liquid-Phase Epitaxial Growth of IngaAsp
*W. T. Tsang*, Molecular Beam Epitaxy for III–V Compound Semiconductors
*G. B. Stringfellow*, Organometallic Vapor-Phase Epitaxial Growth of III–V Semiconductors
*G. Beuchet*, Halide and Chloride Transport Vapor-Phase Deposition of InGaAsP and GaAs
*Manijeh Razeghi*, Low-Pressure Metallo-Organic Chemical Vapor Deposition of $Ga_x in_{1-x} As P_{1-y}$ Alloys
*P. M. Petroff*, Defects in III–V Compound Semiconductors

## Part B

*J. P. van der Ziel*, Mode Locking of Semiconductor Lasers
*Kam Y. Lau and Ammon Yariv*, High-Frequency Current Modulation of Semiconductor Injection Lasers
*Charles H. Henry*, Special Properties of Semiconductor Lasers
*Yasuharu Suematsu, Katsumi Kishino, Shigehisa Arai, and Fumio Koyama*. Dynamic Single-Mode Semiconductor Lasers with a Distributed Reflector
*W. T. Tsang*, The Cleaved-Coupled-Cavity ($C^3$) Laser

## Part C

*R. J. Nelson and N. K. Dutta*, Review of InGaAsP InP Laser Structures and Comparison of Their Performance
*N. Chinone and M. Nakamura*, Mode-Stabilized Semiconductor Lasers for 0.7–0.8- and 1.1–1.6-$\mu$m Regions
*Yoshiji Horikoshi*, Semiconductor Lasers with Wavelengths Exceeding $2 \mu$m
*B. A. Dean and M. Dixon*, The Functional Reliability of Semiconductor Lasers as Optical Transmitters
*R. H. Saul, T. P. Lee, and C. A. Burus*, Light-Emitting Device Design
*C. L. Zipfel*, Light-Emitting Diode-Reliability
*Tien Pei Lee and Tingye Li*, LED-Based Multimode Lightwave Systems
*Kinichiro Ogawa*, Semiconductor Noise-Mode Partition Noise

## Part D

*Federico Capasso*, The Physics of Avalanche Photodiodes
*T. P. Pearsall and M. A. Pollack*, Compound Semiconductor Photodiodes
*Takao Kaneda*, Silicon and Germanium Avalanche Photodiodes
*S. R. Forrest*, Sensitivity of Avalanche Photodetector Receivers for High-Bit-Rate Long-Wavelength Optical Communication Systems
*J. C. Campbell*, Phototransistors for Lightwave Communications

## Part E

*Shyh Wang*, Principles and Characteristics of Integrable Active and Passive Optical Devices
*Shlomo Margalit and Amnon Yariv*, Integrated Electronic and Photonic Devices
*Takaoki Mukai, Yoshihisa Yamamoto, and Tatsuya Kimura*, Optical Amplification by Semiconductor Lasers

## Volume 23 Pulsed Laser Processing of Semiconductors

*R. F. Wood, C. W. White, and R. T. Young*, Laser Processing of Semiconductors: An Overview
*C. W. White*, Segregation, Solute Trapping, and Supersaturated Alloys
*G. E. Jellison, Jr.*, Optical and Electrical Properties of Pulsed Laser-Annealed Silicon
*R. F. Wood and G. E. Jellison, Jr.*, Melting Model of Pulsed Laser Processing
*R. F. Wood and F. W. Young, Jr.*, Nonequilibrium Solidification Following Pulsed Laser Melting
*D. H. Lowndes and G. E. Jellison, Jr.*, Time-Resolved Measurement During Pulsed Laser Irradiation of Silicon
*D. M. Zebner*, Surface Studies of Pulsed Laser Irradiated Semiconductors
*D. H. Lowndes*, Pulsed Beam Processing of Gallium Arsenide
*R. B. James*, Pulsed $CO_2$ Laser Annealing of Semiconductors
*R. T. Young and R. F. Wood*, Applications of Pulsed Laser Processing

## Volume 24 Applications of Multiquantum Wells, Selective Doping, and Superlattices

*C. Weisbuch*, Fundamental Properties of III–V Semiconductor Two-Dimensional Quantized Structures: The Basis for Optical and Electronic Device Applications
*H. Morkoc and H. Unlu*, Factors Affecting the Performance of (Al,Ga)As/GaAs and (Al,Ga)As/InGaAs Modulation-Doped Field-Effect Transistors: Microwave and Digital Applications
*N. T. Linh*, Two-Dimensional Electron Gas FETs: Microwave Applications
*M. Abe et al.*, Ultra-High-Speed HEMT Integrated Circuits
*D. S. Chemla, D. A. B. Miller, and P. W. Smith*, Nonlinear Optical Properties of Multiple Quantum Well Structures for Optical Signal Processing
*F. Capasso*, Graded-Gap and Superlattice Devices by Band-Gap Engineering
*W. T. Tsang*, Quantum Confinement Heterostructure Semiconductor Lasers
*G. C. Osbourn et al.*, Principles and Applications of Semiconductor Strained-Layer Superlattices

## Volume 25  Diluted Magnetic Semiconductors

*W. Giriat and J. K. Furdyna*, Crystal Structure, Composition, and Materials Preparation of Diluted Magnetic Semiconductors

*W. M. Becker*, Band Structure and Optical Properties of Wide-Gap $A^{II}_{1-x}Mn_xB^{IV}$ Alloys at Zero Magnetic Field

*Saul Oseroff and Pieter H. Keesom*, Magnetic Properties: Macroscopic Studies

*Giebultowicz and T. M. Holden*, Neutron Scattering Studies of the Magnetic Structure and Dynamics of Diluted Magnetic Semiconductors

*J. Kossut*, Band Structure and Quantum Transport Phenomena in Narrow-Gap Diluted Magnetic Semiconductors

*C. Riquaux*, Magnetooptical Properties of Large-Gap Diluted Magnetic Semiconductors

*J. A. Gaj*, Magnetooptical Properties of Large-Gap Diluted Magnetic Semiconductors

*J. Mycielski*, Shallow Acceptors in Diluted Magnetic Semiconductors: Splitting, Boil-off, Giant Negative Magnetoresistance

*A. K. Ramadas and R. Rodriquez*, Raman Scattering in Diluted Magnetic Semiconductors

*P. A. Wolff*, Theory of Bound Magnetic Polarons in Semimagnetic Semiconductors

## Volume 26  III–V Compound Semiconductors and Semiconductor Properties of Superionic Materials

*Zou Yuanxi*, III–V Compounds

*H. V. Winston, A. T. Hunter, H. Kimura, and R. E. Lee*, InAs-Alloyed GaAs Substrates for Direct Implantation

*P. K. Bhattachary and S. Dhar*, Deep Levels in III–V Compound Semiconductors Grown by MBE

*Yu. Yu. Gurevich and A. K. Ivanov-Shits*, Semiconductor Properties of Supersonic Materials

## Volume 27  High Conducting Quasi-One-Dimensional Organic Crystals

*E. M. Conwell*, Introduction to Highly Conducting Quasi-One-Dimensional Organic Crystals

*I. A. Howard*, A Reference Guide to the Conducting Quasi-One-Dimensional Organic Molecular Crystals

*J. P. Pouquet*, Structural Instabilities

*E. M. Conwell*, Transport Properties

*C. S. Jacobsen*, Optical Properties

*J. C. Scott*, Magnetic Properties

*L. Zuppiroli*, Irradiation Effects: Perfect Crystals and Real Crystals

## Volume 28  Measurement of High-Speed Signals in Solid State Devices

*J. Frey and D. Ioannou*, Materials and Devices for High-Speed and Optoelectronic Applications

*H. Schumacher and E. Strid*, Electronic Wafer Probing Techniques

*D. H. Auston*, Picosecond Photoconductivity: High-Speed Measurements of Devices and Materials

*J. A. Valdmanis*, Electro-Optic Measurement Techniques for Picosecond Materials, Devices, and Integrated Circuits.

*J. M. Wiesenfeld and R. K. Jain*, Direct Optical Probing of Integrated Circuits and High-Speed Devices

*G. Plows*, Electron-Beam Probing

*A. M. Weiner and R. B. Marcus*, Photoemissive Probing

## Volume 29  Very High Speed Integrated Circuits: Gallium Arsenide LSI

M. Kuzuhara and T. Nazaki, Active Layer Formation by Ion Implantation
H. Hasimoto, Focused Ion Beam Implantation Technology
T. Nozaki and A. Higashisaka, Device Fabrication Process Technology
M. Ino and T. Takada, GaAs LSI Circuit Design
M. Hirayama, M. Ohmori, and K. Yamasaki, GaAs LSI Fabrication and Performance

## Volume 30  Very High Speed Integrated Circuits: Heterostructure

H. Watanabe, T. Mizutani, and A. Usui, Fundamentals of Epitaxial Growth and Atomic Layer Epitaxy
S. Hiyamizu, Characteristics of Two-Dimensional Electron Gas in III–V Compound Heterostructures Grown by MBE
T. Nakanisi, Metalorganic Vapor Phase Epitaxy for High-Quality Active Layers
T. Nimura, High Electron Mobility Transistor and LSI Applications
T. Sugeta and T. Ishibashi, Hetero-Bipolar Transistor and LSI Application
H. Matsueda, T. Tanaka, and M. Nakamura, Optoelectronic Integrated Circuits

## Volume 31  Indium Phosphide: Crystal Growth and Characterization

J. P. Farges, Growth of Discoloration-free InP
M. J. McCollum and G. E. Stillman, High Purity InP Grown by Hydride Vapor Phase Epitaxy
T. Inada and T. Fukuda, Direct Synthesis and Growth of Indium Phosphide by the Liquid Phosphorous Encapsulated Czochralski Method
O. Oda, K. Katagiri, K. Shinohara, S. Katsura, Y. Takahashi, K. Kainosho, K. Kohiro, and R. Hirano, InP Crystal Growth, Substrate Preparation and Evaluation
K. Tada, M. Tatsumi, M. Morioka, T. Araki, and T. Kawase, InP Substrates: Production and Quality Control
M. Razeghi, LP-MOCVD Growth, Characterization, and Application of InP Material
T. A. Kennedy and P. J. Lin-Chung, Stoichiometric Defects in InP

## Volme 32  Strained-Layer Superlattices: Physics

T. P. Pearsall, Strained-Layer Superlattices
Fred H. Pollack, Effects of Homogeneous Strain on the Electronic and Vibrational Levels in Semiconductors
J. Y. Marzin, J. M. Gerárd, P. Voisin, and J. A. Brum, Optical Studies of Strained III–V Heterolayers
R. People and S. A. Jackson, Structurally Induced States from Strain and Confinement
M. Jaros, Microscopic Phenomena in Ordered Suprlattices

## Volume 33  Strained-Layer Superlattices: Materials Science and Technology

R. Hull and J. C. Bean, Principles and Concepts of Strained-Layer Epitaxy
William J. Schaff, Paul J. Tasker, Marc C. Foisy, and Lester F. Eastman, Device Applications of Strained-Layer Epitaxy

*S. T. Picraux, B. L. Doyle, and J. Y. Tsao*, Structure and Characterization of Strained-Layer Superlattices
*E. Kasper and F. Schaffer*, Group IV Compounds
*Dale L. Martin*, Molecular Beam Epitaxy of IV–VI Compounds Heterojunction
*Robert L. Gunshor, Leslie A. Kolodziejski, Arto V. Nurmikko, and Nobuo Otsuka*, Molecular Beam Epitaxy of II–VI Semiconductor Microstructures

## Volume 34   Hydrogen in Semiconductors

*J. I. Pankove and N. M. Johnson*, Introduction to Hydrogen in Semiconductors
*C. H. Seager*, Hydrogenation Methods
*J. I. Pankove*, Hydrogenation of Defects in Crystalline Silicon
*J. W. Corbett, P. Deák, U. V. Desnica, and S. J. Pearton*, Hydrogen Passivation of Damage Centers in Semiconductors
*S. J. Pearton*, Neutralization of Deep Levels in Silicon
*J. I. Pankove*, Neutralization of Shallow Acceptors in Silicon
*N. M. Johnson*, Neutralization of Donor Dopants and Formation of Hydrogen-Induced Defects in $n$-Type Silicon
*M. Stavola and S. J. Pearton*, Vibrational Spectroscopy of Hydrogen-Related Defects in Silicon
*A. D. Marwick*, Hydrogen in Semiconductors: Ion Beam Techniques
*C. Herring and N. M. Johnson*, Hydrogen Migration and Solubility in Silicon
*E. E. Haller*, Hydrogen-Related Phenomena in Crystalline Germanium
*J. Kakalios*, Hydrogen Diffusion in Amorphous Silicon
*J. Chevalier, B. Clerjaud, and B. Pajot*, Neutralization of Defects and Dopants in III–V Semiconductors
*G. G. DeLeo and W. B. Fowler*, Computational Studies of Hydrogen-Containing Complexes in Semiconductors
*R. F. Kiefl and T. L. Estle*, Muonium in Semiconductors
*C. G. Van de Walle*, Theory of Isolated Interstitial Hydrogen and Muonium in Crystalline Semiconductors

## Volume 35   Nanostructured Systems

*Mark Reed*, Introduction
*H. van Houten, C. W. J. Beenakker, and B. J. van Wees*, Quantum Point Contacts
*G. Timp*, When Does a Wire Become an Electron Waveguide?
*M. Büttiker*, The Quantum Hall Effects in Open Conductors
*W. Hansen, J. P. Kotthaus, and U. Merkt*, Electrons in Laterally Periodic Nanostructures

## Volume 36   The Spectroscopy of Semiconductors

*D. Heiman*, Spectroscopy of Semiconductors at Low Temperatures and High Magnetic Fields
*Arto V. Nurmikko*, Transient Spectroscopy by Ultrashort Laser Pulse Techniques
*A. K. Ramdas and S. Rodriguez*, Piezospectroscopy of Semiconductors
*Orest J. Glembocki and Benjamin V. Shanabrook*, Photoreflectance Spectroscopy of Microstructures
*David G. Seiler, Christopher L. Littler, and Margaret H. Wiler*, One- and Two-Photon Magneto-Optical Spectroscopy of InSb and $Hg_{1-x}Cd_xTe$

## Volume 37 The Mechanical Properties of Semiconductors

*A.-B. Chen, Arden Sher and W. T. Yost*, Elastic Constants and Related Properties of Semiconductor Compounds and Their Alloys
*David R. Clarke*, Fracture of Silicon and Other Semiconductors
*Hans Siethoff*, The Plasticity of Elemental and Compound Semiconductors
*Sivaraman Guruswamy, Katherine T. Faber and John P. Hirth*, Mechanical Behavior of Compound Semiconductors
*Subhanh Mahajan*, Deformation Behavior of Compound Semiconductors
*John P. Hirth*, Injection of Dislocations into Strained Multilayer Structures
*Don Kendall, Charles B. Fleddermann, and Kevin J. Malloy*, Critical Technologies for the Micromachining of Silicon
*Ikuo Matsuba and Kinji Mokuya*, Processing and Semiconductor Thermoelastic Behavior

## Volume 38 Imperfections in III/V Materials

*Udo Scherz and Matthias Scheffler*, Density-Functional Theory of sp-Bonded Defects in III/V Semiconductors
*Maria Kaminska and Eicke R. Weber*, El2 Defect in GaAs
*David C. Look*, Defects Relevant for Compensation in Semi-Insulating GaAs
*R. C. Newman*, Local Vibrational Mode Spectroscopy of Defects in III/V Compounds
*Andrzej M. Hennel*, Transition Metals in III/V Compounds
*Kevin J. Malloy and Ken Khachaturyan*, DX and Related Defects in Semiconductors
*V. Swaminathan and Andrew S. Jordan*, Dislocations in III/V Compounds
*Krzysztof W. Nauka*, Deep Level Defects in the Epitaxial III/V Materials

## Volume 39 Minority Carriers in III–V Semiconductors: Physics and Applications

*Niloy K. Dutta*, Radiative Transitions in GaAs and Other III–V Compounds
*Richard K. Ahrenkiel*, Minority-Carrier Lifetime in III–V Semiconductors
*Tomofumi Furuta*, High Field Minority Electron Transport in p-GaAs
*Mark S. Lundstrom*, Minority-Carrier Transport in III–V Semiconductors
*Richard A. Abram*, Effects of Heavy Doping and High Excitation on the Band Structure of GaAs
*David Yevick and Witold Bardyszewski*, An Introduction to Non-Equilibrium Many-Body Analyses of Optical Processes in III–V Semiconductors

## Volume 40 Epitaxial Microstructures

*E. F. Schubert*, Delta-Doping of Semiconductors: Electronic, Optical, and Structural Properties of Materials and Devices
*A. Gossard, M. Sundaram, and P. Hopkins*, Wide Graded Potential Wells
*P. Petroff*, Direct Growth of Nanometer-Size Quantum Wire Superlattices
*E. Kapon*, Lateral Patterning of Quantum Well Heterostructures by Growth of Nonplanar Substrates
*H. Temkin, D. Gershoni, and M. Panish*, Optical Properties of Ga$_{1-x}$In$_x$As/InP Quantum Wells

## Volume 41  High Speed Heterostructure Devices

*F. Capasso, F. Beltram, S. Sen, A. Pahlevi, and A. Y. Cho*, Quantum Electron Devices: Physics and Applications
*P. Solomon, D. J. Frank, S. L. Wright, and F. Canora*, GaAs-Gate Semiconductor–Insulator–Semiconductor FET
*M. H. Hashemi and U. K. Mishra*, Unipolar InP-Based Transistors
*R. Kiehl*, Complementary Heterostructure FET Integrated Circuits
*T. Ishibashi*, GaAs-Based and InP-Based Heterostructure Bipolar Transistors
*H. C. Liu and T. C. L. G. Sollner*, High-Frequency-Tunneling Devices
*H. Ohnishi, T. More, M. Takatsu, K. Imamura, and N. Yokoyama*, Resonant-Tunneling Hot-Electron Transistors and Circuits

## Volume 42  Oxygen in Silicon

*F. Shimura*, Introduction to Oxygen in Silicon
*W. Lin*, The Incorporation of Oxygen into Silicon Crystals
*T. J. Schaffner and D. K. Schroder*, Characterization Techniques for Oxygen in Silicon
*W. M. Bullis*, Oxygen Concentration Measurement
*S. M. Hu*, Intrinsic Point Defects in Silicon
*B. Pajot*, Some Atomic Configurations of Oxygen
*J. Michel and L. C. Kimerling*, Electical Properties of Oxygen in Silicon
*R. C. Newman and R. Jones*, Diffusion of Oxygen in Silicon
*T. Y. Tan and W. J. Taylor*, Mechanisms of Oxygen Precipitation: Some Quantitative Aspects
*M. Schrems*, Simulation of Oxygen Precipitation
*K. Simino and I. Yonenaga*, Oxygen Effect on Mechanical Properties
*W. Bergholz*, Grown-in and Process-Induced Effects
*F. Shimura*, Intrinsic/Internal Gettering
*H. Tsuya*, Oxygen Effect on Electronic Device Performance

## Volume 43  Semiconductors for Room Temperature Nuclear Detector Applications

*R. B. James and T. E. Schlesinger*, Introduction and Overview
*L. S. Darken and C. E. Cox*, High-Purity Germanium Detectors
*A. Burger, D. Nason, L. Van den Berg, and M. Schieber*, Growth of Mercuric Iodide
*X. J. Bao, T. E. Schlesinger, and R. B. James*, Electrical Properties of Mercuric Iodide
*X. J. Bao, R. B. James, and T. E. Schlesinger*, Optical Properties of Red Mercuric Iodide
*M. Hage-Ali and P. Siffert*, Growth Methods of CdTe Nuclear Detector Materials
*M. Hage-Ali and P Siffert*, Characterization of CdTe Nuclear Detector Materials
*M. Hage-Ali and P. Siffert*, CdTe Nuclear Detectors and Applications
*R. B. James, T. E. Schlesinger, J. Lund, and M. Schieber*, $Cd_{1-x}Zn_xTe$ Spectrometers for Gamma and X-Ray Applications
*D. S. McGregor, J. E. Kammeraad*, Gallium Arsenide Radiation Detectors and Spectrometers
*J. C. Lund, F. Olschner, and A. Burger*, Lead Iodide
*M. R. Squillante, and K. S. Shah*, Other Materials: Status and Prospects
*V. M. Gerrish*, Characterization and Quantification of Detector Performance
*J. S. Iwanczyk and B. E. Patt*, Electronics for X-ray and Gamma Ray Spectrometers
*M. Schieber, R. B. James, and T. E. Schlesinger*, Summary and Remaining Issues for Room Temperature Radiation Spectrometers

## Volume 44   II–IV Blue/Green Light Emitters: Device Physics and Epitaxial Growth

*J. Han and R. L. Gunshor*, MBE Growth and Electrical Properties of Wide Bandgap ZnSe-based II–VI Semiconductors

*Shizuo Fujita and Shigeo Fujita*, Growth and Characterization of ZnSe-based II–VI Semiconductors by MOVPE

*Easen Ho and Leslie A. Kolodziejski*, Gaseous Source UHV Epitaxy Technologies for Wide Bandgap II–VI Semiconductors

*Chris G. Van de Walle*, Doping of Wide-Band-Gap II–VI Compounds—Theory

*Roberto Cingolani*, Optical Properties of Excitons in ZnSe-Based Quantum Well Heterostructures

*A. Ishibashi and A. V. Nurmikko*, II–VI Diode Lasers: A Current View of Device Performance and Issues

*Supratik Guha and John Petruzello*, Defects and Degradation in Wide-Gap II–VI-based Structures and Light Emitting Devices

## Volume 45   Effect of Disorder and Defects in Ion-Implanted Semiconductors: Electrical and Physiochemical Characterization

*Heiner Ryssel*, Ion Implantation into Semiconductors: Historical Perspectives

*You-Nian Wang and Teng-Cai Ma*, Electronic Stopping Power for Energetic Ions in Solids

*Sachiko T. Nakagawa*, Solid Effect on the Electronic Stopping of Crystalline Target and Application to Range Estimation

*G. Müller, S. Kalbitzer and G. N. Greaves*, Ion Beams in Amorphous Semiconductor Research

*Jumana Boussey-Said*, Sheet and Spreading Resistance Analysis of Ion Implanted and Annealed Semiconductors

*M. L. Polignano and G. Queirolo*, Studies of the Stripping Hall Effect in Ion-Implanted Silicon

*J. Stoemenos*, Transmission Electron Microscopy Analyses

*Roberta Nipoti and Marco Servidori*, Rutherford Backscattering Studies of Ion Implanted Semiconductors

*P. Zaumseil*, X-ray Diffraction Techniques

## Volume 46   Effect of Disorder and Defects in Ion-Implanted Semiconductors: Optical and Photothermal Characterization

*M. Fried, T. Lohner and J. Gyulai*, Ellipsometric Analysis

*Antonios Seas and Constantinos Christofides*, Transmission and Reflection Spectroscopy on Ion Implanted Semiconductors

*Andreas Othonos and Constantinos Christofides*, Photoluminescence and Raman Scattering of Ion Implanted Semiconductors. Influence of Annealing

*Constantinos Christofides*, Photomodulated Thermoreflectance Investigation of Implanted Wafers. Annealing Kinetics of Defects

*U. Zammit*, Photothermal Deflection Spectroscopy Characterization of Ion-Implanted and Annealed Silicon Films

*Andreas Mandelis, Arief Budiman and Miguel Vargas*, Photothermal Deep-Level Transient Spectroscopy of Impurities and Defects in Semiconductors

*R. Kalish and S. Charbonneau*, Ion Implantation into Quantum-Well Structures

*Alexandre M. Myasnikov and Nikolay N. Gerasimenko*, Ion Implantation and Thermal Annealing of III-V Compound Semiconducting Systems: Some Problems of III-V Narrow Gap Semiconductors

## Volume 47  Uncooled Infrared Imaging Arrays and Systems

R. G. Buser and M. P. Tompsett, Historical Overview
P. W. Kruse, Principles of Uncooled Infrared Focal Plane Arrays
R. A. Wood, Monolithic Silicon Microbolometer Arrays
C. M. Hanson, Hybrid Pyroelectric-Ferroelectric Bolometer Arrays
D. L. Polla and J. R. Choi, Monolithic Pyroelectric Bolometer Arrays
N. Teranishi, Thermoelectric Uncooled Infrared Focal Plane Arrays
M. F. Tompsett, Pyroelectric Vidicon
T. W. Kenny, Tunneling Infrared Sensors
J. R. Vig, R. L. Filler and Y. Kim, Application of Quartz Microresonators to Uncooled Infrared Imaging Arrays
P. W. Kruse, Application of Uncooled Monolithic Thermoelectric Linear Arrays to Imaging Radiometers

## Volume 48  High Brightness Light Emitting Diodes

G. B. Stringfellow, Materials Issues in High-Brightness Light-Emitting Diodes
M. G. Craford, Overview of Device issues in High-Brightness Light-Emitting Diodes
F. M. Steranka, AlGaAs Red Light Emitting Diodes
C. H. Chen, S. A. Stockman, M. J. Peanasky, and C. P. Kuo, OMVPE Growth of AlGaInP for High Efficiency Visible Light-Emitting Diodes
F. A. Kish and R. M. Fletcher, AlGaInP Light-Emitting Diodes
M. W. Hodapp, Applications for High Brightness Light-Emitting Diodes
I. Akasaki and H. Amano, Organometallic Vapor Epitaxy of GaN for High Brightness Blue Light Emitting Diodes
S. Nakamura, Group III-V Nitride Based Ultraviolet-Blue-Green-Yellow Light-Emitting Diodes and Laser Diodes

## Volume 49  Light Emission in Silicon: from Physics to Devices

David J. Lockwood, Light Emission in Silicon
Gerhard Abstreiter, Band Gaps and Light Emission in Si/SiGe Atomic Layer Structures
Thomas G. Brown and Dennis G. Hall, Radiative Isoelectronic Impurities in Silicon and Silicon-Germanium Alloys and Superlattices
J. Michel, L. V. C. Assali, M. T. Morse, and L. C. Kimerling, Erbium in Silicon
Yoshihiko Kanemitsu, Silicon and Germanium Nanoparticles
Philippe M. Fauchet, Porous Silicon: Photoluminescence and Electroluminescent Devices
C. Delerue, G. Allan, and M. Lannoo, Theory of Radiative and Nonradiative Processes in Silicon Nanocrystallites
Louis Brus, Silicon Polymers and Nanocrystals

## Volume 50  Gallium Nitride (GaN)

J. I. Pankove and T. D. Moustakas, Introduction
S. P. DenBaars and S. Keller, Metalorganic Chemical Vapor Deposition (MOCVD) of Group III Nitrides
W. A. Bryden and T. J. Kistenmacher, Growth of Group III-A Nitrides by Reactive Sputtering
N. Newman, Thermochemistry of III-N Semiconductors
S. J. Pearton and R. J. Shul, Etching of III Nitrides
S. M. Bedair, Indium-based Nitride Compounds
A. Trampert, O. Brandt, and K. H. Ploog, Crystal Structure of Group III Nitrides

H. Morkoc, F. Hamdani, and A. Salvador, Electronic and Optical Properties of III–V Nitride based Quantum Wells and Superlattices
K. Doverspike and J. I. Pankove, Doping in the III-Nitrides
T. Suski and P. Perlin, High Pressure Studies of Defects and Impurities in Gallium Nitride
B. Monemar, Optical Properties of GaN
W. R. L. Lambrecht, Band Structure of the Group III Nitrides
N. E. Christensen and P. Perlin, Phonons and Phase Transitions in GaN
S. Nakamura, Applications of LEDs and LDs
I. Akasaki and H. Amano, Lasers
J. A. Cooper, Jr., Nonvolatile Random Access Memories in Wide Bandgap Semiconductors